Divided Spheres

Divided Spheres

Second Edition

Geodesics & the Orderly Subdivision of the Sphere

Edward S. Popko
with Christopher J. Kitrick

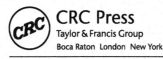

CRC Press
Taylor & Francis Group
Boca Raton London New York

CRC Press is an imprint of the
Taylor & Francis Group, an **informa** business

AN A K PETERS BOOK

Second edition published 2022
by CRC Press
6000 Broken Sound Parkway NW, Suite 300, Boca Raton, FL 33487-2742

and by CRC Press
2 Park Square, Milton Park, Abingdon, Oxon, OX14 4RN

© 2022 Edward S. Popko and Christopher J. Kitrick

First edition published by AK Peters 2012

CRC Press is an imprint of Taylor & Francis Group, LLC

Library of Congress Cataloging-in-Publication Data

Names: Popko, Edward S., author. | Kitrick, Christopher J., author.
Title: Divided spheres : geodesics and the orderly subdivision of the
sphere / Edward S. Popko with Christopher J. Kitrick.
Description: Second edition. | Boca Raton, FL : AK Peters, CRC Press, 2021.
| Includes bibliographical references and index.
Identifiers: LCCN 2021007669 (print) | LCCN 2021007670 (ebook) |
ISBN 9780367680039 (hardback) | ISBN 9780367680749 (paperback) |
ISBN 9781003134114 (ebook)
Subjects: LCSH: Polyhedra. | Geodesic domes. | Spherical projection.
Classification: LCC TA660.P73 P67 2021 (print) | LCC TA660.P73 (ebook) |
DDC 516/.156—dc23
LC record available at https://lccn.loc.gov/2021007669
LC ebook record available at https://lccn.loc.gov/2021007670

ISBN: 9780367680039 (hbk)
ISBN: 9780367680749 (pbk)
ISBN: 9781003134114 (ebk)

Typeset in Times NR MT Pro
by KnowledgeWorks Global Ltd.

To my loving family for all their support—Geraldine, Ellen, Gerald, and Amy.

Edward S. Popko

To my wife Tomoko, who has always been an integral part of the adventure.

Christopher J. Kitrick

Table of Contents

Foreword

I wonder

As one flips through each page from cover to cover of Edward Popko and Chris Kitrick's book *Divided Spheres: Geodesics and the Orderly Subdivision of the Sphere*, one will immediately realize the book will assume that rare location on your bookshelf reserved for books that will be visited most often. It has been wonderfully written for inquiring minds who are interested in discovering connections across many disciplines. It opens the mind to seeking answers to the *I wonder* questions all have from time to time. For those approaching the subject with little knowledge about *Divided Spheres*, introductory material is carefully explained when appropriate. For those seeking in-depth information on the different subjects and their connections, many threads run throughout the visual as well as the written text.

While growing up, one learns many lessons about spheres. There are evenings when one can point to twinkling points in the sky that seemed close enough to touch. Often those points of the Milky Way became indistinguishable from the points of light from the fireflies emerging from the ground. That point, it's *Venus*, a planet, or that one; it's a *photinus pyralis (Firefly)*, that group of points is "*the Boys*," the Cherokee name for the *Pleides*. A geologist may explain the chemistry of the growth of the minerals introducing the atomic structure as related to crystal lattices of points of the atomic elements and lines of force. In the physical reality points are real, containable and measurable. The stars in the sky were spheres. The comets were described as lines as the spheres streaked across the sky. The spiral galaxies were explained as being planer made up of clusters of particles and gases. And, we live on a planet that is a volumetric spherical form. One would see them, touch them, and alter them. They exist. Those points, lines and planes are real. They are not imaginary.

However, in school it is taught that the point that was illustrated on the chalk board is imaginary and does not exist. It has zero dimensions. The teacher would place a series of points on the chalk board in a row and explain they represented a line that likewise did not exist (one dimension). Following that, the teacher would place a series of parallel lines on the chalk board and name it as a plane having two dimensions. It still did not exist. Finally, it was explained that by placing the non-existing planes on many parallel chalk boards, a solid having volume (three dimensions) would be generated.

Point, line, plane, and solid—
metaphysical to physical

How is something made from nothing? It is all dependent on one's point-of- view which is constantly changing both individually and collectively.

A subtle lesson is learned throughout the book. As one looks more closely at the written text and the visual illustrations, separations of the two basic elements of Universe become apparent. One is the (metaphysical), the other is the physical. Both are required to make up the whole; Universe.

The points, lines, planes, and solids appear separated into two groups, one being abstract (metaphysical), the other real (physical).[1] Ways of communicating ideas are metaphysical and objects made up of concentrated energy are physical.

Therefore, the metaphysical points can be a location in space; lines can denote direction and distance, planes define area and solids volume. Lines, points, planes, and solids can all be at a common location at the same time.

However, physical lines, points, planes, and solids cannot be at a common location at the same time. They must accommodate each other by altering their physical form in order to share a common space.

Metaphysical at common location   *Physical at common location*

Early in the book is the introduction of Fuller's comprehensive *anticipatory design science* method, a unique process of combining metaphysical and physical with the scientific method.

Searching for principles operating in *Universe* crosses many disciplines but it is common to all. *Universe* is plural; the physical being *energy* and the metaphysical being *the rules of behavior.*

Without the whole the parts would not be.
   Without the parts the whole could not be.

As the physical and metaphysical make connection, their relationship initiates a transformation of energy to take on a form that can be measured.

The physical has need for the metaphysical for instruction.
   The metaphysical needs the physical to give instruction.

---

[1] All images and photographs courtesy of Joseph D. Clinton.

While exploring the *anticipatory design science* method, (discussed in Chapter 2,) through the eyes of atomic physics, an *I Wonder* view of the fractal patterns of atoms might present itself. Instead of the common atomic model used to describe the atom one may ask; is it possible that a *fractal-link/knot* model may explain it more completely? [2]

Three-link knot of a spherical tetrahedron

One of the exciting aspects of *Divided Spheres* is the complimentary use of metaphysical visual language and written language to describe physical reality.

These two languages emerged from the need to explain individual points-of-view of the physical and metaphysical aspects of one's environment. The meaning of the elements of language evolved from a personal point-of-view to a collective point-of-view common to the understanding of the individual point-of-view. Yet the true meaning of a symbol used in our language today is still only known by the individuals using the word. And, this changing understanding comes from the experiences one draws from exposure to unique observations of one's environment plus one's understanding of others experiences. As Edwin A. Abbotts 1884 book *Flatland: A Romance of Many Dimensions* so aptly described. *"It all depends on one's-point-of-view."*[3]

After the initial spellbinding *I wonder* experience one is faced with the decision to read *Divided Spheres* from cover to cover or to follow the threads provided from the beginning. That is where the choice must be made. For those who are new to the subject, start at the beginning. But, don't be afraid to follow a subject thread and deviate from your original path. The *I wonder* path will be greatly rewarding. For those who wander through the book by starting in the middle, there will be a return to the beginning, looking for inspiration many more times in the future. *Divided Spheres* is a resource book that will help satisfy the yearning for answers to *I wonder* questions regardless of discipline or inquiry. The following examples may come to mind while traveling through the book.

In the early 1970s, an *I wonder* showed itself again. I began exploring polyhedral forms from the intersection of cylinders based on the symmetry axes of the regular and semi-regular polyhedra. During the same period, Charles E. Peck became interested in the forms.[4] After many discussions, we both expanded our ideas and began

---

[2] (Briddell 2012)
[3] (Abbott 1884)
[4] (Peck 1995)

creating computer drawings of new polyhedron forms. In December 2003 Paul Bourke published many of the same forms on his website.[5] In 2004, I published a paper expanding the family of forms to two additional classes: the dual forms and the stellated forms.[6] These forms have taken on the name *Polycylinderhedron*.

Class I, Class II, Class III polycylinderhedron

An art expression, like those discussed in Chapter 3, emerged from a series of thoughts revolving around the geometry of the *Polycylinderhedron*.

The *Polycylinderhedron* exemplifies the transformation from the metaphysical to the physical for the axis of the polyhedron chosen to become a *Polycylinderhedron*. The metaphysical axis, as a line, is given a diameter and becomes a physical reality, a cylinder. The metaphysical line of accommodation of two adjacent cylinders is an elliptical path of an energy transformation of adjacent metaphysical planes. The art form evolved after many iterations of applying metaphysical rules of behavior to the physical realities of energy and the *"point of origin of all things"* emerged as the *Radix Universum*.

Radix Universum sculpture

[5] paulbourke.net/geometry/.
[6] (Clinton 2004)

Divided Spheres, the first edition, was a primary *I Wonder* inspiration to add to the collection of new polyhedron forms: *Polyconehedrons.* The *Polyconehedrons* are an extension of the family of *Polycylinderhedrons.* The symmetry axis of any polyhedron can define a cone axis with its base circle being a great circle of the sphere. The vertex of the cone is located on the surface of the sphere. It is proposed that by using the same mathematical operations defined for the family of *Polycylinderhedrons,* three unique classes of forms will emerge. Gary Doskas describes them as *Polyconix.*[7]

Twelve-cone polyconehedron

After discussing the regular and semi-polyhedron, an early *I wonder* project was introduced to a group of students in Mr. Boles class at Joplin Junior College, Joplin, Missouri in 1959. He assigned to the class to use the polyhedron as inspiration, to do something with the objects discussed.

Growing up in a mineral rich part of the Midwest, there was a fascination for the crystals collected and their polyhedron forms. Instead of decorating models of the polyhedron, a different approach was taken.

Weaving lesser circles of a sphere, using spring steel wire, the symmetry of the faces of any spherical form of polyhedron could be illustrated. By penetrating the sphere with a circle of the same material and diameter representing the edges of the polyhedron, something unusual occurred. The structures can collapse into a single circle.[8]

Over the years a family of new forms has emerged. *Divided Spheres* could become a source for continuous inspiration that could lead one to occasionally add new forms to the collection.

Twelve collapsible rings inside of a spherical Hexahedron

---

[7] (Doskas 2011), see Section 1.7 Other Ways to Use Polyhedra.
[8] (Clinton 1999)

Chapters 2 and 10 are major additions to this second edition of Edward Popko and Chris Kitrick's book *Divided Spheres*. They are excellent presentations, on Chris Kitrick's developments stemming from the following two *I wonder* thoughts.

In 1998, a change was taking place in the simulator field. The cathode ray tube (CRT) projectors were beginning to be phased out and replaced by solid state (LED) video projectors. In order to reshape an image for projection onto a spherical simulator screen with the CRT projectors, the image array was reshaped in the CRT. The image array was preserved. It was reshaped and there was little loss in image brightness and resolution. With the new LED projectors, however, the horizontal and vertical image array would be cropped with image reshaping. There was loss in the image array and both brightness and resolution was diminished. Being familiar with tessellations on a sphere, the *I Wonder* impulse was put to the test. Could a spherical surface be divided into diamonds having equal edges? If so, the image might be able to be reshaped optically for each area of the spherical projection screen. The result was a patent *A Method of Tessellating a Surface.*[9]

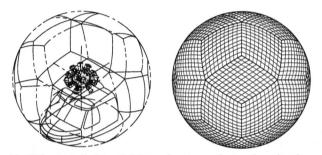

Rhombic Triacontahedron LED projector optics and projection screen

While studying Buckminster Fuller's geodesic domes and tensegrity structures, Dick Boyt had an *I Wonder* vision. As a professor at Crowder College, Neosho, Missouri, he wrote a paper describing his calculations for *Rotegrity* spheres.[10]

At the *Soft Energy Fair*[11] *in 1978,* while demonstrating his Rotegrity spheres, Dick Boyt was introduced to Fuller. The two began a discussion making Bucky thirty minutes late in giving his keynote lecture. Dick discovered the "*Rotegrity*" geometry based on Bucky's "*turbining*" effect at the tensegrity vertexes, see Figure 10.5. *Fuller's Central Angle Turbining.*

Dick Boyt and rotegrity sphere at the 1978 Soft Energy Fair

[9] (Clinton 2004).
[10] (Boyt 1991).
[11] Workshop and conference held at Amherst, MA, 1978.

The book is a remarkable collection of history, methods for tessellating the sphere and applications across disciplines. It is an inspiration to others to discover new forms and applications to add to the collection in the future editions. It is a must have, must use, must share addition to one's very special location on the bookshelf.

A sphere has no beginning, has no end. It is a duality. It is both an abstraction and a reality. It can be perceived as position lacking size, only location. Or, it is a totality as the whole of all things. It has a duality of inside and outside or both neither. If perceived as position, it must have another sphere or location cannot be measured. But to be measured, it must have size. If it has size it is not an abstraction, but a reality.

A sphere can be an imagination or thought as part of the world of abstractions where points are only positions in space. The lines can intersect at those positions in space, and planes that only exist to describe areas that can blossom into physical reality. Or, a sphere can take on form in the world of physical reality where micro points, lines, planes, and macro points all have dimensions where they are not allowed to share a common space, but they can only accommodate each other.

Joseph D. Clinton

Preface

This book summarizes the key spherical subdivision techniques that have evolved over 70 years to help today's designers, engineers, and scientists use them to solve new problems.

I became enthusiastic about geodesic domes in the mid-1960s, through my association with Buckminster Fuller and his colleagues. I was an architectural intern at Geometrics, Inc., a Cambridge, Massachusetts, firm whose principals, William Ahern and William Wainwright, had pioneered geodesic radome designs. During my internship, Ahern and Wainwright were collaborating with Fuller and Shoji Sadao on the US Pavilion dome at Expo'67 in Montreal. I was immediately drawn to the beauty and efficiency of designs based on geodesic principles.

I first met Fuller at Geometrics. He surprised me with his stature and energy, his easy rapport with audiences, and his discourse on design. It was, however, a struggle to understand him. He had his own language, which combined geometry with physics and design into a personal philosophy he called *Synergetics*. Later, I visited the original geodesics "*skunk works*"—Synergetics, Inc. in Raleigh, North Carolina. At Synergetics, Jim Fitzgibbon, T. C. Howard, and others showed me projects from their start-up years in the early 1950s, as well as their latest work. They were making history with their innovative spherical designs. Duncan Stuart, another early member of the firm, had made his own history in the early 1950s, when he invented a method (*triacon subdivision*) that made geodesic domes practical to build. The spherical grids this method produced had the fewest number of different parts of any subdivision technique. It is still one of the best gridding methods and it is detailed in this book.

More recently, I met Manuel Bromberg and Chizuko Kojima. In the late 1940s, Bromberg, along with Jim Fitzgibbon, Duncan Stuart, and other faculty members at North Carolina State University, formed the first geodesic start-up company, Carolina Skybreak. When architectural commissions materialized, the original company evolved into two others, Synergetics, Inc. and Geodesics, Inc., and Raleigh quickly became the epicenter of geodesic design. Kojima was a "*computer*" for Geodesics in the late-1950s. In those days, "*computer*" was a job title for someone who calculated. Kojima's job was to calculate hundreds of angles and grid coordinates for early dome projects—by hand.

Many years ago, I met Magnus Wenninger, a monk, teacher, mathematician, and polyhedral model builder *par excellence*. Wenninger has built thousands of models, and his classic books on polyhedral and spherical models have inspired generations of schoolchildren, artisans, and mathematicians. He showed me how to work with some lesser-known skewed spherical subdivisions; the main techniques are covered in this book.

My professional work with Computer Aided Design (CAD) systems for architects and engineers led me to more applications of geodesics. My hobbies in sailing and celestial navigation with a sextant sharpened my understanding of spherical trigonometry and improved my skills in geodesics. I had always been interested in world maps and their varied graphic projections as well as the delicate spherical structures found in microorganism exoskeletons. But what surprised me most were spherical designs in things like ocean-bobbing fish pens, panoramic photography, underground neutrino observatories, and virtual-reality simulators for the military—all of which used geodesic geometry. Geodesics appeared again in virus research, astronomy catalogs, weather forecasting, and kids' toys. But the most unexpected geodesic application I found was in the innovative design and layout of golf ball dimples. Dozens of different dimple patterns resembled small geodesic domes. These applications and many more are detailed in a later chapter.

Divided Spheres

For designers, the principles of spherical design are, at first, counterintuitive and somewhat obscured by a unique vocabulary. Chapter 1, "*Divided Spheres*," highlights the major challenges and approaches to spherical subdivision. The chapter states the design objectives that many designers use. We will meet them in later chapters. Key concepts and terms are introduced.

Buckminster Fuller's pioneering work in the late 1940s and the research and development of his colleagues in the 1950s led to many of the techniques we use today. Chapter 2, "*Bucky's Dome*," examines how Fuller's design cosmology, *Synergetic Geometry*, was first applied to cartography and then to geodesic domes. The interplay between Fuller and key associates who worked out practical solutions to geometry and construction problems is particularly important. Many of today's subdivision techniques were developed at this time. This chapter contains the first commentary on how geodesic domes were originally calculated.

Chapter 3, "*Putting Spheres to Work*," provides a brief glimpse of the wide diversity of today's spherical applications in fields like biology, astronomy, virtual-reality gaming, climate modeling, aquaculture, supercomputers, photography, children's games, and sports balls. If you are a golfer, you will enjoy seeing how manufacturers use spherical design and unique dimple patterns to maximize player performance.

Spherical geometry is quite different from Euclidean geometry, though they share common principles. Chapter 4, "*Circular Reasoning*," develops circular reasoning with points, circles, spherical arcs, and polygons. *Spherical triangles* are the most common *polygon* created when subdividing spheres. We look closely at their properties to establish an understanding of their areas, centers, type, and orientation.

It is easy to evenly divide the circumference of a circle on a computer to any practical level of precision. It's not so easy to evenly subdivide spheres, computer or not. Chapter 5, "*Distributing Points*," focuses on the challenge of evenly distributing points on a sphere and, in so doing, defines specific design conditions that this book will develop.

Spherical polyhedra offer a convenient starting point for subdivision. Chapter 6, "*Polyhedral Frameworks*," describes useful Platonic and Archimedean solids. We are particularly interested in their *symmetry* properties and their spherical versions.

Many readers will be surprised by the rich and varied spherical subdivisions that golf ball dimple patterns demonstrate. Chapter 7, "*Golf Ball Dimples*," references some of the famous US government patents, showing the golf ball industry's amazing diversity of designs and illustrating why a seemingly small thing, such as dimple patterns have become a key part of a multimillion-dollar sports ball industry.

Divided Spheres presents six classic subdivision techniques grouped into three classes. In Chapter 8, "*Subdivision Schemas*," each technique is presented in the same step-by-step format. These techniques are flexible and apply to many different spherical polyhedra. Depending on the designer's requirements, certain combinations of spherical polyhedra and subdivision techniques may be more appropriate to use than others. Thus, one objective of this chapter is to show how various techniques affect the final subdivision.

With so many design choices, it is natural to ask which combination is best. Of course, the answer depends on the application. Chapter 9, "*Comparing Results*," shows how to cut through all the design variables and select a combination that best fits your design requirements. This chapter relies extensively on graphics rather than taking a statistical approach. This book is also the first to use graphical analysis, such as Euler lines and stereographics, to highlight the subtle differences between subdivision techniques.

Chapter 10, authored by Christopher Kitrick, introduces a new concept in spherical subdivision—self-organizing grids. These grids include honeycomb hexagonal grids and grids that are a combination of hexagons and triangles called rotegrities or nexorades. The self-organizing concept is appealing because it can precisely solve some very challenging configurations that traditional subdivision techniques outlined in Chapter 8 cannot.

Three primers are included in the appendices: Stereographic Projection, Coordinate Rotations, and Geodesic Math. Stereographic projection is a graphical technique for making 2D drawings of the surface of the sphere. Stereograms appear in several sections of the book; we use them extensively in Chapter 9 to compare the results of subdivision methods in Chapter 8. Appendix A explains the theory behind this classic graphical technique. Each of the subdivision techniques explained in this book create grids or point distributions that covers only a small part of the sphere. This geometry must be replicated and rotated locally and then about the entire sphere to cover it without overlaps or gaps. Appendix B explains how this is done. Appendix C—Geodesic Math shows useful computer algorithms for each of the subdivision and optimization methods presented. Small computer code snippets suggest how they might be implemented.

Graphic Conventions

This book is about 3D spherical geometry. The foundations for the most uniform subdivisions are based on the Platonic and Archimedean solids—forms discovered by the Greeks and made from combinations of *regular polygons* (polygons in which all edges and angles are equal), such as equilateral triangles, squares, pentagons, or hexagons. These 3D forms have pure "*theoretic*" definitions, but we use a number of graphic conventions in this book to illustrate and explain their features.

Figure 1 shows six different graphic conventions for the common *dodecahedron* (12 faces, 30 edges, and 20 vertices). Each graphic is the same scale and is shown from the same viewpoint. If placed on top of one another, the vertices of all six figures would be coincident. Each convention emphasizes a different aspect of the solid.

The wireframe (a) and planar versions (d) are the most familiar and come closest to the pure definition of a planar dodecahedron. Its spherical cousin (b) makes the differences in volumes and face angles apparent. The holes in version (c) reveal the relationships of interior faces, show what lies behind the form, and draw attention to

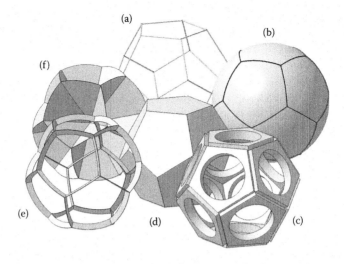

Figure 1. Polyhedron graphic conventions used in this book.

the centers of the faces (there is more than one type of center). Thick faces emphasize the angles (dihedral) between adjacent faces. Graphic convention (e) emphasizes the spherical arcs and planes of the dodecahedron's edges, while the graphic convention in (f) is useful when examining the surface angles where arcs meet. The planes and arcs are part of the volume encompassed by the form. These six graphic conventions appear throughout this book.

You have already seen another graphic convention used in this book - emphasized key words in italic and blue font. They appear near where they are defined and are important terms about spherical geometry and subdivision. You will find them throughout this book and in most of the references in the Bibliography.

I have covered the basics of what I have learned over the years in my own spherical work. I hope that *Divided Spheres* makes it easier for you to understand the principles of spherical subdivision and to develop your own designs. Spherical applications are as limitless as they are beautiful.

—Edward S. Popko
Woodstock, New York

Acknowledgments

Ed Popko would like to thank the following persons and institutions.

This book is a tribute to the work of many people. In a way, it does not seem fair that only two names are one on the cover. It is a pleasure to recognize those who contributed so much and supported me along the way.

I want to give a special thanks to Chris Kitrick for being part of this edition of Divided Spheres. In the years since the first edition, new fronters have opened for spherical subdivision and Chris is one of the pioneers. It's with great pride that he shares his research and pioneering work in self-organizing grids (Chapter 10).

No one has been more patient or taught me more about putting a book together than Jim Morrison. I cannot say enough about his enthusiastic support. Jim was my sounding board for new ideas, an instructor *par excellence* on PostScript graphics programming, and my mentor on manuscript preparation. His book, *The Astrolabe,* paved the way for mine, and I am grateful for all the advice and experience he shared with me.

I have been fortunate to have known and worked with experts in polyhedral geometry and spherical design. The late Father Magnus Wenninger, OSB, stands out. He has built thousands of models over the years and his books, including *Polyhedron Models, Spherical Models,* and *Dual Models,* have made these subjects accessible. I am grateful for his friendship and for the advice he offered throughout this project. The polyhedron chapter and skewed grid sections owe a great deal to him.

Buckminster Fuller's invention of the geodesic dome was the biggest stimulus for spherical subdivision research and development. This book samples some of his amazing work and the work he inspired in others. I was fortunate to have met him on several occasions and to have worked in one of his affiliate offices in the 1960s. Those experiences stayed with me and greatly enriched my life. I also met many of Fuller's associates and early geodesic pioneers. I am particularly indebted to Manuel Bromberg, Jim Fitzgibbon, T. C. Howard, Bruno Leon, Chizuko Kojima, Shoji Sadao, William Ahern, William Wainwright, and Duncan Stuart. All of them were part of early Fuller start-ups like Skybreak Carolina, Synergetics, Geodesics, and Geometrics. Each taught me different aspects of geodesics and the early history of the dome development. Manuel Bromberg and Bruno Leon shared their personal experiences with Fuller in the late

1940s early 1950s. These were the early days of research and development and Fuller's first major dome customer, the Marine Corps. I am fortunate to be able to include some of their accounts in this book. Duncan Stuart revolutionized spherical subdivision with his *triacon method*. He taught me how the method worked and encouraged me to write one of the first computer graphics programs based on it in the 1960s. A section of this book outlines Stuart's method and shows how it made domes practical to build. Stuart's method is used today in many applications unrelated to geodesic domes. Chizuko Kojima was a "*computer*" (a job title) at Geodesics in the 1950s. She gave me a sense of how exacting and tedious it was to calculate the geometry of Fuller's early domes *by hand*. William Ahern and William Wainwright pioneered radomes. They taught me the principles of *random subdivision*. Both have unique patented subdivision systems. I am grateful for every lesson.

Important records of Fuller's life and geodesic dome projects are cataloged in the R. Buckminster Fuller Papers collection at Stanford University's Department of Special Collections and University Archives. Staff librarians were particularly helpful in guiding me through this enormous collection and in securing rights to use and reference material. John Ferry, collections curator for the Estate of Buckminster Fuller, has been a constant supporter of my research. I am grateful for his help.

Joseph Clinton wrote many of the early primers on geodesic geometry and he developed several unique subdivision methods. He also created the class system we use today to characterize different spherical grid systems. I'm much indebted to him for his pioneering work and personal support in making this book happen. It's a privilege to have him present this work in the *Foreword*.

I'm grateful to Christopher Fearnley (aka CJ) for his help in content proofing Divided Spheres. His leadership of the *Synergetics Collaborative* (SNEC)[1], his activism in promoting the ideas of Buckminster Fuller and his professional consulting in Information Technology make him uniquely qualified to review every topic of *Divided Spheres*.

Mathematics is precise and clarity is important. I am grateful to Professors David W. Henderson, Cornell University Mathematics Department; Gregory Hartman, Virginia Military Institute Mathematics and Computer Science Department; and to Taylor and Francis reviewers for their advice and suggestions in clarifying terms and improving the text.

Many of the CAD images in this book were created with SOLIDWORKS and CATIA, both powerful 3D modelers from the French multinational Dassault Systèmes. Ricardo Gerardi taught me how to write CATIA macros that automatically create 3D geometry. Without his instruction, I would never have been able to create the number of detailed illustrations I have created for this book. Dassault Systèmes and Alain Houard facilitated my access to CATIA, without which I could not have created many of the illustrations.

Good editing clarifies a subject and makes it easier for readers to understand. This is particularly important in a technical primer like this one. Bernadette Hearne did an amazing job of editing. She tolerated many rough drafts and is an artist with scissors and paste pot. Valerie Havas and Lynne Morrison picked up where my high school English teacher left off and showed me the value of consistent style. Michael Christian provided invaluable advice about publishing. I greatly appreciate his positive support along the way.

Like most writers, I greatly underestimated how much time and effort it would take to write this book. At times, domestic projects and family were ignored. My family

[1] SNEC is a non-profit organization dedicated to bringing together a diverse group of people with an interest in Buckminster Fuller's Synergetics, See www.synergeticscollaborative.org

was patient and tactful; never once did they ask if I was done yet. My wife, Geraldine, daughter Ellen (aka Turtle), and son Gerald acted as a fan club that kept me going. Clearly, this book would not have happened without their support. Love to them all.

Christopher Kitrick would like to thank the following persons and institutions.

First and foremost, I want to thank Ed Popko for inviting me to contribute new material for the second edition of his wonderful book, *Divided Spheres*. I still remember being inspired when I first saw Ed's first book, *Geodesics*, in the late 1970s with its straightforward attention to design and construction details. Still a wonderful primer on the subject. The second edition of *Divided Spheres* has been a very fruitful collaboration and the results are a product of Ed's helpful feedback and editorial support. I first met Ed in the winter of 2014 at a Synergetics Collaborative conference in Rhode Island after having been personally invited by Joe Clinton to present some of my work and where Ed was also presenting so thanks to Joe for setting this in motion.

Buckminster Fuller was a pivotal influence on my work on spherical subdivision. Having had the rare opportunity to being a college intern for three years in his office in the late 1970s working on a myriad of projects and interacting directly with him for three years was all that was needed to keep me motivated for almost 40 years to research geometric challenges. Being in the presence of someone, already in their 80s, so motivated to continue to explore and explain all that he knew and learned cannot be overestimated. Being in that office exposed me to many who were in Bucky's circle and gave me the opportunity to develop long-term relationships with Joe Clinton, Fr. Magnus Wenninger, Prof. Koji Miyazaki, Shoji Sadao, Ed Applewhite, and Prof. Tibor Tarnai. In addition, I was able to work hand in hand with another intern Robert Grip, whose brilliance and tenacious model building provided a special working environment that forced me stay on my toes to learn and keep up.

While books like *Domebook* exposed the public to geodesic domes, it was Joe Clinton's NASA funded work that laid the foundation for detailed mathematical approaches to tessellations and gave me the impetus for expanding many of the original concepts that I built upon to further my own research. Joe's work over many years continues to provide a foundation for new ideas and explorations and I owe him a debt of gratitude. He has been kind for many years.

Fr. Magnus Wenninger and Tibor Tarnai are both important figures in my own development, both being key figures in modern work in polyhedra, geodesics, and structural topology. They had worked together on a number of projects and I was fortunate to have met them both separately and together. Fr. Magnus follows in the footsteps of so many monks who have throughout history done so much to enlarge our horizons on beauty of the elements of the real world. Prof. Tarnai has brought engineering and mathematical rigor to structural topology as well a keen historical perspective on polyhedra that and has been someone who has reviewed my work over the years.

My family has been steadfast in their support of my efforts both during the work on this book and for so many years before. I have been so fortunate not only to have met my wife, Tomoko, while working for Bucky, but to have had her meet Bucky and be directly involved with some of the projects that ensued. My children Francis, Ian, and Eileen have also been supportive of someone who continues quixotic pursuits in geometry.

The rendering software, *DisplaySphere*, used for all 3-dimensional images in Chapter 10, was developed specifically to view and capture images of spherical geometry, is available through GitHub as open source.

① Divided Spheres

We owe the word *sphere* to the Greeks; *sphaira* means ball or globe. The Greeks saw the sphere as the purest expression of form, equal in all ways, and placed it at the pinnacle of their mathematics. Today, the sphere remains a focus of astronomers, mathematicians, artisans, and engineers because it is one of nature's most recurring forms, and it's one of man's most useful.

The sphere, so simple and yet so complex, is a paradox. A featureless, spinning sphere does not even appear to be rotating; all views remain the same. There is no top, front, or side view. All views look the same. The Dutch artist M. C. Escher (1898–1972) ably illustrated this truth in his 1935 lithograph where the artist holds a reflective sphere in his hand as he sits in his Rome studio.[1] It's easy to see that even if Escher were to rotate the sphere, the reflected image would not change. His centered eyes could never look elsewhere; the view of the artist and his studio would be unchanging. His sphere has no orientation, only a position in space.

The sphere is a surface and easy to define with the simplest of equations, yet this surface is difficult to manage.[2] A sphere is a closed surface with every location on its

---

[2] The equation of a sphere is very simple. For a sphere whose three-dimensional Cartesian origin is (0, 0, 0), a point on the sphere must satisfy the equation $r = \sqrt{x^2 + y^2 + z^2}$.

surface equidistant from an infinitely small center point. A mathematician might go one step further and say the sphere is an unbounded surface with no singularities, which means there are no places where it cannot be defined. There are no exceptions.

The sphere is unusual, it has no edges, and it is *undevelopable*. By undevelopable, we mean that you cannot flatten it out onto a two-dimensional plane without stretching, tearing, squeezing, or otherwise distorting it.[3] You can test this yourself by trying to flatten an orange without distorting it. You can't without stretching or tearing it no matter how small or large the orange is.

Any plane through the center of the sphere intersects the sphere in a *great circle*. For any two points, not opposite each other on the sphere (opposite points are called *antipodal points*), the shortest path joining the points is the shorter arc of the great circle through the points, and this arc is called a *geodesic*. Geodesics are the straightest lines joining the points; they are the best we can do since we can't use straight lines or *chords* in three dimensions when we measure distance along the sphere. The length of a geodesic arc is defined as the distance between the two points on the sphere.[4]

Like the *equator* on a globe, any great circle separates the sphere into two *hemispheres*. A plane not passing through the origin that intersects the sphere either meets the sphere in a single point or it intersects the sphere in a *small circle*, also called a *lesser circle*. Small circles separate the sphere into two *caps*, one of which is smaller than a hemisphere and resembles a contact lens. We use spherical caps later in the book to compare spherical subdivisions.

Great circles play a dual role when subdividing spheres. Points define great circles and great circle intersections create more points. When we start to subdivide a sphere, we typically start with just a few points that define a relatively small number of great circles. We define more points by intersecting various combinations of the great circles. The new points derived from intersections can now be used to define more great circles, and the cycle repeats. You can see already that we are going to make great use of the dual role of great circles when we describe the various techniques and their resulting grids in Chapter 8. The difference in techniques is primarily how we define the initial set of points and great circles, and what combinations of great circles we select to intersect to define additional points.

Spherical polygons are polygons created on the surface of a sphere by segments of intersecting great circles. Spherical polygons demonstrate other differences between spherical and plane Euclidean geometry. The sides of spherical polygons are always *great circle arcs*. As a result, two-sided polygons are possible. Just look at a beach ball or slices of an orange or apple for examples. These two-sided polygons are called *lunes* or *bigons* (bi instead of polygons). Spherical triangles are also different. They can have one, two, or three right angles. And one of the oddest differences between spherical and Euclidean geometry is that there are no *similar triangles* on a sphere! They are either *congruent* or they are different. In plane Euclidean geometry, three angles define an infinite number of triangles differing only by the proportional length of their sides. But on a sphere, triangles cannot be similar unless they are actually congruent.

---

[3] The Swiss mathematician and physicist Leonhard Euler (1707–1783) did intense research on mathematical cartography (mapmaking). In a technical paper, "*On the geographic projection of the surface of a sphere*," published by the St. Petersburg Academy in 1777, he proved that it was not possible to represent a spherical surface exactly (preserving all distances and angles) on a plane. The surface of a sphere is undevelopable.

[4] The word geodesic comes from geodesy, the science or measuring the size and shape of the Earth. A geodesic was the shortest distance between two points on Earth's surface, but today, it is used in other contexts such as geodesic domes.

The sphere is a challenging but fascinating place to work. With all of these differences, how do we work with spheres and what Euclidean principles, if any, can we use?

1.1 Working with Spheres

A sphere can be any size at all. Its radius, r, could be any distance and range from subatomic dimensions, to the size of a playground dome, to light-years across the observable universe. To make spheres easier to work with when their radius could be anything, we treat r as a positive real number and make it equal to one unit. A sphere with a *unit radius* is called a *unit sphere*. One what? Are we talking about 1 mile, 1 foot, 1 inch, or one anything? Yes, to all these questions. Unit spheres, ones where $r = 1$, are easy to calculate, and any spherical result is easily converted to an actual dimension such as miles, feet, inches, or whatever. Angles do not have to be converted; they are used as-is no matter the size of the sphere. However, for distances, lengths, and areas, we need to convert unit sphere dimension into the true radius of the sphere our application needs.

1.2 Making a Point

Although the sphere is an infinite set of points all equally distant from its center, practical design applications require us to locate specific points on the surface that relate to the design we have in mind. Locating points requires us to define a reference system and orientation for our work. In the simplest case, placing the sphere's center at the center of the *Cartesian axis system*, the familiar *xyz*-coordinate system we use most often, we have defined at least six special reference points on the sphere's surface, one for the positive and negative points where each coordinate axis intersects the sphere's surface. In so doing, we have also adopted standard design conventions where we can refer to a top, bottom, side, or front, if we need to. We are off to a good start. Out next challenge is to define points on the surface that help us with our design. So how do we do this?

It's natural to think of points on a sphere like points around a circle. While it is easy to *evenly* distribute any number of points around the circumference of a circle, doing so on the surface of a sphere is actually quite difficult and certain numbers of points are impossible. Figure 1.1 gives us a sense of the problem. In Figure 1.1(a), an equal number of points are arranged around rings, or lesser circles, similar to the lines of latitude on Earth. Small circles surround each point to give a visual indicator of their spacing. We see that as the lesser circles get closer and closer to the sphere's two poles,

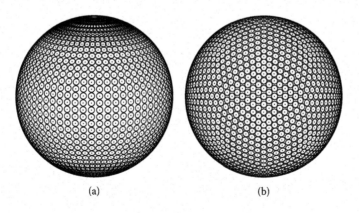

(a)   (b)

Figure 1.1. Distributing points on a sphere.

the points around them are getting closer and closer together; the circles surrounding them are overlapping. At the two poles, dozens of points are nearly superimposed. In this subdivision, the points are not uniformly distributed at all.

In Figure 1.1(b), we see a dramatically different distribution for the same number of points as in (a). Clearly, they are more uniformly spaced with a consistent symmetry and appearance. Notice that while some circles touch, none overlap. A grid connecting each point with its nearest neighbors would produce consistently shaped triangles, whereas in the first layout they won't. These two layouts illustrate our geometric challenge when we subdivide a sphere—how do we define arrangements like the one in (b) that give us freedom to have more or fewer points that allow us to make different shapes on the sphere's surface that meet our design requirements?

We know that on a circle, we can evenly space an arbitrary number of points—say, 256, or 2,011, 9, or 37 points. In a sense, this logic is what was used in Figure 1.1(a). So why can't we evenly space an arbitrary number of points and get the result shown in (b)? The answer lies in the number of points we try to distribute and how symmetrical we want their arrangement to be. Let's look at each of these related design issues and how we intend to address them.

1.3   An Arbitrary Number

Physicists and mathematicians have theorized how to distribute an arbitrary number of points on a sphere for some time. And today, with the help of computers, there are some partial solutions. One approach, familiar to physicists, sees the point distribution problem much the way they see the way particles interact. One technique uses particle repulsion to distribute points. It beautifully illustrates the problem of trying to evenly space an arbitrary number of points on a sphere.

Particles with the same charge (positive or negative) repel each other with strength inversely proportional to the square of their distance apart. The closer they are together, the stronger they repulse each other. This natural law, called the *inverse-square law*, can be exploited to subdivide a sphere when particles represent points that can be connected to form spherical grids.[5]

If you cannot relate to charged particles, think of hermits instead of particles. Each hermit is standing on the Earth and each wants to get as far away from everyone else as they can.

Imagine a given number of particles randomly distributed over the surface of the sphere, say 252 of them. Figure 1.2(a) shows them surrounded by the same diameter sphere (white), each centered on the surface of an inner sphere (transparent gold). One particle's position is fixed, but all others can move on the surface relative to the fixed one. A computer program simulates the effect of the particles repulsing each other. For each particle, the strength and direction of repulsion of every other particle acting on it is found using the inverse-square law. After every particle has been evaluated, they are allowed to move a little in the direction found so that the repulsion forces acting on them are lessened. Although the particles move a little at this stage, they always remain on the surface of the sphere. The computer once again simulates the effect of the particles repulsing each other and again, they are allowed to reposition a little to decrease the forces acting on them. In each cycle, all the particles are closer and closer

---

[5] Inverse-square law—a physical quantity like the intensity of light on a surface or the repulsion strength of two like-charged particles is inversely proportional to the square of the distance from the source of that quantity. For example, if the distance to a source of light is doubled, only one quarter of the light now reaches the subject.

Figure 1.2. Distributing an arbitrary number of points on a sphere.

to maximizing their distance to their neighbors; that is, each one is trying to get away from its neighbors and yet remain on the sphere's surface.

Figure 1.2(b) and (c) shows their progress. At each stage, the particles are progressively more evenly distributed; the spaces between them are more uniform and fewer small spheres are intersected. After a number of iterations, a program parameter, the particles reach an equilibrium state. Figure 1.2(d) shows that within tolerances, another program parameter, they have established their final position on the sphere and the repositioning cycles end. None of the spheres surrounding the particles interfere now, though some touch each other. We have a visual check that the particles are as equally spaced from one another as they can be. A grid connecting neighboring particles (points) defines a triangular subdivision grid.[6]

The particle repulsion technique looks promising. The result in Figure 1.2(d) looks quite good and we have the advantage of distributing any number of points our design requires. So why not go with this approach? No need for great circles here!

As attractive as this approach is, there are some serious drawbacks that eliminate this approach for all but the most specialized applications. First, there are an infinite number of final arrangements for the same number of particles. Resprinkle the same number of them again, let them find their equilibrium positions again, and for sure, they will settle down in a different arrangement. This means you cannot repeat the process for the same number of points and get the same design outcome. Randomization and the sequence of rebalancing in each cycle guarantees that no two simulations for the same number of particles will produce the same end result. Second, for certain numbers of particles and initial placements, it is possible they will jostle around forever and never find their equilibrium state. Forever is not a good timeframe to wait if you are trying to design something. This is one reason we fixed one point's position before we started letting them rearrange themselves. Without one fixed point, each cycle would

---

[6] Particle repulsion simulation program Diffuse, courtesy of Jonathan D. Lettvin.

simply continue to move particles around and they would never balance. Third, and this is a very big drawback, there is no symmetry in the final arrangement of points. Except for an occasional lucky number of points, there may be no stable pattern. The final arrangement of points depends on the initial random distribution, how forces are simulated and the iterative effects of relaxation and re-positioning. There is no way to anticipate if multiple points will lie on the same great circle, form antipodal points (two points on opposites sides of the sphere), or create any symmetric arrangements at all. This also means there is no way to prove whether a final solution is unique or not.

None of the above outcomes are good, especially if the points are to be used in designs we intend to manufacture. What we are really looking for is a way to subdivide a sphere by

- distributing points evenly and define grids as course or fine as needed;

- minimizing the variation within the grid (chords and areas);

- creating grids where some members form continuous great circles for applications that require them;

- maximizing symmetry and reuse of local grids;

- defining coordinates that uniquely define any point on the grid;

- developing simple ways to convert from one coordinate system to another; and

- defining metrics for comparing one subdivision method with another.

So how can we do this? The Greeks will show us.

1.4   Symmetry and Polyhedral Designs

Like many things in geometry, the Greeks seem to have gotten there first. Although the five *Platonic solids* have been known since prehistoric times, the Greeks were the first to recognize their properties and relationships to one another.[7] Figure 1.3 shows

Figure 1.3. Platonic solids.

---

[7] Stone figures resembling polyhedra, found on the islands of northeastern Scotland, have been dated to Neolithic times, between 2000 and 3000 BC. These stone figures are about 2 inches in diameter and many are carved into rounded forms resembling regular polyhedra such as the cube, tetrahedron, octahedron, and dodecahedron. By 400 BC, the time of Plato, all five regular polyhedra were known.

them with holes in their faces to make it easier to see how their parts relate. Starting in the back and going clockwise, they are the tetrahedron, cube, dodecahedron, icosahedron, and octahedron. The Platonics are one family of polyhedra (there are dozens). *Polyhedra* are three-dimensional (3D) forms with flat faces and straight edges and they take on an amazing variety of forms, many of which are very beautiful in their symmetry and spatial design. They are one of the most intensely studied forms in mathematics.

The Platonics are unique among polyhedra. Every Platonic solid has regular faces (equilateral triangles, squares, or pentagons) and every *edge* on a Platonic face has the same length. All *face angles* are the same and the angle between faces that meet at an edge are the same.

Every Platonic solid is highly symmetrical. This means that they can be orientated in many ways and still retain their appearance. Certain pairs of Platonics can be placed inside one another and this characteristic demonstrates how they are interrelated. The Platonics and the properties just mentioned are so important in subdividing spheres, we devote an entire chapter to them and explain them in detail.

Abraham Sharp (1653–1742)

Sharp was an English mathematician, astronomer, and geometrist. He was assistant to the famous astronomer Flamsteed from 1689 to 1696. Demands for highly accurate sky references clearly influenced his work. In 1717, Sharp published a significant treatise on geometry and logarithms entitled *Geometry Improv'd*.[8]

Sharp's tables and polyhedron calculations were unsurpassed for accuracy, some to 15 and 20 decimal places. Numerous illustrations show polyhedra for which he calculated their geometric properties and the sequence of cuts needed to make 3D wooden models of them. For those engaged in spherical subdivision today, Sharp's illustrations resemble reference polyhedra for geodesic spheres where face areas that might be used for small-area grid development and projection to an enclosing circumsphere.[9]

---

[8] (Sharp 1717).

[9] In Section 1 of "Solid Bodies" p. 71, Sharp describes a "very elegant geometrical solid defined from the dodecahedron or icosahedron with planes upon all the edges of either." Today, this polyhedron is called the Rhombic Triacontahedron (32 vertices, 60 edges, and 30 faces) and it's a member of the Catalan family of polyhedra. It is also a dual to the icosahedron. A spherical version of it was used by Jeffrey Lindsay in 1950 as design reference for the triangular gridding system of the first full-scale geodesic dome, Weatherbreak, built in Montreal, Canada. See Section 2.7.3.

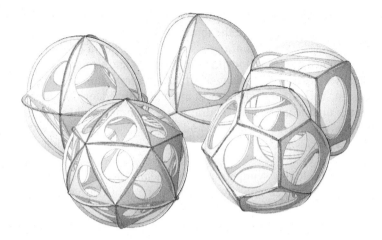

Figure 1.4. Spherical Platonics.

Another key property of the Platonics, particularly useful in spherical subdivision, is their vertices or points (we use the two words interchangeably). They are evenly spaced and lay on the surface of sphere that surrounds them and share the same center point as the polyhedron. This circumscribing sphere is called the polyhedron's *circumsphere*. Figure 1.4 shows the spherical versions of the five Platonics. The planar versions are shown inside to make the association between planar edges and spherical great circle arcs easier to see. Notice that in each spherical version, some pairs of vertices are on opposite sides of the sphere; they are antipodal points. Notice also that every spherical *face* in each Platonic is bounded by arcs of great circles and that a single Platonic has but one spherical face type (equilateral triangle, square, or pentagon). This is important because any design we develop on one face can be replicated to cover the others, thus covering the entire sphere with a pattern that has no overlaps or gaps.

1.5   Spherical Workbenches

Spherical Platonics give the designer a huge head start in subdividing the sphere because the polyhedron's vertices are already evenly distributed on its circumsphere and we can use them immediately to define great circles and reference points for further subdividing. Let's take a quick look at how we are going to develop a design starting with a simple polyhedron. Later chapters in the book will explain the details of how this is done.

Referring to Figure 1.5, we start the subdivision process by first selecting one of the Platonic solids that best meets our design requirements. In this example, we select the icosahedron (a). It is the most used in spherical work, by far, but we could have used any of the other Platonics. We define the icosahedron's spherical version (b); it's an easy step given the planar one. The spherical version is now our subdivision workbench. We pick a conveniently positioned face to work with, or some symmetrical area within it (c). In this example, a complete icosahedral face is selected; one of its vertices is at the sphere's zenith (top); this can simplify our calculations. Most of our efforts will be spent here, subdividing this single face. This face is not our only working area option, but it is an obvious choice. Within this face, we can use any one of several gridding techniques. In Chapter 8, we describe six techniques and Chapter 10 describes several more. Each technique produces a different grid and each grid's set of points offer its own benefits, depending on your application. Once the subdivision is

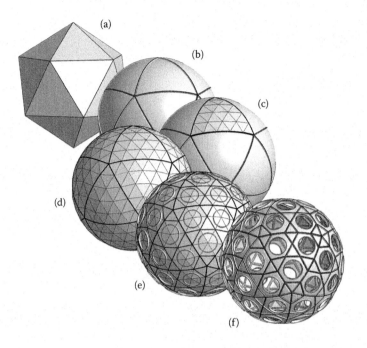

(a)

(b)

(c)

(d)

(e)

(f)

Figure 1.5. Progression of spherical subdivision to basic design application.

complete, we replicate the resulting points and grid to cover the rest of the sphere as shown in (d).

We have considerable flexibility in (a) through (c). We can use any of the five Platonics as our base, we can pick different standard areas to subdivide, and we can use different techniques to subdivide this area, making the grid as course or fine as we wish. Once we have a set of points, we have more flexibility in how we join combinations of points to make triangles, hexagons, pentagons (all shown in Figure 1.5(f)), diamonds, or any other spherical shape. The design possibilities are limitless. With so many choices, one might ask, which is best? The answer, of course, depends on the application you have in mind. In Chapter 9, we will describe a series of metrics that you can use to evaluate the appropriateness of one layout or another to your needs. Various metrics show the differences between subdivision methods and this helps you decide, which combination best fits your requirements.

For many users, the subdivision grid in Figure 1.5(d) is just the starting point for further refinement. It must be developed into a physical design. The grid might define locations for openings, panels, struts, or even positions of dimples on a golf ball, and not every subdivision point or grid member may be part of the final design.

Figure 1.5(e) shows one of many possible designs for this particular grid. Triangles might be combined into diamonds, hexagonal, or pentagonal patterns. Such refinements call for other programs such as Computer Aided Design (CAD) where points, chords, and face definitions generate, manually or automatically, prismatic shapes, structural elements, surfaces, or geometric elements, as shown in a simple design application (f). Grids can be combined or layered to create more intricate patterns or trusslike structures. If the design is for a geodesic dome, part of the subdivision may be cut off at the ground level to allow for foundations or supports. The design flexibility is limitless, and CAD provides powerful visualization and analysis capabilities as well. If the design is to be manufactured, the 3D geometry can be used to automate cutting, welding, molding, or bending equipment during production. CAD is an accessible technology and a powerful tool for the *spherist*.

1.6   Detailed Designs

The grid shown in Figure 1.5(d) is quite modest. The surface of the sphere is covered with 262 points. However, from the same starting point, the initial 12 vertices of the spherical icosahedron, shown in (b), we can distribute fewer or more points to make our grid. Figure 1.6 shows how. In each figure, the *frequency* of points, a term explained later, steadily increases from the spherical icosahedron's 12 vertices (top left figure) to 10,242 (bottom right figure) by progressively placing points between the points of the previous layout. We could continue the process indefinitely, but most spherical applications do not need more points. Notice also, as more points are added, we can generate more subdivision triangles by connecting points to their nearest neighbors.[10]

In this series, the number of points from left to right, top to bottom, is 12, 42, 162, 642, 2562, and 10,242, respectively. The radius of all accent spheres is the same in all illustrations. In the last illustration, some accent spheres touch, but none interfere. In later chapters, we will use accent spheres such as these to visualize and analyze the point distributions that result from the spherical subdivision techniques we present.

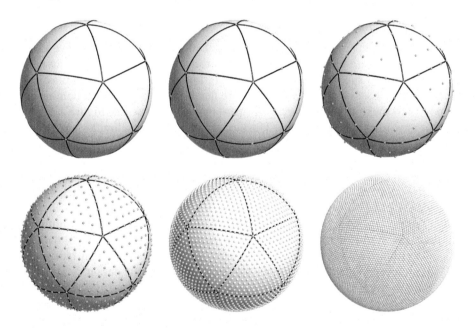

Figure 1.6.   Increasing the number of points over the sphere.

---

[10] The total number of points in any one of these spherical triangles is $n(n+1)/2$. A single face in each of the images in Figure 1.6 contains 3, 6, 15, 45, 153, and 561 vertices, respectively. In Section 8.2, we will discuss the triangulation number formula which tells how many points result from any spherical subdivision.

1.7   Other Ways to Use Polyhedra

The polyhedral subdivisions we have been discussing are based on points defined by intersecting great circles. Great circle techniques are the ones we emphasize in this book, but it is possible to use polyhedra to define points on a sphere by intersecting lesser circles instead. Here's how.

We will use the icosahedron again but instead of focusing on its spherical faces, we will use its vertex axes to develop our grid. Every one of the icosahedron's six vertex axes passes through its center and they are perpendicular to the polyhedron's circumsphere. A cylinder is positioned around each axis. Figure 1.7(a) represents the icosahedron's axes with arrows; a single cylinder is positioned around one of them. If the cylinder radius is large enough, every cylinder will intersect its neighbor's cylinder, as shown in (b). Each cylinder will also intersect the polyhedron's circumsphere and define a lesser circle, as shown in (c). A point can be defined at each lesser circle intersection, as shown in both (c) and (d).

The lesser circle arcs between the points in (d) are not geodesics. Only an arc of a great circle, the shortest distance, can be a geodesic. The lesser circle arcs, technically speaking, do not define spherical polygons either. Again, only great circle arcs define the sides of spherical polygons. It is possible, however, to define great circles between these points by passing a plane through pairs of them (and the origin).

There are many variations on this technique. For example, one can substitute cones for cylinders. The cone's apex is at the center of the sphere and the cone's surface intersects the sphere defining a lesser circle. Equations can systematically generate lesser circles on the sphere as well, and where circles intersect, subdivision grid points can be defined. Gary Doskas has done extensive research in this area; the patterns he explores

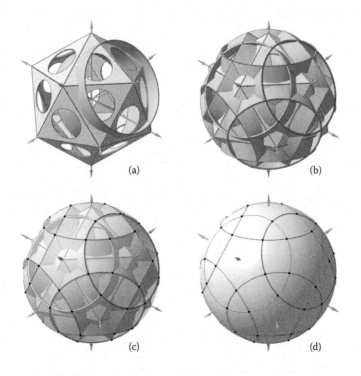

(a)   (b)

(c)   (d)

Figure 1.7.   Lesser circle spherical subdivisions.

and their symmetry are beautiful and many are utilitarian.[11] We mention lesser circle subdivision of spheres for completeness; they are not developed further in this book.[12]

1.8 Summary

In this chapter, we looked at the challenges of subdividing spheres and the fact that spheres have unique geometric properties. We have seen that it is impossible to evenly distribute an arbitrary number of points on a sphere and achieve symmetric and predictable results. Spherical polyhedra, particularly the Platonics, offer a symmetrical framework and jump start the subdivision process. The vertices of spherical Platonics are evenly spaced reference points on a sphere that surrounds the polyhedron. A small work area can be defined and grids developed there, as course or fine as a design requires. The grid can then be replicated to cover the rest of the sphere without overlaps or gaps.

Designers can use any of the spherical Platonics as a design framework, define working areas on its faces, and use any one of a number of different subdivision techniques to develop a grid over that face. The variety is limitless.

In the next chapter, we will look at the fascinating history of spherical subdivision. Buckminster Fuller, the inventor of the geodesic dome, was the first person to recognize the value of spherical polyhedra and subdivision grids to general architectural construction. Most of the subdivision techniques we use today were developed in the late 1940s and 1950s by Fuller and his associates to build geodesic domes. In their day, some domes were the largest free-span structures on Earth. Fuller's work is particularly important because he achieved highly creative results while leveraging manufacturing techniques. The spherical subdivision techniques that evolved from the 1950s are as relevant today as they were then and are applied in a wide range of science and industrial applications that have nothing to do with geodesic domes. We survey some of these applications in Chapter 3.

Additional Resources

Doskas, Gary. *Spherical Harmony—A Journey of Geometric Discovery*. LuLu Marketplace: Hedron Designs, 2011.

Kitrick, Christopher J. "Geodesic Domes." *Structural Topology* 11 (1985):15–20.

Messer, Peter W. "Polyhedra in Building." *Beyond the Cube: The Architecture of Space Frames and Polyhedra* ed. J. Francois Gabriel. New York: John Wiley & Sons, 1997.

Stuart, Duncan R. "The Orderly Subdivision of Spheres." *The Student Publications of the School of Design*. Raleigh, NC: North Carolina State University, 1963.

---

[11] See (Doskas 2011) for more information.
[12] Polyhedra created from cones and cylinders have been explored and a general typology developed. See (Pecks 1995) and (Clinton 2004).

Bucky's Dome

Buckminster Fuller (1895–1983), seen in the painting above,[1] was a true American polymath. Bucky, as he was called, was a philosopher, designer, engineer, architect, author, futurist, and prolific inventor. His earliest inventions include building blocks, temporary shelters, and an environmentally sensitive bathroom where you could take a "fog shower" utilizing pressurized mist and less than a gallon of water. He even invented a bullet-shaped, three-wheeled car that could turn complete circles in its own length. However, he is best remembered for the invention of the geodesic dome and designs that do more with less.[2]

Millions of visitors have seen the US Pavilion at Expo'67 in Montreal, Epcot Spaceship Earth at Disney World in Orlando, or the La Géode Theater in Paris. Countless others live, shop, or worship in geodesic domes. Geodesic domes enclose radar equipment at airports and air defense stations in remote polar regions. They corral fish in ocean-bobbing pens and record neutrinos from outer space, as they zip through underground spherical observatories. Geodesic tents shelter Boy Scouts and Mount

---

[1] Painting of inventor and philosopher, R. Buckminster Fuller, Jr., by Boris Artzybasheff (1899–1965). The media is tempera on board, 21.5 × 17 inch, circa 1964. *Time Magazine* cover, January 10, 1964, and gift of Time, Inc. to the National Portrait Gallery, Smithsonian Institution, Washington, DC, USA. Used with permission of the National Portrait Gallery and the Estate of R. Buckminster Fuller.

[2] Buckminster Fuller patents include Prefabricated Bathroom (Patent 2,220,482, November 5, 1940) and Motor Vehicle (Patent 2,101,057, December 7, 1937).

Everest climbers alike. A nearby playground might even have a geodesic jungle gym for kids to climb on. No single construction system has been built in so many sizes and of such diverse materials – wood, pipes, sheets of plastic and metal, foam panels, cardboard, plywood, bamboo, fiberglass, concrete, and even bicycle wheels and the tops of junked cars.[3]

These applications and many others use spherical techniques that Buckminster Fuller, or one of his associates, developed in the late 1940s and 1950s. They are as useful today, as they were then, because they continue to solve new problems in fields that have nothing to do with domes, such as astronomy, weather prediction, materials science, virology, product design, and PC game development. Even the dimple patterns on golf balls owe a debt to Buckminster Fuller.

The concept of geodesics is not entirely new. For mathematicians, they are the shortest path between two points on a curved surface, *any* curved surface, not just a sphere's. Spherical triangles and grids are hardly new, and neither is geodesic construction. The earliest example dates to 1922 when Walther Bauersfeld, an engineer for Carl Zeiss optical company, developed the world's first reinforced concrete dome in Jena, Germany,[4,5] for Zeiss' planetarium. The dome's steel reinforcing grid resembles the lattice we associate with today's geodesic dome.

Bauersfeld's structure was highly innovative at the time. However, unlike Fuller's domes, Bauersfeld's dome was never developed into a generalized construction system or used elsewhere. Fuller was the first to establish geodesics in a framework he called *energetic Synergetic Geometry,* or Synergetics for short. For sure, Fuller's relentless promotion of geodesics had a lot to do with the success of the geodesic dome, but this success was also the result of Synergetics. Synergetics acted as a vehicle for moving concepts in physics (spin, charge, attraction), mathematics (plane and spherical geometry), materials science (tension and compression), natural building processes (triangulate and space-filling forms), and design intent (conservation of energy, high strength-to-weight structures, and industrial processes) to different settings, such as mapping, long-span truss systems, and, most important, geodesic domes. The fact that these innovations were based on a broader framework for design increased their appeal and encouraged their application in new fields.

Fuller did not invent the geodesic dome in isolation. In the late 1940s, he taught at several colleges where he came into contact with highly creative and intelligent people. His style, somewhat non-academic for the times, mixed seminars, workshops, and hands-on projects, in which his ideas were freely associated with those of students and colleagues. He often made their ideas his own. This style, borrowing ideas, characterized his business life as well. There, designers, architects, engineers, and artisans, such as Kenneth Snelson, Duncan Stuart, Jeffrey Lindsay, Donald Richter, T. C. Howard,[6] William Wainwright, Jim Fitzgibbon, Shoji Sadao, and William Ahern, also had creative skills and energy. Some were inventors and would be awarded their own geodesic patents. They were often the designers and architects on projects that only Fuller received credit for. There is no doubt that Fuller's free-association inspired others,

---

[3] Drop City, a hippy commune in the 1960s, was an assemblage of geodesic panel domes made from the sheet metal of automobile roofs and other inexpensive materials.
[4] (Kahn 1989).
[5] (Fernández-Serrano 2019).
[6] Thomas C. Howard (aka T.C.) (1931) was principal designer, architect, and engineer for Synergetics, Inc. from 1955 until 2006 and lead designer on many award winning geodesic dome projects. The T. C. Howard Papers on Synergetics, housed at NC State University Libraries, archives significant architectural drawings and related documentation to worldwide projects and showcase geodesic domes, octetrusses, and Charter-Sphere Domes. Several of T. C. Howard's projects appear in the book.

increased the flow of ideas, and hastened the development of geodesics. And while most relationships were symbiotic, some were not. This book recognizes the contributions of others, as often as possible.

Buckminster Fuller is a fascinating study of innovation, entrepreneurship, self-belief, and opportunism. He was a prolific inventor and writer. While some writings are razor-sharp, others read like stream of consciousness. He was always concerned about being misunderstood and, given his elliptical style and self-made vocabulary, it was a reasonable worry. Many authors and historians have tried to place this complex personality and his inventions within the broader context of twentieth-century events. Anyone interested in the genesis of geodesic domes should attempt to understand the man, Buckminster Fuller. Those wanting to know more about him can consult the additional resources at the end of this chapter.

2.1   Synergetic Geometry

Volumes have been written by and about Fuller's Synergetic Geometry. This book examines only the very small part of it that built the foundation for dividing spheres. By the late-1940s, Fuller was shifting his attention from developing affordable housing to exploring new ways of thinking about design. He kept a sketchbook called *Noah's Ark II*, where he diagrammed geometric relationships and spherical grids.[7] Fuller had always been interested in natural forms: rocks, crystals, shells, and so forth. He sensed that nature always found the most efficient solution to problems and he was taken by the idea that nature is in constant motion and that motion itself is relative. He looked for a way to unify fundamental laws of physics (atoms, orbits, spin, energy, charges, and bonding) with geometry (polyhedra, space-filling lattices, great circles, symmetry, and maximum packing) into a design cosmology he would call Synergetic Geometry. Synergetics was Fuller's summary of natural phenomena and his framework for design. He drew his principles from natural systems and used mathematics, particularly solid geometry, to show relationships – *synergy* – between systems. "The whole is always more than the sum of its parts," he would say. In his cosmology, mathematics, observation, and the analysis of natural systems were all telling us how to make the most efficient design. Fuller personalizes Synergetics with his own language, which is often difficult to understand. But his basic dictum was clear: man is part of a natural system; he must learn from it, respect it, and use its principles in his work.[8]

He added, *"I would not suggest that it is the role of the individual to add something to the universe. Individuals can only discover the principles and then employ them to move forward to greater understanding."*

Fuller had a keen sense of how geometry increased strength and stability, and how some polyhedra could be packed together to fill space without leaving voids. He recognized the inherent stability and rigidity of the tetrahedron, octahedron, and icosahedron, three solids with only equilateral triangular faces. The cuboctahedron has both square and triangular faces, but can be constructed from just eight tetrahedra. As a result, it is highly stable and rigid. These solids were particularly important in Synergetics and would appear and reappear in different combinations in his future inventions.

---

[7] (Fuller 1950).
[8] (Edmondson 1987, 2007).

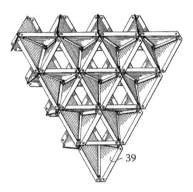

Figure 2.1. Synergetic building construction.

The tetrahedron (four equilateral triangular faces, six edges, and four vertices) is the only polyhedron where every vertex is equidistant from every other one. Fuller visualized its geometry by clustering three equal-diameter spheres into a triangle and nestling a forth on top, as seen in the figure to the left. All four spheres are packed in their closest three-dimensional (3D) configuration; their centers define the vertices of a tetrahedron. The tetrahedron is the most basic form in Fuller's Synergetic Geometry. It appears in numerous configurations in his later work.

The octahedron (8 equilateral triangular faces, 12 edges, and 6 vertices) can be visualized as two back-to-back square pyramids. Octahedra and tetrahedra together can be placed side by side and perfectly fill 3D space. Figure 2.1 shows the octahedral-tetrahedral truss system Fuller later called the *octet truss*. You can see how these two solids pack together to fill all the space.[9]

Fuller describes the relationship of the tetrahedron-octahedron: "Nature's simplest structural system in the universe is the tetrahedron. The regular tetrahedron does not fill all space by itself. The octahedron and tetrahedron complement one another to fill all space. Together they produce the simplest, most powerful structural system in the universe."[10]

The icosahedron (20 equilateral triangular faces, 30 edges, and 12 vertices) has the highest number of identical regular faces of any regular polyhedron. Its faces can be subdivided into six right triangles; thus, the overall solid could be made from 120 right triangles. The subdivision of any one of them can be replicated over its surface without overlaps or gaps. It is the most used polyhedron for spherical work by far. We discuss the icosahedron in great detail later in the chapter.

The cuboctahedron (14 faces, 24 edges, and 12 vertices) is the only polyhedron out of hundreds where every edge is equal in length *and* this length is the distance between every vertex and the center of the polyhedron. You can visualize the cuboctahedron as a polyhedron made of eight tetrahedra, each sharing one vertex at the center of the polyhedron and their other vertices tangent to one another. Like the tetrahedron, the cuboctahedron can be created by close-packing spheres. A total of 12 spheres

---

[9] (Fuller 1961, sheet 7/7, fig. 14).
[10] (Fuller 1983, 168).

nested around a 13th, central sphere, as shown in the figure to the right. Fuller thought this configuration and angular relationship was so special he called it the *vector equilibrium* (VE), claiming that it was nature's coordinate system as opposed to normal *Cartesian coordinates* man uses.[11] Cuboctahedron and VE all refer to the same polyhedron and we use the terms interchangeably.

2.2 Dymaxion Projection

Perhaps due to his prior Navy service and the importance of charts and navigation, Fuller began his spherical work with cartography. Map projection, one of the longest standing geometric challenges, involves figuring out how to accurately represent features of a round Earth on a flat plane, such as a piece of paper. The surface of a sphere is undevelopable, which means it cannot be rolled out or flattened without distortion. This centuries-old problem has attracted scores of geographers and mathematicians and led to numerous projection schemes. Some projections and maps preserve the shape of spherical shapes and angles, while others preserve great circle distances or compass headings between points on the sphere; others represent the area of shapes accurately. No two-dimensional (2D) projection preserves all these characteristics. In the end, all 2D maps of a sphere have some form of distortion.

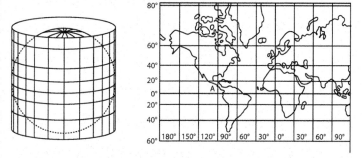

At the time, and it's still true today, the most common map projection was the *Mercator projection*. In this projection, the Earth is placed inside a cylinder (a curled-up map that wraps around the Earth), and rays from the center of the Earth paint the outline or land masses onto the cylinder. Only the equator, which is tangent to the cylinder, is accurately projected. The land masses distort more and more as the projection nears the poles. Greenland and the arctic regions appear huge in comparison to their true size. The poles cannot be projected at all, since rays from the Earth's center to the poles would be parallel to the surface of the cylinder and never project at all.

In 1944, Fuller took a totally different approach and based his map projection on the spherical subdivision of the cuboctahedron, shown in Figure 2.2(a). Essentially, Fuller developed a grid over the faces of the cuboctahedron and then projected the grid onto a sphere that surrounded the solid. The grid he developed essentially runs parallel to the edges of the cuboctahedron's regular square and triangular faces. The resulting spherical projection undeniably resembles a geodesic dome, though Fuller would not make this connection for a few more years.

In Figure 2.2(b), the projection is unfolded into a 2D view, sometimes called a *net*, and various combinations of land and ocean areas are displayed on a series of contiguous squares and triangles.[12] Some displays show a continuous ocean, others continuous land. Each serves its own purpose. Few thought it possible to invent anything new in a field so exhaustively studied and thoroughly developed as cartography

---

[11] The Cartesian coordinate system uses three mutually orthogonal axes; Fuller believed that nature's coordinate system is based on a 60° coordinate system. See also Williams (1979, 164).

[12] Map image courtesy of the Estate of R. Buckminster Fuller.

Figure 2.2. World maps with Dymaxion Projection.

but Fuller did just that. In his patent, he briefly describes the problem he solved: "The Earth is a spherical body, so the only true cartographic representation of its surface must be spherical. All flat surface maps are compromises with truth."[13]

Fuller called this configuration a *Dymaxion Projection*.[14] The projection is based on a subdivision grid that runs either parallel or perpendicular to the edges of the square and triangular faces.

After pointing out the disadvantages of current projections, including the popular Mercator projection, Fuller goes on to say: "Another expedient has been to resolve the Earth's surface into a polyhedron, projecting gnomonically to the facets of the polyhedron, the idea being that the sections of the polyhedron can be assembled on a flat surface to give a truer picture of the Earth's surface and of directions and distance."[15]

In essence, Fuller claims that a gnomonically projected[16] map placed onto the faces of a polyhedron, the cuboctahedron in this case, provides a truer representation of areas, boundaries, directions, and distances than any plane surface map heretofore known. By using the cuboctahedron as the base polyhedron for his projection, all vertices lie on one of four great circles.

Fuller makes three invention claims in his patent. The Dymaxion map is

- a projected map of square and triangular sections where edges are represented by projected great circles with a uniform cartographic scale;

---

[13] Fuller (1946, p. 1, col. 1, para. 1).

[14] In the 1930s, Fuller was advised to find a better name for his four-dimensional (4D) house invention, and with the help of an advertising wordsmith, Waldo Warren, the word "Dymaxion" was coined by selecting suitable syllables from Fuller's account of his design ideas. The roots of this name are "dynamism," "maximum," and "ions" (Marks 1973).

[15] Fuller (1946, p.1, col. 1, para. 2).

[16] Gnomonic or gnomic projection is a map projection obtained by projecting points on the surface of a sphere from a sphere's center to a plane that is tangent to the sphere or a projection from the center of a sphere to the surface that surrounds an object (such as a sphere around a polyhedron). In a gnomonic projection, great circles are mapped to straight lines.

- a map with two or more sets of matching sections where equilateral triangles and square sections are constructed on a two-way and three-way great circle grid; or

- a world map comprising six equilateral square sections and eight equilateral triangular sections matching along their edges.[17]

Fuller believed that his map projection minimized distortion, provided uniform scales, and did not have a special orientation (there was no "up"). Although the context is mapping, we clearly see an initial step towards creating a spherical version of a planar polyhedron and subdividing its faces with a grid. The same method Fuller uses here for mapping would lead him to invent the geodesic dome in less than five years.[18]

By 1954, Fuller had extended the Dymaxion Projection concept to the icosahedron. The Air ocean World Map, shown in Figure 2.3, is perhaps the most recognized of Fuller's maps.[19] It shows all land masses contiguous across the faces of an unfolded version of the icosahedron. The color scheme represents the mean low annual temperature (blue regions are colder than warm red ones).[20] While Fuller positions the icosahedron so the land masses are continuous, he does not grid the icosahedron's equilateral faces. He retains the latitude and longitude reference grid instead. Later, we will see some recent mapping applications where latitude and longitude grids are eliminated altogether in favor of geodesic grids.[21]

Figure 2.3. Dymaxion Airocean World projection.

[17] Fuller (1946, p. 3, col. 2).
[18] In 1943, Irving Fisher published a technique for gnomonic projection of world maps onto a regular icosahedron. He subsequently patented the technique (Fisher 1943, 1948).
[19] Map image, courtesy of the Estate of R. Buckminster Fuller.
[20] Shoji Sadao had enrolled as a student in Cornell University's School of Architecture. Fuller was a visiting professor in 1952 and began to construct a 20-foot-diameter "miniature Earth," a sphere with the continental land masses marked on its surface. Shoji's prior experience in the topographic maps made him the project leader. Years later, in a life-time association with Fuller, he drafted the early versions of Fuller's Dymaxion Maps.
[21] See also Sadao (1954).

The World Map Dymaxion and Airocean projection proved that practical grids can be made by subdividing the faces of spherical polyhedron and that these grids can be generated from great circles.[22] Without recognizing it at the time, Fuller was already establishing the future methodology behind geodesic domes. He studied the cubocta-hedron and icosahedron in detail, but at the time, he did not recognize the applicability of these grids and polyhedral frameworks to construction systems. It would be five years before they would be used in geodesic domes.

2.3   Cahill and Waterman Projections

Fuller was not the first person to use polyhedra for map projections, but he was the first to use them with grids that generalized the subdivision of spheres, that he later adapted to his famous geodesic dome construction system. Bernard Joseph Stan-islaus Cahill (1866–1944), cartographer and architect, proposed using polyhedra for projections as early as 1909.[23] Cahill patented his projection in 1913, explaining his goals as follows:

> *None of the methods of projection and development at present used in forming a map of the Earth furnishes a representation of its land surface with any considerable de-gree of accuracy in regard to relative dimensions of the various parts, or is free from noticeable distortion and exaggeration, and also free from discontinuities of said land surface. The object of the present invention is to provide a map of the Earth, which will avoid the above objections.*[24]

Steve Waterman extended Cahill's concept to *Waterman polyhedra*, producing a low-distortion map based on a truncated octahedron that minimizes landmass separations. Applying a *gnomonic projection*, the map is neither *conformal* nor *equal-area*;[25] it does, however, have constant scale at the equator (a fold line). The four boundary meridians at 45° east and west of each lobe's central meridian are standard meridian lines.

Figure 2.4 shows Cahill's original 1909 octahedral world map projection (a)[26] alongside Carlos A. Furuti's display of Waterman's butterfly projection using a Waterman polyhedron (b).[27] Cahill's projection represents all land masses on con-tiguous faces in the 2D foldout (called a net) of an octahedron. The Waterman polyhedron in (b) is named W5 and is a regular truncated octahedron in which vertex truncation results in side ratios of 1:2.[28] For the most part, Waterman's projection represents all land masses on contiguous faces, though a part of Siberia and some Canadian islands are divided. Antarctica is treated separately, otherwise parts of it would be separated over four other map sections, as they are in Cahill's projection.

---

[22] The 1980 edition of the Dymaxion map, produced in Fuller's office at the University of Pennsylvania, was the first introduction of a mathematical transformation algorithm used to produce, up to that time, the largest and most accurate global presentation. See Kitrick (2019) for details.

[23] Cahill (1909). See Keyes (2009) for a comparison of Cahill's map projections to Fuller's Dymaxion map.

[24] Cahill (February 25, 1913, col. 1, reference lines 9–20, sheet 1). Restored image courtesy of Gene Keyes.

[25] Conformal means the fidelity of shape and the angle between any two lines on a sphere are the same as between their projected counterparts on the map. Equal-area means that any spherical area projects ap-proximately the same area, regardless of where the area is on the sphere.

[26] "Figure 11 – Map with four lobes, modified from Fig. 10." (Cahill 1909, 464).

[27] Image courtesy of Carlos A. Furuti.

[28] All sides are equal on the Archimedean truncated octahedron.

(a)   (b)

Figure 2.4. Cahill octahedral world map and Waterman's butterfly world map.

2.4 Vector Equilibrium

The Dymaxion Projection is based on a subdivision of a spherical cuboctahedron. When projected on to the surface of a circumscribing sphere, the cuboctahedron's edges define a set of great circles. These great circles intersect one another, defining areas on the sphere's surface that can be subdivided into an endless variety of grids.

Figure 2.5 shows the cuboctahedron's master set of 25 great circles, the result of overlaying four sets of great circles.[29] The edges of a spherical cuboctahedron themselves define the first set of four great circles, shown in (a). Six more great circles mark

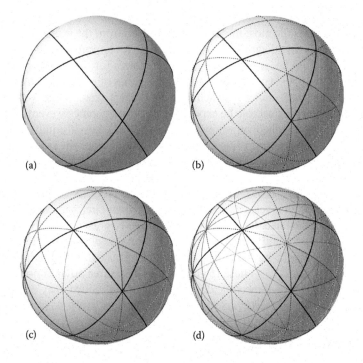

(a)   (b)

(c)   (d)

Figure 2.5. Evolution of the Vector equilibrium's 25 great circles.

---

[29] Robert Gray has done an extensive analysis of the primary and secondary great circles of spherical vector equilibrium and icosahedron; see www.rwgrayprojects.com.

the heights of each equilateral and square face; they are shown dashed in (b) and define the second set. These great circles are the principal grid axes in the world map Dymaxion Projection [Figure 2.2(a)].

The diagonals of the cuboctahedron's square faces define the third set of great circles. They are mutually *orthogonal* planes aligned to the Cartesian *x*-, *y*-, and *z*-planes. They are shown as light lines in (c). And finally, 12 more great circles pass through vertices of square faces and through the centers of surrounding equilateral faces that are not adjacent to these vertices. This fourth set of great circles is shown together with the other sets in (d).

In Figure 2.5(d), the whole sphere is covered by a set of spherical right triangles – 24 right-handed and 24 left-handed triangles for a total of 48; one is highlighted. Fuller called this single right triangle the *least common denominator (LCD)*, because the entire surface of the sphere can be described by this single common triangle. Any pattern or subdivision developed within this small area can be replicated to cover the entire sphere without leaving gaps or causing overlaps. Mathematicians call the LCD a *Schwarz triangle* after Hermann Amandus Schwarz (1843–1921), who first discovered it (and many triangles like it) while investigating polygons that cover spheres. Schwarz triangles are very important in spherical subdivision. Chapter 4 discusses them in detail.[30]

Vector equilibrium was a pivotal discovery for Fuller, forming the basis of several of his inventions. We will return to it in a later chapter when we build 3D models of his famous octet truss construction system.

2.5   Icosa's 31

The icosahedron is the most important polyhedron from the standpoint of spherical subdivision. Fuller's first diagrams of its 31 great circles appeared in 1947. Combinations of these circles would become the geometric references for every geodesic gridding system invented in the 1950s. James Ward, editor of a series of books on Fuller, said, "The materialization of great circles was achieved in Fuller's sketchbook *Noah's Ark II*, which alludes to the teleological role he expected his applied geometry eventually would play."[31]

The spherical icosahedron, shown in Figure 2.6(a), has 20 equilateral faces, 30 edges, and 12 vertices. Each vertex has a companion on the opposite side of the sphere. Thus, edges between vertices also have a corresponding edge on the opposite side. The pairs of opposing edges lie on the same plane and define the icosahedron's first set of 15 great circles. Figure 2.6(b) shows them as dashed lines. Note that five of these great circles always pass through a vertex and that the same great circles intersect every face edge at its center point. Just like the cuboctahedron, these 15 great circles divide every one of the icosahedron's equilateral faces into six LCDs. For a whole sphere, there are 60 left-handed LCDs and 60 right-handed ones. A typical LCD is colored gold in Figure 2.6(b)–(d).

These 15 great circles have three other important relationships: they coincide with the height of a face; they define the face's three vertex axes; and they intersect every edge perpendicular at its midpoint. If any of these great circles are treated as a ground plane, the icosahedron will be oriented so that one of its edges is in the uppermost position. This so-called *edge-zenith* orientation is shown in the figure to the left. The points where great circles intersect edges define the next two sets of great circles.

[30] See also Wenninger (1971, 132–134).
[31] Fuller (1985, 31).

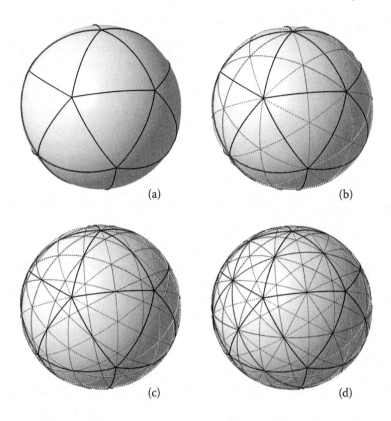

(a)   (b)

(c)   (d)

Figure 2.6. Icosahedron's 31 great circles.

Ten more great circles pass through the mid-edge points of two adjacent icosahedral faces. They are shown dotted in Figure 2.6(c). They also define 60 more points along icosahedral edges and pass perpendicular to one edge of each face. Unlike the first 15 great circles, none in this set pass through any icosahedral vertices. If any one of these great circles is treated as a ground plane, one of the icosahedron's faces is in the uppermost position. The orientation is called face-zenith and is shown in the figure to the right.

Fuller identified one more important set of great circles, a set of six that span each icosahedral face from mid-edge to mid-edge, with none of them passing through any of the icosahedron's vertices. Figure 2.6(d) shows them as solid lines along with the previous set of 25 great circles. Arcs from these great circles run parallel to the edges of the icosahedron. This grid-edge orientation will be the basis of an entire family of subdivisions called Class I. We will detail them in a later chapter. If any one of these great circles is used as a ground plane, the icosahedron will be oriented vertex-zenith, as shown in the figure to the right. Most geodesic domes are subdivided spherical icosahedra oriented vertex-zenith.

This last set of great circles illustrates an odd fact about spherical geometry. When a spherical equilateral triangle is subdivided around its mid-edges, as shown in the last example of icosahedral great circles, a curious thing happens. At first glance, it appears the grid divides the overall equilateral triangle into four smaller equilateral triangles. And in two dimensions, this is precisely what happens – but not on a sphere. On a sphere, only the center triangle remains equilateral and the other three around it are isosceles. Furthermore, the face angles for the inner and outermost equilaterals are not the same. Spherical oddities such as these are covered in Chapter 4.

Fuller's development of the 31 great circle sphere was pivotal. First, it led to the invention of the geodesic dome. Almost all domes and many other spherical applications today are based on the icosahedron. Second, two out of the three sets of 31 great circles would become reference frameworks for two different classes of subdivision grids. Each class offers distinct advantages, depending on design requirements. These developments would take place in the bucolic mountains of North Carolina.

2.6   The First Dome

Black Mountain College was an ideal place for experimenting with new ideas such as geodesic domes. The avant-garde interdisciplinary college located in Asheville, North Carolina, was founded in 1933 on the principle that the arts are central to learning. Although the school lasted only 24 years, it became the model for many future liberal arts colleges. Over the years, its faculty and lecturers would include some of the most famous artisans and thinkers of the era, including John Cage, Willem de Kooning, Ben Shahn, Robert Motherwell, Harry Callahan, Aaron Siskind, Merce Cunningham, and Buckminster Fuller.

Encouraged by others, Josef Albers invited Fuller to join the 1948 Summer Institute Program as a substitute for an architecture professor who withdrew at the last minute. It was Fuller's first teaching job and upon arrival, his seminars captivated faculty and students alike with his vision of how technology would solve world problems like hunger and the lack of adequate housing. He lectured on Synergetic Geometry and prepared his students to build the first geodesic dome.

Fuller, who regarded the geodesic dome as a three-way grid of great circles,[32] naturally chose to base his first dome on a hemispherical version of the *icosahedron's 31 great circles* [see Figure 2.6(d)]. Bucky was not well off and was only making ends meet through his lectureships. He bought a large quantity of inexpensive aluminum slat material, a common product used in manufacturing window venetian blinds. The material seemed stiff enough to work in compression (it is slightly cupped and has some longitudinal rigidity) and is very strong in tension. The students set about cutting and marking long strips of aluminum stock using the 31 great circle model as a reference and calculations by Donald Richter. The strips were labeled and marked where they overlapped one another. Then the strips were laid out on the ground in the order they would assume in the final dome. Figure 2.7 shows Fuller in his Black Mountain studio (a) and in the field (b) supervising the dome's construction. In (b), Fuller is in the background dressed in a white shirt, the painter Elaine de Kooning is to his right, and Albert Lanier, an architecture student, is to his left.[33] The great circle reference model is clearly visible in both photographs. Notice the arrangement of strips in (b) and how they approximate their position in the final dome.

To no one's surprise, the dome collapsed as soon as it was lifted from the ground! It simply could not support its own weight. Though strong in tension, the thin-gauged slats were not very stiff in compression. They just buckled under the load. Fuller attributed the collapse to poor-quality material, but the LCD's geometry wasn't helping, either. The *arc lengths* within

---

[32] Marks (1973, 60).

[33] Left photograph of Buckminster Fuller by Hazel-Frieda Larsen Archer, who documented her years at Black Mountain College in the 1940s and early 1950s. Both photographs courtesy of the Estate of R. Buckminster Fuller.

(a)   (b)

Figure 2.7. Fuller's first geodesic dome construction.

the LCD triangle varied in length by as much as 2:1. A different grid with shorter and more equal arc lengths as well as a smaller radius dome might have worked.

Albert Lanier said, "We worked like the devil all summer and waited for the dome to rise like the second coming of Moses, but it laid there like a bowl of wet spaghetti."[34] The class and school were deflated, but Bucky got all excited. He told the students, "Failure is a part of the process of inventing, and success is achieved when one stops failing."[35] Elaine de Kooning later dubbed it the "supine dome" (lying on the back, face up). The setback was only temporary, however; Fuller knew he was on the right track.

That winter, Fuller continued his geodesic work with students at the Institute of Design in Chicago but returned to Black Mountain's 1949 Summer Institute Program. This time "twelve disciples" followed him back. Among them was Donald Richter, a skilled mathematician and designer who had done the calculations on the supine dome. Fuller intended to apply the 31 great circle layout to new design, the Autonomous Dwelling facility, it would be a turning point in his approach to both design and constructing domes.

2.7 Dome Development

2.7.1 Tensegrity

Black Mountain would lead to another and very different discovery whose benefits in dome development would not be realized for another ten years. In 1948, Kenneth Snelson, an art student at the same time Fuller was lecturing, created a sculpture called *X-Piece*; it's shown below. It was the first structural demonstration of discontinues compression members (wood paddles) suspended in a network of tensioned wires.[36] When Fuller

---

[34] Thomas Zung (2011, pers. comm.).

[35] Marks (1973, 71).

[36] X-Piece reconstructed from the original, photograph by Kenneth Snelson, image courtesy of State Agency Records, N.C. Museum of Art, Black Mountain College Research Project, Visual Materials, Kenneth Snelson "Wooden Floating Compression Column," Winter 1948, 11-inch-long by 5-inch-wide, materials wood and wire.

saw the sculpture, he immediately realized the benefits of separating compression and tension members and how such an arrangement could be applied to towers and geodesic domes. Fuller was so impressed with *X-Piece* he commissioned Snelson to build a small tower using the same principles. Years later, Fuller patented its application in construction and called it *Tensile-Integrity Structures*;[37] he later blended these terms calling it *Tensegrity*.

X-Piece by Kenneth Snelson
Tensegrity is a wonderful example of collaboration (even if it wasn't formal or agreed upon) and how an idea can move from one setting (sculpture) to another (construction and long-span structures) creating its own identity and value in the transition. Clearly, tensegrity was the product of two minds – Snelson's and Fuller's.

In tensegrity structures, dis-continuous compression members like struts are suspended in a continuous network of tensile members such as cables, wire or rope where no two compression members touch. Tensile forces in cables transfer their loads to compression members and vice versa. In concert, they form an efficient force-transfer system. Fuller felt tensegrity structures used materials to their best advantage; they were either loaded in pure compression or pure tension. Members could be selected based on their best structural property. Under load, tensegrity structures tend to deflect somewhat as members take on more and more of the load. At the same time, the entire system becomes stiffer and stiffer. In some applications, this elasticity is an advantage in absorbing impact or seismic loads.

Fuller's initial patent shows the fundamental framing concept of tensegrity systems and how it can be applied to his prior geodesic dome invention. Figure 2.8(a) shows a simple structure resembling a tensile-dodecahedron and (b) tensegrity adapted to a geodesic dome.[38]

Relatively few tensegrity domes have been built owing to the difficulty of enclosing them; however, tensegrity principles have found their way into large scale bridge and tower design and are often found in sculpture, crafts, and furniture making.

From the perspective of divided spheres, tensegrity opened the door to entirely new spherical layouts such as Rotegrity and Nexorades. Both are described in Chapter 10.

2.7.2 Autonomous Dwellings and Fly's Eye

The Autonomous Dwelling was a breakthrough project. Under Fuller's direction, Black Mountain students built a collapsible 31 great circle dome made from a network of cables running through aircraft tubing and connectors. It was covered with a heat-sealed, double-walled, inflatable plastic translucent skin that controlled the environment inside, see Figure 2.9.[39,40] It was lightweight, very rigid and used standard sized members and connectors. It could be mass produced, shipped anywhere, quickly erected or packed up, and moved elsewhere if needed.

[37] Fuller (1962).
[38] (a) Fuller (1975, sheet 1/8, fig. 1) and (b) Fuller (1962, sheet 7/13, fig. 11).
[39] Photo courtesy of State Archives of North Carolina, Black Mountain College Research Project Visual Materials.
[40] The Autonomous Dwelling Facility was also called the Skybreak Dwelling, Black Mountain College, NC, Summer 1949.

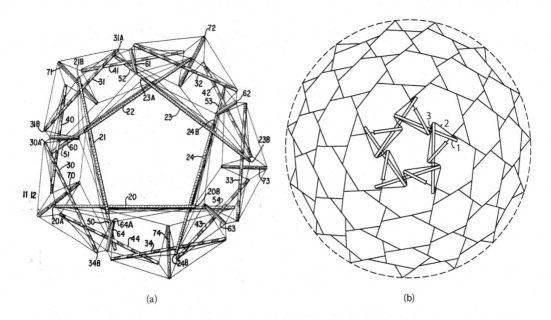

(a)   (b)

Figure 2.8. Tensile-integrity structures.

A year later, Fuller expanded the Autonomous Dwelling concept. He found a new application for Synergetic grids and way to applying composite materials. He called it the *Mono-Hex Dome* or, more popularly, the *Fly's Eye*. It was the first dome to use *stressed skin construction*.

Figure 2.9. The Autonomous Dwelling facility dome.

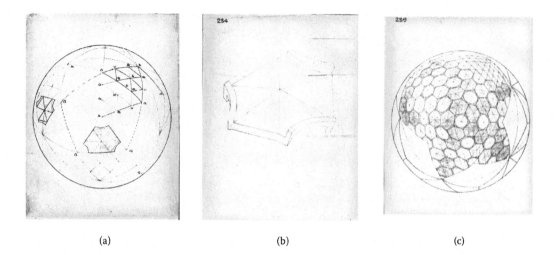

(a)   (b)   (c)

Figure 2.10. Fuller's notebook sketches of Fly's Eye Synergetic Geometry.

Fuller's first sketches of Fly's Eye appeared in his 1950s workshop notes. Figure 2.10(a) and (c) clearly shows a novel adaptation of Class I and II geodesic grids (detailed in Sections 8.5 and 8.6).[41] The triangular reference grid produced by subdividing the spherical icosahedron were no longer treated as struts and vertex connectors as before. Instead, a new spherical *hexagonal-pentagonal panel* design, vertex centered, became the new layout. Fuller saw that certain grid frequencies created large enough hex-pent panels that they could accommodate circular or bubble-like windows and service penetrations. Unlike prior domes where forces were transferred tension and compression loads though struts, these new panels would transfer forces through stressed skin, laminated fiberglass in this case. What's more, these composite panels would be molded.

Stressed skin panels were a totally new way to build a dome. Panels could be joined mid-chord as shown in Figure 2.10(b). The panels could easily be nested, stored, and transported. No other dome had ever been designed, manufactured, shipped, or assembled this way. In 1965, Fuller was awarded a design patent for Fly's Eye, see Figure 2.11(a) and (b).[42]

Fly's Eye was a significant departure from previous concepts. It was a stressed skin geometry where previous domes were based on vertices, not faces. In stressed skin structures, forces transferer through surface materials, such as composites or sheet material and not through frames (struts, beams, pipes,...). Fiberglass, a relatively new and inexpensive composite material was common in boat building but not often found in architectural construction at that time. No one knew how forces flowed through sculpted forms with compound curves and circular openings. The design had curious sound effects too. In traditional domes, sounds focus in the center; in Fly's Eye, they focused at each dome bubble. The bubbles also had a lensing effect affecting both sound and optics. They could potentially be solar capture elements as well.

Most of Fly's Eye's advanced features were not explored at the time. Bucky was distracted by other projects and the concept lie dormant for nearly 25 years. Then in 1981,

---

[41] Workshop notebook sketches from: Hall, William L., Stuart, Duncan R., Buff, Glenn A., et al., *An Introduction to the Energetic-Synergetic Geometry of Buckminster Fuller*. The School of Design, North Carolina State College, Raleigh, NC. May 1955. Notebook images by permission of the Estate of R. Buckminster Fuller.

[42] Fuller (1965, sheets 1 and 2/6, figs. 1 and 2).

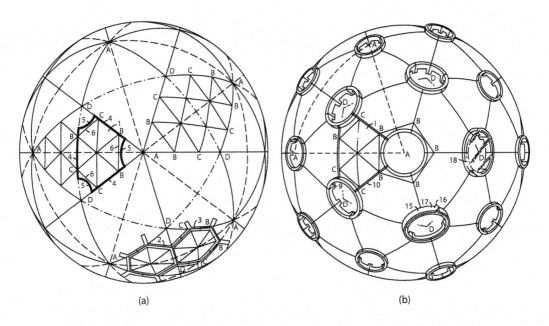

(a)   (b)

Figure 2.11. Fly's Eye dome patent illustrations. (a) Fly's Eye original concept patent 1. (b) Fly's Eye original concept patent 2.

three were built – a 12-foot, a 24-foot, and a 50-foot version. The largest was 36-feet tall, 44 feet in diameter at ground level, and 50 feet in diameter at its widest point. Figure 2.12 shows its stressed skin panels, large openings (the result of a low frequency spherical subdivision) and the novel mid-chord panel joinery.[43]

The 50-foot dome was originally displayed at Pershing Square in Los Angeles as part of the city's 1981 Bicentennial celebration. After the event, the dome was disassembled and lay abandoned in a warehouse for many years. In 2013, it was rediscovered, extensively refurbished by architectural historian Robert Rubin, and acquired in 2015 by Crystal Bridges Museum of American Art, Bentonville, AR for permanent exhibit.

Figure 2.12. 50-foot-diameter Fly's Eye dome prototype.

---

[43] Fly's Eye dome images by permission of the Estate of R. Buckminster Fuller.

Although other Fly's Eye domes have been built, none fulfilled Fuller's dream of a mass-produced Autonomous Dwelling. However, his legacy of innovation, Synergetic Geometry, and the commitment to sustainable design for low-cost housing still influences architecture today.

2.7.3   A Full-Scale Project

1949 was a pivotal year for Fuller. After several Black Mountain workshops and field experiments, he understood the 31 great circle framework better, and he had completed enough tests to know what materials and joinery worked best. He was ready for a full-scale project. This time it would be built in Montreal.

With the help of three former Institute of Design/Black Mountain students – Jeffrey Lindsay, Donald Richter, and Ted Pope – Fuller built his first large-size geodesic dome, a 49-feet diameter and slightly more than a hemisphere design.

To implement the project, Lindsay, a Canadian, set up and directed a satellite office in Montreal in 1949 and called it the *Fuller Research Foundation Canadian Division*.[44] Canada was an ideal location.

> *After the war, Canada didn't have the same rations on materials as the US, and there were possibilities here with the aluminum industry that was booming in Québec. Fuller gave his students in that first class [at Black Mountain] a "Dymaxion License," where they were all given the right to use these ideas.*[45]

Figure 2.13 shows the dome they built, it was called "*Weatherbreak*" and erected in Baie d'Urfé, QC, on the outskirts of Montreal during the fall and winter of 1950.[46] *Weatherbreak* was a new way to build geodesic domes. It was the first dome calculated with this many members, the first self-supporting and largest span geodesic dome ever built and just as important, it was the first successful practical demonstration of Fuller's theories.

The dome's layout used a spherical *rhombic triacontahedron* as reference geometry. More popularly called the *triacontahedron*, this 3D solid has 30 diamond faces, 32 vertices, and 60 edges; it's the dual solid of the icosidodecahedron and related to the icosahedron Fuller had been working on at Black Mountain, see Section 6.3.2. Because of its dual relationship, a spherical triacontahedron can overlaid a spherical icosahedron and the long axis of the triacontahedron's diamonds align with every edge of the icosahedron. Every triacontahedron diamond was the same and symmetrical about its major and minor axes; it was the standard design, calculation, and construction unit.

Lindsay divided the triacontahedron diamond into four symmetrical right triangle quadrants. He calculated a triangular grid for one of those right triangles and replicated it to the other three thus creating 32 total triangles in an 8^v or 4,4 grid, (see Section 8.6).[47] The chord factors and dihedral angles for each grid member were calculated (see Section 4.4). The chord factors determined the length of each strut.

---

[44] Jeffrey Lindsay (1924–1984) was founder and director of the Fuller Research Foundation Canadian Division based in Montréal between 1949 and 1956. The University of Calgary, Archives, and Special Collections, houses the fonds of Jeffrey Lindsay. It contains materials relating to his architectural work primarily focused on geodesic domes and early work with the Fuller Research Foundation.

[45] Comments of architectural historian Cammie McAtee in Hyperallergic, an online publication about contemporary perspectives on art and culture. Interview by Allison Meier, January 2018.

[46] Photo of *Weatherbreak* dome by Jeffrey Lindsay. Courtesy of the Jeffrey Lindsay at the Canadian Architectural Archives, Archives, and Special Collections, the University of Calgary. Gift of Jasmine Lindsay Forman. A print of this photograph was displayed as part of a 1952 exhibition at the Museum of Modern Art in New York.

[47] Fuller, Ward (editor) (1985, 44–45).

Figure 2.13. *Weatherbreak*, the first full-scale dome project.

This subdivision method – dividing the sphere into equal and symmetrical diamonds, dividing a typical diamond into four right triangle quadrants[48] and triangulating the quadrant – became known as the *Triacon Method* or Class II subdivision (see Section 8.6). *Weatherbreak* was likely the first time this gridding method was used. A few years after *Weatherbreak* was erected, Duncan Stuart, a staff member in Fuller's Skybreak Carolina office in Raleigh, reexamined Lindsay's method, simplified calculations, and found a more optimal Triacon geometry. Stuart's technique would be used on many future projects. The Triacon gridding system also appears on Fuller's first geodesic dome patent.

Weatherbreak was made of aluminum pipe struts; there were 20 different lengths, most were a little more than 4-feet long. The pipe truss was further braced by wires threaded through short aluminum pipes protruding from each grid connector. Orlon™ fabric[49] provided an enclosure, it was inside the truss and suspended from the frame. The dome took 48 man-hours to erect and weighed approximately 1140 pounds. It was remarkably stiff and lightweight for such a large span.

> *"Fuller could not have been happier [with Weatherbreak]," says McAtee. "It gave credence to all of his theories. It was a famous structure, featured on a cover of Architectural Forum and put into a show at MoMA [Museum of Modern Art]."*[50]

---

[48] This symmetrical quadrant was a spherical triangle, left- and right-handed versions covered the entire sphere without gaps or overlaps. Fuller called this triangle the Least Common Denominator (see Sections 2.5 and 4.16.9).

[49] Orlon (a trademark of E. I. Du Pont de Nemours and Company) is an acrylic fiber and woven cloth resistant to sunlight and atmospheric gases and commonly used for awnings and other outdoor uses.

[50] Mark Byrnes interview with architectural historian Cammie McAtee entitled "In Quebec, Buckminster Fuller's Domed Dreams Live On." Bloomberg CityLab (2017).

Lindsay and Richter would go on to design other domes; both were skilled geodesic designers, even in these early days.[51] We will see other Richter projects when he designs domes for Kaiser Aluminum and Chemical Company, see Section 2.7.10.

2.7.4   NC State and Skybreak Carolina

In late 1948, North Carolina College of Agriculture and Engineering in Raleigh, later renamed North Carolina State University, was in the process of establishing a fully accredited architectural program. Henry Kamphoefner, an architectural history professor at the University of Oklahoma, was hired to be the new dean of the program, which would later be called the School of Design (SOD). Kamphoefner, in turn, recruited two others from the University of Oklahoma, James Fitzgibbon and Duncan Stuart. At NC State, Kamphoefner was to build one of the most influential school of design (SOD) in America offering accredited programs in architecture, landscape architecture, and graphic and industrial design. Fitzgibbon and Stuart were part of that dynamic faculty and would teach classes in structural concepts and geometry for design, among others. Both would make major contributions to the development of geodesics.

Fitzgibbon was building a house for a client in Knoxville in the summer of 1949. He detoured going to Knoxville from Raleigh and went to see what Black Mountain College was all about. He met Bucky, saw the projects he was working on, and was much impressed with his vision and geodesic experiments. He told Bucky about the new SOD in Raleigh.[52] As part of his plan to modernize NC State's curriculum, Kamphoefner planned to invite esteemed guest lecturers. Fitzgibbon, along with Duncan Stuart and Manuel Bromberg, recommended they invite Fuller. Fuller's first lecture at NC State was in March 1949.

Fuller was not well-known at the time. Some faculty members felt that he was not really an architect, and questioned whether he was qualified to lecture their students on design. The engineering faculty had their doubts about Fuller too. The prevailing philosophy of design, best characterized by the writings of renowned architectural historian, Lewis Mumford (another SOD faculty member at that time), was that architecture created spaces where people would be brought together in secure and permanent settings that were "of and fixed to the Earth" as it were. Fuller believed that buildings should be lightweight, portable, and use materials in tension. They should be fabricated as pop-up tinker toys. They could be manufactured elsewhere and plunked down anywhere. These ideas were not part of the design vernacular at the time. Fuller's lectures on autonomous living packages showed that you could go anywhere you wanted *and* take your house with you; Synergetic Geometry and geodesic domes were literal examples of thinking outside the box.

Fuller's marathon eight-hour, run-on lecture style did not impress the entire faculty. Some characterized his lectures as semi-coherent, unedited streams of thought. Fuller freely associated his ideas and moralized about the future of man and the destiny of designers. He mixed incomplete concepts in geometry with spinning particles, packed spheres, and natural laws, and delivered them in long, run-on sentences. For him, there were design imperatives such as housing mankind in the most equitable and efficient

---

[51] Extensive archival material about Jeffrey Lindsay's dome projects can be found at search archives. ucalgary.ca/jeffrey-lindsay-fonds. See also McAtee (2017).

[52] Fuller's work appealed to Fitzgibbon on many levels. In addition to Fuller's fresh approach to synergetic design and geodesic construction, Fitzgibbon had a life-long interest in "ephemeral" architecture, portable structures built by nomadic tribes or used by fairs or circuses. The portability of geodesic domes perfectly fit his interests. Fitzgibbon later wrote several books on the subject (Brook 2005, 38).

way possible. Manuel Bromberg, a faculty supporter who attended these sessions, describes him this way: "Bucky was highly intuitive and seductive with an audience. With thick glasses and fleeting stares, he seemed to quickly find the soul of people he was with. He seemed to be a thinker and con artist at the same time, creating an atmosphere of 'we're off to see the wizard.'"[53]

Although some faculty members had their doubts about Fuller, he was an instant celebrity with the architecture students. Bucky's lectures quickly took over the school. Students quit attending other classes just to hear him speak. Faculty attitudes would change as Fuller's ideas gained acceptance slowly over time. So much so that NC State University would award him his first honorary degree in design in 1954.[54]

Fitzgibbon was convinced that Fuller's basic design philosophy was correct, and became one of Fuller's business partners in several companies. In 1951, along with Stuart, Bromberg, and several other SOD faculty members, he formed *Skybreak Carolina Corp.*,[55] as a way to develop new ideas about the built environment and keeping contact with Fuller (who only occasionally visited Raleigh). There were no paying customers and no company mission per se, but Skybreak founders saw many possibilities with Synergetic Geometry, and the company gave them a forum for their creativity. This unusual collective of talented architects, painters, sculptors, and engineers was essentially the first geodesic start-up company. Its interdisciplinary mix would become the norm at many of Fuller's future offices.

Fitzgibbon, a highly talented architect in his own right, wanted to build a 30-foot-diameter prototype dome. He and Duncan Stuart saw the prototype as a way to test geometric layouts, understand the tolerances required when building a dome, try different ways of prefabricating, and experiment with ways of weatherizing domes. They also saw value in having a demonstration model they could show prospective clients, if they ever had a client.

With the help of students from NC State's SOD,[56] a dome was built in Fitzgibbon's backyard. It was made of inexpensive 3/8-inch by 2-inch Carolina pine wood slats bolted together in large triangulated sections they called rafts.[57] Figure 2.14 shows the lattice dome before it was enclosed.[58] The structure was a 16-frequency subdivision of a spherical icosahedron and it was surprisingly rigid, yet lightweight. The cluster of semi-complete triangles in the center of the photograph is where the apex of five rafts meet and form a pentagonal pattern; all others are hexagonal shapes around any vertex. Once the frame was completed, Fitzgibbon considered ways to weatherize it. He wanted to experiment with "cocooning," a spray process developed by the US Navy to preserve moth-balled ships after World War II. It encased objects in a rubber-plastic neoprene material making them weatherproof. His idea was to apply this neoprene material over chicken wire nailed to the slats. The process worked well, and the dome, when it was completed in January 1952, weighed only 1000 pounds, a featherweight by any construction standard.

---

[53] Bromberg (2004, pers. comm.).

[54] Buckminster Fuller was awarded 47 honorary degrees in his lifetime.

[55] Skybreak Carolina Corp., Raleigh, NC, 1951, founders: James Fitzgibbon, Duncan Stuart, Manuel Bromberg, Roy Gusso, George Matsumotoa, and William Parkhurst.

[56] Two students would become notable architects. T. C. Howard became the principal architect and geodesic dome designer in Synergetics, Inc., and Bruno Leon became a practicing architect and dean of the School of Architecture, University of Detroit.

[57] A raft corresponded to a large diamond symmetrical to each edge of the icosahedron. All rafts had the same triangular geometry except for those rafts that were cut off at the ground.

[58] Image used with permission, NC State School of Design. Photo of wood lattice dome built by Fitzgibbon and students from School of Design taken by Mark Mills in 1951. Appearing from left to right: James Fitzgibbon and Duncan Stuart. See Student Publications of the School of Design, 1952.

Figure 2.14. Lattice dome.

Stuart was not satisfied with the design or prefabrication of the lattice dome. The lattice grid was based on modifications he had made to Fuller's (1948) three-way regular grid, but it was not ideal. Each raft required too many different and individually cut triangles and many arcs did not meet exactly at a common grid point. Bucky was aware of the regular grid problems and had discussed it with Stuart. Shoji Sadao, an early pioneer in geodesics and a longtime business associate of Fuller's, described the grid problem this way:

> There had always been a nagging problem with the regular grid – it had windows. That is to say, at certain vertices the three great circles did not go through a common point. Calculations were checked and rechecked, but the windows persisted to the point where Fuller thought it was a message from God that the trigonometric tables had an error in them! A copy of "Tables of Sines and Cosines to Fifteen Decimal Places at Hundredths of a Degree," published by the US Department of Commerce, National Bureau of Standards,[59] was used, to no avail.[60]

Stuart was a skilled mathematician with an extraordinary sense of 3D relationships. He could visualize and render the most complex spatial arrangements (one of Stuart's freehand sketches appears in Section 3.24). Stuart's dissatisfaction with the lattice dome and the regular grid made him rethink the whole approach to spherical subdivision. There was no problem with Bucky's 15 great circle icosahedron reference. The problem

[59] US Department of Commerce (1949).
[60] Interview "A Brief History of Geodesic Domes" by Shoji Sadao, (Zung 2001, 25).

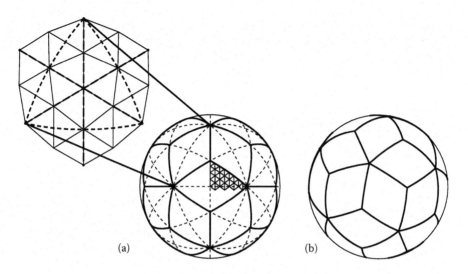

Figure 2.15. Triacon subdivision or typical diamond (a) and reference spherical rhombic triacontahedron (b).

was how to achieve consistent gridding between these reference great circles. He was looking for a geometric layout that would minimize the number of different *chord lengths* and angles between members. Up to this time, every dome prototype started with Bucky's 15 great circle icosahedron reference but further subdivided it in slightly different ways. No two domes used the same method and even with high precision trigonometric tables, some grid members did not intersect at exactly the same point, creating small triangular windows, the ones that Sadao commented on above. Stuart noticed that if the edge of the spherical icosahedron was evenly subdivided [see Figure 2.15(a)], a set of repetitive and nested spherical right triangles could be defined and when those triangles were further subdivided, they produced a very uniform grid. Stuart worked out this new geometry for a series of increasingly dense grids and compared the results to previous methods like the one used for the Weatherbreak dome in Montreal. His analysis showed conclusively that the denser the subdivision grid, the more advantageous the new system was. Stuart wrote to Bucky in June 1952 and described his new system.[61] He also sent him a report detailing the new method and comparing its results to the previous method. His new system offered[62]

- a higher order of symmetrical relationships – rendering the system more advantageous structurally;

- a great simplification in terms of the number and kinds of parts, a greater simplicity in determining the relationships or the various parts;

- more advantageous diameter-frequency relationships; and

- more favorable erection conditions when the system is applied to structural problems.

---

[61] Stuart (1952), unpublished letter to Buckminster Fuller, R. Buckminster Fuller archives at Stanford University.
[62] Stuart (1953), unpublished technical Skybreak report, R. Buckminster Fuller archives at Stanford University.

Stuart named his new breakdown *triacon* because the diamond pattern resembles the diamond faces of the rhombic triacontahedron (see Section 8.6), shown in Figure 2.15(b), another polyhedron closely related to the icosahedron.[63] Stuart's triacon was the breakthrough Fuller needed. Although the triacon was related to earlier gridding systems, it was unique in its approach and made geodesic domes practical to build. Domes using this new method would be much easier to calculate (right triangles are the easiest to solve), easier to prefabricate, and cheaper to build. Stuart saw that the repetitive diamond pattern offered additional possibilities for prefabrication. The triacon method was so key it would be employed on almost every major project in the 1950s. Today, the triacon subdivision method is one of several breakdowns where the resulting grid members essentially run perpendicular to the edges of the spherical polyhedron on which they are based. Grids such as this are categorized as Class II grids. But Stuart's triacon method is the only Class II grid that minimizes the number of different parts; it is detailed later in Chapter 8. Stuart was one of Fuller's indispensable resources and made many more contributions to dome projects in the 1950s. Although Fuller depended on his skills, he never recognized or credited Stuart for his work.[64]

Based on his geodesic experiments at Black Mountain, the successful Weatherbreatk dome in Canada, and exhibitions at the Pentagon Garden dome in Washington, DC, and in Lawrence, New York, Fuller filed for US patent protection in 1951. His building construction patent describes a general method for subdividing spheres and creating a structural truss work. In 1954, he was awarded patent 2,682,235, shown in Figure 2.16. Patent claims 3, 6, and 7 state the essence of the invention:[65]

3.   A framework of generally spherical form, in which the main structural elements form a substantially uniform overall pattern of great circle arcs intersecting in a three-way grid.

6.   A building framework of generally spherical form, in which the longitudinal centerlines of the main structural elements lie substantially in great circle planes whose intersections with a common sphere form grids comprising substantially equilateral spherical triangles.

7.   A building framework constructed in accordance with claim 6, in which the main structural elements are interconnected to form a truss, the outermost points of which lie substantially in a common spherical surface.

It is interesting to note that in its application review, the US Patent Office found only one remotely related prior patent by another inventor.[66] This suggests just how novel and unprecedented Fuller's basic geodesic dome invention really was.

Fuller's patent announced the geodesic concept to the world and ensured Fuller the commercial rights to the idea, if it ever caught on. Stuart's triacon system gave designers at *Fuller Research Foundation* and Skybreak the confidence that geodesic building construction was practical. It also made Raleigh the epicenter for geodesic development.

---

[63] The rhombic triacontahedron is the dual polyhedron of the icosidodecahedron. See Section 6.4.

[64] Stuart's contributions were not the only ones overlooked by Fuller. Press coverage often attributed projects to Fuller when, in fact, it was the independent work done at one of Fuller's franchised offices. See Chu et al. (2009) for an excellent analysis of the R. Buckminster Fuller archives at Stanford University.

[65] Fuller (1954, sheet 1/6, figs. 1-2).

[66] Pantke (1930).

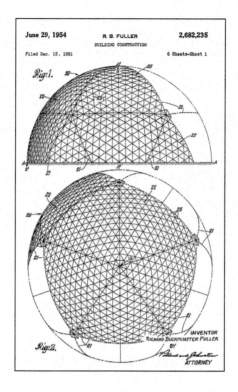

Figure 2.16. Buckminster Fuller's geodesic dome patent.

2.7.5 Ford Rotunda Dome

The Ford Rotunda dome was Fuller's first real commercial project and his patron was none other than Henry Ford himself. The Rotunda was originally built as an exhibit building for the 1933 Chicago World's Fair. After the close of the fair, the building was taken apart, shipped to Dearborn, Michigan, and reassembled on Rotunda Drive, across from Ford's Central Office Building, his world headquarters at the time. After remodeling the building, it was opened to the public, and was a huge tourist attraction until the outbreak of World War II. After the war, and in preparation for Ford Motor Company's fiftieth anniversary celebration in 1953, the Rotunda again would be re-modeled and reopened to the public. Plans called for covering the interior courtyard.

Various roofing schemes were considered, but all proved impractical. The Rotunda walls were not strong enough to support the weight of long span steel roofing systems estimated to weigh as much as 160 tons. Ford wasn't satisfied, and asked his engineers to contact Buckminster Fuller. Bucky considered the requirements and proposed covering the 93-foot circular opening with a lightweight aluminum truss, a geodesic dome, which he said would weigh only eight tons. The dome would incorporate a new light-weight truss system he called *octet truss*, which combined tetrahedral and octahedral structural units made of light extruded 5-ounce aluminum struts into a rigid frame. Ford liked the idea and signed a contract in 1952.

Fuller's associate, T. C. Howard, designed a hemispheric dome that required almost 20,000 yard-long struts. Each one was cut and marked with color codes. Workers matched the ends of struts with the same color codes and preassembled them into 160 large triangular octet truss subassemblies; each subassembly was about 12 feet on a side. In Figure 2.17(a), Fuller holds Howard's model of the dome; the subassemblies are clearly visible as darker triangles inside the model. A truss subassembly is made in

(a)   (b)

Figure 2.17. Fuller demonstrating concept dome and typical truss section
(a) completed Rotunda dome (b).

the following way: three struts formed a triangle, and four triangles made an octahedral unit such as the one pictured behind Fuller in Figure 2.17(a). Ten octahedral units form a single truss subassembly. Each weighs only about four pounds and is easily handled by a single person.[67] The dome was assembled from the top down. A central mast elevated it much like an umbrella as workers added sections to its edges. They worked from a wide bridge platform above the court. In Figure 2.17(b), workers apply transparent polyester fiberglass panels to weatherize the dome.[68] The completed dome required 720 man-hours to erect and weighed approximately 17,000 pounds, or about 2.5 pounds for every square foot covered.[69]

The entire Rotunda dome project was designed, manufactured, preassembled, and installed in just four months, two days ahead of schedule. The Rotunda was open to the public again in June 1953 and quickly became one of the top ten tourist destinations in the country at the time. And Bucky's dome was in plain sight.[70]

The Ford project marked another turning point for Fuller and his associates. It was their first commercial project, and it was a resounding success. There was another first: it was the first large dome project to use a new spherical breakdown based on the icosahedron's mid-edge great circles, shown dashed in Figure 2.18. Unlike the triacon, in which grid members run perpendicular to face edges, the Rotunda's grid would run parallel to the edges. The Rotunda dome divided each face of the spherical icosahedron into 16 subfaces, and each subface was covered by a planar octet truss structure. The subdivision technique was aptly called the *Ford* or *alternate* (alternate to the triacon breakdown). Today the technique is grouped with similar ones and they are collectively known as *Class I* breakdowns. They are detailed in Chapter 8.

---

[67] LIFE (1953, 67–70).
[68] Both images courtesy of the Estate of R. Buckminster Fuller.
[69] For an excellent collection of architectural drawings detailing the entire design, see Fuller (1985, 96–151).
[70] These roofing panels were a constant source of leaks and eventually led to the demise of the dome. Workman heating a tar mixture for patching inadvertently set fire to panels, destroying the entire structure, November 1962.

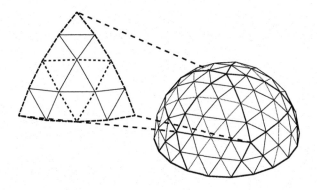

Figure 2.18. Ford or alternate breakdown.

2.7.6   Marines in Raleigh

The late 1940s marked the resurgence of the Soviet Union as a post-World War II super-power and the beginning of a Cold War with the United States that would last more than 40 years. In the years to come, battles would take place on many fronts – propaganda, espionage, weapons development, industrial advances, and even in outer space, where the two superpowers aimed to prove their technical superiority in a "space race." In the 1950s, unrestrained nuclear testing and increased proliferation made an all-out nuclear war between the United States and Russia appear inevitable. Every branch of the US government was engaged to counter the threat. The US Marine Corps and Air Force were no exception—and they were taking notice of geodesic domes.

In 1953, Colonel Henry C. Lane, USMC Head of Aviation Logistics, visited NC State to interview James Fitzgibbon about the new geodesic construction system. He was look-ing for solutions to a wide variety of rapid deployment shelters – personnel shelters, aircraft hangars, and storage facilities, to name a few. Lane noticed a model of the lattice dome (Figure 2.14) in Fitzgibbon's office and asked about it.[71] Fitzgibbon told him how strong and light geodesic domes were, that they could span large distances without column sup-port, and that they were very portable. He went on to explain how they can be built of many materials, such as metal, wood, cardboard, or plastic, and that they can be weatherized in many ways – through cocooning, for example. They had very high strength-to-weight ratio, which meant that even a large dome was light enough that it could be built off-site and airlifted into place. They were easily prefabricated in pieces and could be shipped anywhere.

Lane was completely sold on the concept, its potential for prefabrication, simple construction, and its lightweight. But he needed proof. He needed to test it and see how it met the Corps' many logistical needs. Shortly after, Lane arranged for a Marine Corps helicopter to lift the lattice dome out of Fitzgibbon's backyard and fly it across a nearby Catholic church. This made front-page headlines as far north as New York, where the *New York Times* took note. So convincing was this demonstration that Lane authorized a contract for further development.

Lane's contract with Skybreak was a turning point in geodesic dome development, but the next step wasn't so clear. Lane knew what he needed to know, but Skybreak did not have enough resources to develop all of the prototypes he wanted. Fuller rec-ognized this too, so parallel efforts were started that involved a number of universities. In 1954, the same years Fuller's geodesic dome patent was issued, Fitzgibbon and T. C. Howard (a graduate of NC State who had worked in Fuller's New York office)

---

teamed with Fuller to form Geodesics, Inc. A year later, the same team along with others formed a second company, Synergetics, Inc. The former developed military projects; the latter did commercial work. Fitzgibbon acted as office manager for both companies, and Howard became the principal architect and designer.

2.7.7 Plydome

Fuller experimented constantly with new materials and techniques for building domes. Fly's Eye introduced the concept of stressed skin and molded composite panels; Radomes made extensive use of them. Plydome would explore yet another panel material fast becoming a standard in house construction – plywood.

Normally, plywood and similar planar materials such as cardboard or particleboard are designed and manufactured to be stiff and flat. At most, one can form gentle curves with them but they resist forming compound curves – bends where more than one fold-axis intersect – unless they are formed in pressure molds or stamping machines, such as sheet metal. In the 1970s, Fuller found a unique way to arrange sheet material to form the compound curves of a geodesic dome and in so doing, created the first membrane, thin-shell dome.

Fuller found that it was possible to shingle planar sheets in such a way that slight corner bends and rows of fasteners produced a triangular pattern by virtue of the way sheets overlap. Rows of fasteners between sheets could be aligned and act much like struts in traditional construction. Together, the folds and fasteners formed a triangulated geodesic grid and an enclosing surface at the same time. Fuller called this a *self-strutted geodesic dome*, but more popularly, a *Plydome*. At the time, this was a radical way to build; no other structure with compound curves has been made of planar material. In 1957, Fuller patented the invention.

Figure 2.19(a) and (b), from his patent drawings, clearly shows how shingling overlaps and sheet's fastener patterns follow a regular triangulated grid.[72] Here the reference grid is Class II Triacon 6^v subdivision of an icosahedral spherical triangle face, see Section 8.6.

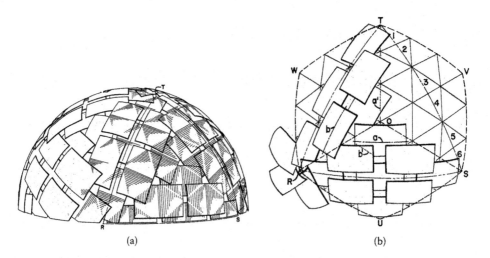

(a)　　　　　　　　　　　　　　(b)

Figure 2.19. Plydome overall (a) and shingling-fastener relationship to a reference geodesic grid (b).

---

[72] Fuller (1957, sheet 1, figs. 1 and 2).

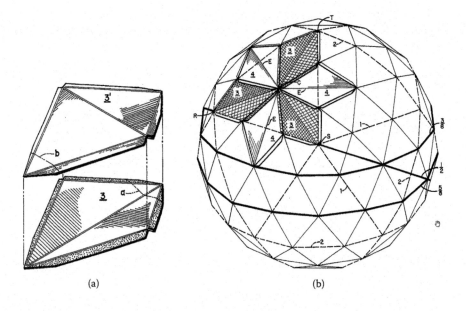

(a)   (b)

Figure 2.20. Laminar geodesic dome sandwich panel (a) and overall arrangement (b).

In addition to an innovative use of planar materials, the Plydome's three-way grid creates a monolithic membrane, which absorbs and distributing loads. Unlike previous geodesic structures where tension and compression forces are transmitted axially through a system of struts (pipes, beams, etc.), the Plydome transmits forces through the stressed skin plywood sheets.

Fuller continued to explore stressed skin and membrane construction and soon developed a double sandwich diamond panel of cardboard or similar lightweight material. The standard unit was a diamond with a fold creating two back-to-back triangles. Two diamond panels with opposing fold-axes completed the sandwich and provided double-bracing. Overlapping diamond edges created a rigid frame as well. The overall layout is shown in Figure 2.20.[73]

Richard Fischbeck continued to explore the shingling concept and developed a system based on vertex-centered conical elements. His patent drawing, Figure 2.21 shows

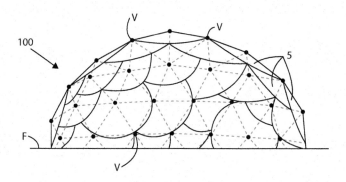

Figure 2.21. *RanDome* shingled conical forms and reference geodesic grid.

---

[73] Fuller (1965, sheets 1 and 2, figs. 1 and 7).

how shingled elements arranged vertex-centered and fastened together along overlapping chords define a geodesic. Although Fischbeck's illustrations shows a Class I 4^v icosahedral grid, many other layouts are possible.[74]

Fischbeck describes his *RanDome* concept this way:

> The RanDome structure is not an assemblage of edges and faces, as are traditional shell structures. It is assembled from a number of overlapping cones called vertex elements. A RanDome has all the structural benefits of a traditional geodesic shell structure but without the complexity of design, manufacture and construction. That all vertex elements are identical leads directly to manufacturing efficiency. When one is fabricating a large number of identical elements, the economies of scale and ease of process are best taken advantage of. The basic building component of a RanDome is a vertex element and this makes all the difference. The cones can be mass produced and stacked efficiently for shipping.[75]

The concept of stressed skin has been used in many variations since; see Kaiser's Dome and Union Tank Car in Sections 2.7.10 and 2.7.11.

2.7.8 University Circuit

In the late 1940s and early 1950s, Fuller traveled extensively and lectured at many universities. His reputation was spreading, and his network of associates rapidly increased. Under pressure from the Marine Corps contract, he reached out to his university contacts and engaged a number of them in research and development. His format was similar at each institution. He would use the same seminar and hands-on approach that had proved successful at Black Mountain College. He taught the basics of Synergetic Geometry and then defined a dome problem the university team was to address. Projects differed in size, materials, joinery, and enclosing method. After starting up a project, he would move onto the next one, picking up ideas from one and dropping them off elsewhere. Some said he often left the impression the borrowed ideas were his own.

The cost of materials in most cases was quite low. Many prototypes were built in corrugated paperboard reinforced and water proofed with polyester resin, or used simple wood struts with wire tension members. Stuart's triacon was the favored spherical subdivision method. By February 1954, Tulane University had completed a one-sixth scale model of a six-plane hangar dome 18 feet in diameter; the full scale would have been 108 feet in diameter. In rapid succession, prototypes were completed at the University of Michigan, Virginia Polytechnic Institute, North Carolina State College, and the Marines Corps Schools in Quantico, Virginia. Domes ranged in size from a small 14-foot-diameter, six-person personal shelter made of cardboard to a large 50-foot-diameter hangar dome with a magnesium frame that was light enough to be moved 50 miles in 60 minutes without disassembly. Figure 2.22 shows its erection sequence. It could hangar any plane or helicopter the Marine Corps owned except conventional transport aircraft.[76] These pilot projects tested a wide variety of sizes, materials, and techniques. Some newspapers were carrying stories that Fuller's cardboard domes would soon replace the army's standard-issue tent.[77] Although some domes encountered problems, especially in the detailing, these demonstrations proved that the geodesic principle was essentially sound.

At the same time the universities were developing prototypes, Geodesics, Inc. was designing and building three 30-foot-diameter domes made of cardboard. The Marines erected them at their Quantico post in Virginia, next to the Potomac River. Geodesics' project manager, Bruno Leon, recalls that the marines disregarded instructions

---

[74] Fischbeck (2008, sheet 4, figure 4).

[75] Paraphrase from Richard Fischbeck, *The RanDome*, randomeshelter.blogspot.com.

[76] Lane (1955, 51–52, 54).

[77] New York Times (July 18, 1954).

Figure 2.22. Marine Corps hangar assembly.

to anchor the domes. The wind blew them into the Potomac River, but the enclosure concept was successfully demonstrated.[78]

By the end of the year, Lane had his proof points and enough backup data to recommend that the Marine Corps replace conventional tent structures with geodesic dome-type shelters and use the 55-foot dome for aircraft shelters wherever there was no available substitute.[79] He also recommended that the Marine Corps continue its research and development in geodesic domes.

The Marine Corps project made beachheads out of Fitzgibbon's backyard experiment. The Marine Corps went on to deploy hundreds of domes worldwide. Geodesics, Inc. would go on to design many more military projects, not all of which were domes.

2.7.9 Radomes

The Marines were not the only ones fighting the Cold War with geodesic domes. In 1954, President Eisenhower signed a bill authorizing the construction of the Distant Early Warning Line (*DEW Line*) system, an "over the pole" line of radar stations operated by the US Air Force to detect airspace intrusions. In one of the most hostile environments on Earth, with temperatures as low as –30°F and winds as high as 150 miles per hour, the government built 58 radar stations across a 3000-mile stretch of Alaska and northern Canada between 1955 and 1957. The Air Force required radar enclosures to be knocked down, flown to sites, and set up in 20-hour windows of predictable good weather.[80] They had to withstand a 210-mile-per-hour wind safety factor and not use any metal, since metal reflects the radar's beam. Fuller's fiberglass-reinforced plastic domes were feasible solutions.

In 1954, William Ahern, Bernard Kirschenbaum, Sherry Proctor, and William Wainwright of Geodesics, Inc., an office affiliated with Fuller in Cambridge, Massachu-

---

[78] Leon (2004, pers. comm.).
[79] Lane (1955, 114).
[80] Marks (1973, 61).

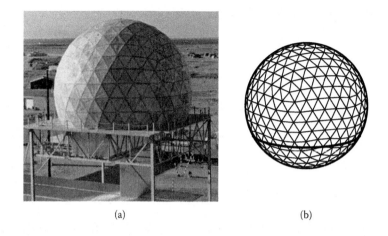

(a)   (b)

Figure 2.23. DEW line radome (a) and reference geometry (b).

setts, designed a 55-foot-diameter 3/4 sphere that stood 40-feet high.[81] It was made of fiberglass-reinforced plastic and was based on prototypes developed earlier at the Massachusetts Institute of Technology. This dome weighed 12,000 pounds and was tested to withstand 150 mile-per-hour winds. The structure consisted of 363 glass-reinforced plastic components (polyester resin) that resembled giant diamonds and circle cake pans. Cake pan flanges were molded with epoxy resin molds and pans were bolted together to form the sphere and all parts were interchangeable. An outer perimeter edge of both diamond and circular pans were recessed to receive rubber gaskets, which sealed the joints between the diamonds and circular pieces.[82] A single radar dome, or *radome* as they were called, could be assembled in only 14 hours. At the time, radomes were the largest plastic structures ever built. Figure 2.23 shows a variation of the Geodesics, Inc. dome in service along the DEW Line in Alaska[83] alongside the icosahedral reference geometry that was used.

DEW Line radomes used the Ford breakdown method again, but instead of dividing the icosahedron's edge into four segments – a four-frequency subdivision – DEW Line domes, such as the one shown in Figure 2.23(a) divided them into six. Other frequencies were also used. The dome support geometry was also different. Unlike the hemispheric Ford Rotunda dome, DEW Line radomes were 3/4 domes, thus the support structure's ground plane intersects the sphere as a lesser circle, not a great circle, as can be seen in the diagrammatic view of a six-frequency grid in Figure 2.23(b). William Wainwright later modified this grid to eliminate the need for the numerous small non-standard panels around the base. His modified grid is called the parallel truncatable grid and it greatly reduced fabrication and erection costs of radomes.

DEW Line domes were strong, weather resistant, and remained in service for many years, demonstrating, once again, the adaptability of geodesics to extreme conditions, innovative construction, and new materials, such as fiberglass.

Radome spherical breakdowns would evolve very differently than methods used for traditional domes. Two types of radomes became commonplace: air-supported radomes and self-supported radomes. Air-supported radomes are simple structures with fewer frame members. Additional support comes from having a pressurized interior, something like a

---

[81] The Geodesics, Inc. Cambridge, Massachusetts, office would later reincorporate and become Geometrics, Inc.

[82] Geodesics Inc. (1955).

[83] FOX-Main Radome 1962, photo courtesy of Brian Jeffrey, www.VE3UU.com.

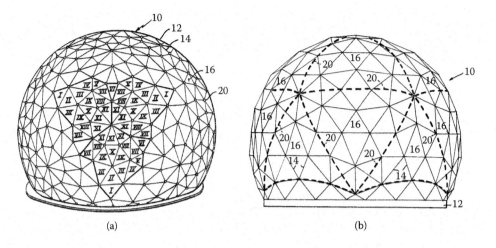

Figure 2.24. Radome showing which panels are the same (a) and typical symmetry (b).

slightly inflated ball. Self-supporting radomes, such as those used on DEW Line, have more structural frame elements because their frame supports the entire weight of the structure.[84] Self-supporting radomes were becoming larger and larger, and radar itself was changing. The higher frequency of subdivision and the inherent regularity of geodesic geometry were interfering with the radar's signal. A new breakdown approach was needed to eliminate long sequences of chords that approximated large great circle arcs. To overcome the interference problem, designers developed "randomized" layouts. So-called random domes are somewhat of a misnomer, as many random domes are actually based on highly regular spherical grids similar to the ones we have already seen. But instead of defining chords between closest neighboring grid points (the usual practice), triangles or diamonds are defined by selecting other combinations of vertices. Some domes also mixed isosceles and equilateral panels in a set pattern. Either way, the near-equilateral pattern we associate with geodesic domes is partially or totally eliminated, yet enough symmetry remains in the overall pattern that sets of identical panels are still used to build the dome. Figure 2.24 shows two approaches to randomization. Despite their appearance, they can be prefabricated from standard, if irregular, components. A 1950s patent by Albert Cohen, Figure 2.24(a), uses 11 different panels for a 150-foot-diameter radome. Another design from the 1960s, Figure 2.24(b), by William Ahern and William Wainwright (who were part of Fuller's early prototype work at the Massachusetts Institute of Technology in the early 1950s) is based on four different panels.[85] In 1977, Ahern and Birgit Mathe would again refine this layout. They discovered that for moderately sized domes, only four panel sizes are required, whether or not the interior surface was the same or different from the exterior surface. Their layout was ingenious and an amazing demonstration of spherical trigonometry.[86]

In the early days of radome development, designs patented by companies or individuals not licensed to use Fuller's basic geodesic dome patent 2,682,235 presented potential intellectual property violations. Their "randomness" somewhat disguised their underlying geometry and staff at Fuller's Geodesics, Inc. office reviewed patent claims to see if the radome actually fell within Fuller's patent or not.[87]

---

[84] Ahern (1968, 1).
[85] Figure (a) Cohen et al. (1961, sheet 2/4, fig. 2); Figure (b) Ahern and Wainwright (1968, sheet 1/4, fig. 1).
[86] Ahern and Mathe (1977, sheet 1/5, fig. 1).
[87] Howard (1958).

Randomized radomes are highly sophisticated geodesic applications. Despite their appearance, many use a basic polyhedral framework and generate their grids by intersecting great circles. In most cases, the result is still a highly symmetrical subdivision of the sphere.

2.7.10   Kaiser's Domes

Donald Richter had met Fuller at the Institute of Design in Chicago and did calculations for the supine dome at Black Mountain as well as the first full-scale dome in Montreal in 1950 (see Section 2.7.3). These projects, and several others like them, led Richter to develop a another gridding technique he called the alternate grid. We will explain how it works in a later chapter. He was now working as a designer at Kaiser Aluminum and Chemical Company, and once again, a chance encounter with a geodesic dome model would alter the course of geodesics.

As the story goes, the famous industrialist Henry Kaiser was passing through his company's offices and noticed a model of a dome in Richter's office.[88] His first impression was that it was one of Kaiser's new products. But when the industrialist asked Richter what it was, Richter explained that it was a geodesic dome. Richter then proceeded to point out the virtues of the dome, i.e., high strength-to-weight ratio, suitability for prefabrication, and lightweight aluminum as an ideal material to build them. Kaiser, just as Colonel Lane had been, was completely sold on the idea. He saw enormous potential in producing aluminum domes, and immediately engaged company lawyers to make Kaiser Aluminum one of the first licensees of Fuller's geodesic patent.[89]

Kaiser was one of the earliest and biggest boosters of the Hawaiian tourist industry. He asked Richter to design a dome that would cover the auditorium for his new Hawaiian Village Hotel in Honolulu. Richter's design was an ingenious 49.5-foot-high, 149-foot-diameter dome that combined lightweight curved aluminum panels with aluminum struts. It was shipped to the resort and reported to have been erected in just 20 hours in 1957. Figure 2.25 shows Richter's basic design.[90]

Richter, like Duncan Stuart, had a keen sense of 3D geometry and the mathematical skills for the required calculations. His geodesic design was unusual in many ways. In most geodesic domes, struts alone distribute the compression and tension loads from the dome's own weight and external sources (wind, snow, rain, etc.) to the dome's

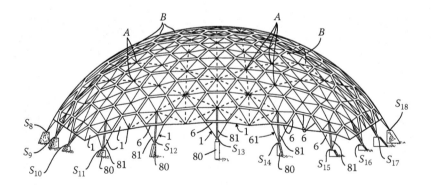

Figure 2.25. Elevation view of the basic Kaiser dome.

[88] Fuller and Zung (2002, 36).

[89] Richter recounts "Working with Buckminster Fuller" in Fuller (1985, 381–394).

[90] Richter (1962, sheet 4/8, fig. 5A).

Figure 2.26. Typical Kaiser tetrahedral membrane unit (a) with details of long and short axis unit connectors (b).

foundations. But Richter also transferred loads through a membrane made of aluminum sheets, the same ones that weatherized the dome. The Kaiser dome and another dome, the Union Tank Car dome, which we will see next, both used the same structural membrane technique. But unlike Union Tank Car's inner layer of steel plates, outer layer of pipes, and in-between layer of rod bracing, Richter's design bent diamond-shaped sheets of aluminum and added a crossbar, also of aluminum, to form a 3D unit that resembled a tetrahedron. A typical unit is shown in Figure 2.26(a).[91]

The sheets of each diamond panel were lightly impressed with creases radiating from the vertices at the ends of the long diamond axes when sheets were cut and curving. These creases did not stretch or weaken the metal. Instead, they stabilized it, preventing the so-called "oil canning" effect that sometimes happens to sheet metal when it flexes or warps after a temperature change or the application of other loads. An aluminum cross-bar spanning the narrow axis of the diamond is riveted to the sheet metal and completes the tetrahedral pattern. When the panels are assembled to make the dome, cast aluminum hubs, as shown in Figure 2.26(b), join the corners of adjacent panels. The fact that panel edges were bent, making them even stiffer, and riveting adjacent panels together weatherized them as well. A construction worker could easily handle a panel without using any equipment. The combination of metal membrane and lattice of struts formed an extremely rigid, weather-tight integral structural system when the dome was completed.

2.7.11   Union Tank Car

By far, the most technically complex and highest risk geodesic project that Fuller and the Synergetics team (lead by T. C. Howard) faced in the 1950s was the Union Tank Car Company dome built in Baton Rouge, Louisiana. The dome was 120 feet at its zenith and 384 feet in diameter, and would cover 110,000 square feet. A 200-foot paint

---

[91] Richter (1962, sheets 2 and 6/8, figs. 3A, 10, and 11).

(a)   (b)

Figure 2.27. Union Tank Car dome under construction with prefabrication fixtures in foreground (a) and detail showing outer pipe truss and rod bracing (b).

tunnel, 20-feet high and 40-feet wide, was also part of the structure. When completed in 1958, it was the world's largest dome and free-span structure.[92] Figure 2.27(a)[93] gives an overall view of the massive structure.

The dome was approximately a one-quarter sphere, constructed of 321, 11-gauge (0.125-inch or 1/8-inch) steel plates with an outer truss of 4-inch-diameter pipe struts braced with 3/4-inch-diameter tension rods. The plates were edge-welded to weatherize the dome's interior. Together, the outer and inner layers acted as a huge double-layer spherical truss system approximately 4-feet deep. Figure 2.27(b)[94] shows how the plates, pipe struts, and tension rods form a double-layer truss structure.

What is particularly remarkable about the dome is that all fabrication and subassembly was done on site. Large sections were built on the ground. Steel plates, pipes, and tension rods were preassembled on fixtures to ensure accurate layup. Some fixtures are visible in the bottom left of Figure 2.27(a). The sections, some weighing as much as 4000 pounds, were then hoisted into place by mobile cranes. Some cranes are visible in the background of (a). The structure was self-scaffolding; that is, throughout its construction, it always formed a structurally stable ring and did not call for any bracing. Temporary ground supports positioned subassemblies at the moment subassemblies were joined to already completed sections. The general construction sequence was from the ground up. Subassemblies were continuously added to the free edge as the dome progressively reached its final 384-foot span. This structure was a triumph of design and construction excellence for Synergetics, Inc. T. C. Howard was the principle architect. It was the first major dome to use the triacon subdivision method to create a double layer of hexagonal-octagonal grids, as seen in (b). The inner layer was made of steel plates, with the outer layer being constructed of compression pipes. Steel tension rods were placed between the outer and inner dome layers. The two layers, combined with the tension rods, formed an octahedral pattern. Recognizing the enormous strength and lightweight this configuration offered, Howard would use variations of this octahedral design on future projects.

More than 20 domes were built in the 1950s, and many more followed in the 1960s. We have only touched on a couple of key ones long enough to see the origins of the

---

[92] The domes at St. Peter's Basilica in Rome, Pantheon of Ancient Rome and the Duomo in Florence are 136, 142, and 130 feet in diameter.

[93] Photo courtesy of RIBA Library Photographs Collection.

[94] Photo courtesy of Tai Ray-Jones.

alternate and triacon subdivision techniques; they would join other techniques in a formal classification system presented in Chapter 8.

The Union Tank Car project was not only the most challenging dome at the time, it was the first dome project to use a computer to calculate the geometry. In the spring of 1957, the hexagon wheel octahedral truss conceptual geometry for the dome had been approved by the client[95] and design detail work at Synergetics, Inc. had begun. At this point, every calculation was performed by hand. Synergetics had developed standard office forms for computing spherical triangles and they had implemented a system of double checks. They also made detailed 1-inch to 1-foot scale paper and wood models of critical jigs and assembly models to verify dimensions. By the summer of that year, Synergetics' staff was under tremendous pressure to accelerate design work and provide the shop drawings needed for start construction. It was at this point that Synergetics turned to a new technology - computers. We discuss this transition shortly.

2.7.12 Spaceship Earth

Viewed by millions annually at the Walt Disney World theme park in Buena Vista, Florida, Epcot Spaceship Earth, is perhaps the most recognized structure in the world.[96] Built in 1981, at 165 feet in diameter and 180 feet to the top, it is the largest complete geodesic sphere in the world.[97] It is supported 15 feet off the ground by three steel legs. The structure appears to float; passing beneath it, one feels as if they are going up into spaceship rather than just walking through a doorway. It's a memorable experience for any visitor.

Designed by Architect Peter Floyd of Geometrics, Inc. Cambridge, Massachusetts and heavily influenced by Buckminster Fuller and prior projects using stressed skin construction, this dome was one of the most innovative designs to date.

Geometrically, the sphere is a subdivided face-zenith spherical icosahedron. Although not immediately apparent, Spaceship Earth is actually a double layered sphere. The inner sphere is a steel frame structure, Class 2 Triacon 8^v triangular grid of 1450 structural steel members (see Sections 8.1 and 8.6). This sphere is climate controlled and weatherproof to protect interior rides and exhibits.

Construction-wise, the sphere it is split in two. The upper portion sits on a structural ring. A grid of trusses inside the ring supports two helical structures of the interior's ride and display systems. The lower portion of the sphere is hung from the bottom of the ring, completing the shape.

EPCOT Spaceship Earth sphere

The outer sphere is approximately 2 feet away from the inner core and supported by aluminum hubs and standoffs attached to the grid connectors of the inner-sphere. Clusters of three aluminum pyramids cover each inner-sphere grid triangle. The visual effect is a doubling of the apparent spherical subdivision from 8^v to 16^v (see Section 8.6.2). The spherical edges of the reference icosahedral geometry are not apparent physical features in the final sphere.

Although in plain sight, the 954 exterior Alucobond pyramidal panels[98] hide three of the most innovative aspects of the sphere's design. A detail is shown in Figure 2.28.[99]

---

[95] The hexagonal wheel octahedral truss design was first proposed in 1956 but rejected for use in Kaiser Aluminum Company domes (Fitzgibbon 1957).

[96] EPCOT is an acronym for Experimental Prototype Community of Tomorrow. Walt Disney originally planned EPCOT to be a model utopian city, but doubts about its acceptance gave way to themes of humanity, cultural development, traditions, food and innovation. The named also changed from EPCOT Center to simply Epcot.

[97] EPCOT image courtesy of Benjamin D. Esham/Wikimedia Commons.

[98] Alucobond panels are anodized aluminum on both sides heat-bonded with a polyethylene core in the center.

[99] Photo courtesy of Ellen S. Popko.

Figure 2.28. Detail view of Alucobond panels.

They are anodized to resist corrosion and reflect the sun; they reduce the solar heat gain of the interior (there is a 2-foot airspace between the outer pyramidal cover and the sealed inner sphere); and drainage.

An inch-gap between panel edges allow water to flow through and onto the inner sphere where it flow into an internal gutter system. Figure 2.29 shows Spaceship Earth during construction (circa 1980). The relationship between the inner geodesic steel frame, Alucobond outer panels on the right with their support standoffs and the rainwater drains on the left of the sphere are clearly visible.[100]

Disney's Epcot Spaceship Earth is recognized worldwide as an architectural and engineering achievement. It was the apex of Fuller's life-time pursuit of geodesic structures. Sadly, he passed away in 1983, shortly after Epcot was completed. Fuller never saw the huge public acceptance and the success Spaceship Earth had combining education with entertainment.

Figure 2.29. Epcot Spaceship Earth partial construction.

---

[100] Photo credit: Wikimedia Commons.

2.8   Covering Every Angle

Most of the domes presented thus far were built before computers were available. To-day's PCs and scientific pocket calculators make short work of geodesic calculations, but calculations were a major challenge in the 1950s, 1960s, and 1970s. Every one of the early geodesics domes was calculated by hand – one triangle at time! Office forms, developed by staff at the Fuller Research Foundation, Skybreak, Geodesics, or Synergetics, ensured that correct procedures were followed when using spherical trigonometry. A completed set of calculation forms for a modest-sized dome might be a stack of paper an inch thick. Forms included classic equations such as the law of cosines or sines or *Napier's rule* and a circle diagram of triangle parts. Forms also indicated where calculated values were filled in, added, subtracted, or multiplied (adding logarithms). Often, the form indicated the number of digits required for a calculation. To add to the tedium, most angles were represented in sexagesimal (base 60) notation; that is, degrees, minutes, and seconds. Seconds were sometimes carried out to six or seven decimal places.

Computers (a job title at the time) would follow the sequence of steps on forms developed by Stuart, Richter, T. C. Howard, Joe Stinborn, Richard Lewontin, and others. A typical form, when completed, would list the triangle's reference number, its central and surface angles, chord and arc factors, and sometimes its area. Some forms indicated how to check the results. This was particularly important because angles from one triangle were often used to start the calculations for an adjacent triangle, thus, any error that crept in would simply spread from triangle to triangle. In some cases, calculations were repeated by another person using different methods when possible. Often, the alternative method checks took more time than the original calculations that tended to use the most efficient method. Stuart was particularly careful in his calculations. He relied on high precision 15-place US Government Printing Office sine tables for trigonometry (see Figure 2.30).[101] He also computed his own supplementary tables to nine decimal places for converting angles in sexagesimal to decimal degrees and vice versa.[102]

By the late 1950s, dome projects such as Union Tank Car, were becoming very large and more numerous. They were capturing the public's imagination and numerous

Figure 2.30. Monro-Matic calculator (a) and 15-place sine and cosine tables (b) and (c).

---

[101] US Department of Commerce (1948).
[102] Stuart (1951).

companies wanted to use the technology. It's said that it took Don Richter more than a year to work out the calculations for Fuller's Black Mountain 31 great circle model. The Union Tank Car project with its double truss dome would involve significantly more calculation than any previous dome project and it had to be done quickly. The success of geodesics not only depended on proving they were safe, but that they could be built to required tolerances. Fuller was vulnerable now. A project failure, especially in these start-up years, could discredit the entire geodesic concept. On the other hand, continued success would lead to more and more patent licensing, and this required design methodologies other companies could follow. Fast and accurate computation was now essential; better tools and procedures were needed.

High-end Post and Keuffel and Esser slide rules and the Chemical Rubber Company (CRC) standard mathematical tables and formulae[103] were used for some geodesic work, but slide rule accuracy, typically two significant digits or a maximum of three, was not adequate. In 1957, the Synergetics office started using Friden and Monro-Matic electromechanical calculators to compute spherical geometry. Some machines weighed more than 30 pounds.[104] Figure 2.30 shows the calculator and an often-used US Government 15-place sine and cosine tables used for calculation.[105] For sure, the calculator and reference tables did speed things up and improved precision. However, the bulk of calculations still fell to human computers. The real breakthrough took place only when computation passed from human to electronic computers.

In 1955, NC State received one of the first IBM mainframe computers, the 650 magnetic drum calculator. A typical installation is shown in Figure 2.31.[106] The operator in (a) is coding instruction with console switches. In the background is a computer card reader and punching unit. In (b), a tabulating machine for listing punched cards onto office print forms is shown in the foreground. Initially, the 650 was used by farm management experts in the agricultural school to find the most profitable combinations of crops and livestock for a particular farm. But designers in Fuller's offices wanted to have it programmed to make geodesic reference tables. The idea was to have a desktop reference for spherical breakdowns, where the values of angles, chords, and coordinates could be looked up when needed. This was similar to the way engineers used the CRC standard mathematical tables.[107]

(a)   (b)

Figure 2.31. IBM 650 magnetic drum calculator (a) and printer/tabulator in left foreground (b).

---

[103] Chemical Rubber Company (1955).

[104] Fitzgibbon (October 1957). The Monro-Matic (Model 8N-213 shown) was fully automatic with ten keyboards, 11+10 counter, and 21-place accumulator. Photograph courtesy of John Wolff's Web Museum.

[105] US Department of Commerce (1949).

[106] Photographs courtesy of IBM Corporate Archives.

[107] Chemical Rubber Company (1953).

The IBM 650 was the first mass-produced digital computer sold to non-governmental customers. It was inexpensive (only half a million dollars), small (it fit in a single room), and user-friendly (it was programmed in decimal rather than binary).[108] Most important, it understood an English-like programming language called FORTRAN-SIT (FORTRAN Internal Translator), a subset of what was later called FORTRAN (FORmula TRANslation). The first geodesic computer program produced tables for Stuart's triacon breakdown.[109] It is based on the subdivision of a single spherical right triangle, the one shown in Figure 2.15. The angles between each grid member vary somewhat within this triangle, but once they were calculated, all other triangles were the same. Chord and arc lengths depended on the radius of the final sphere, but chord and arc factors can be calculated just once for a unit radius sphere. Actual arcs and chord lengths are found by multiplying each factor by the final dome's radius.

Perhaps the most tedious calculations involved grid point coordinates. Every point has unique coordinates, and even modest spherical subdivisions involve scores of angles and hundreds of coordinates. Having the computer calculate them would be a major step forward, improving accuracy and eliminating tedious, time-consuming work.

In the summer of 1957, Richard Lewontin was in charge of writing a FORTRAN program for the IBM 650 to do just that.[110] His program could be considered the first application of computers aiding design (CAD) in architecture. For a given subdivision frequency, his program computed the angles, chord, and arc lengths and coordinates for all parts of the triacon right triangle for a spherical icosahedron and punched the results onto paper cards. Tabulating machines listed the card's numeric values on large paper forms. All angles were listed in decimal and sexagesimal (base 60) along with their sine and cosine values. Under considerable pressure to compute all panel dimensions and prefabrication fixtures for the Union Tank Car dome project, Lewontin's program initially produced tables for just 24- and 36-frequency subdivisions. Thanks to the symmetry of the triacon grid, the combination of these two tables also contained the angles and coordinates for 2-, 4-, 6-, 8-, 12-, and 18-frequency grids as well; they are subgrids within the 24- or 36-frequency grids, see Section 8.3.[111] Finding subgrid angles in a table designed for a larger grid is less convenient because it involves careful look up and additional hand calculations, but computing only a few high-frequency tables kept computing costs down and produced the highest priority tables first. Lewontin's program had another advantage: it computed tables for a unit-radius sphere. One table could be used for many different-sized domes, even for domes where there were multiple layers of grids and trussing, such as the Union Tank Car. Angles do not change, no matter how large or small the dome is, but chord and arc lengths, as well as vertex coordinates, do change, as the radius of the dome changes and must be rescaled to the actual radius of the sphere being built. Lewontin's program was so successful that by the fall of 1957, tables were being produced for a wide range of frequencies for both icosahedral and octahedral subdivisions and the Synergetics office was offering them or custom computer runs to other Fuller companies as proprietary and copyrighted data.[112]

---

[108] Art Miller, professor emeritus of mathematics and computer science at Mount Allison University, made this characterization.

[109] In 1954, John Backus headed the IBM team of researchers at the Watson Scientific Laboratory that invented FORTRAN, or FORmula TRANslation. FORTRAN was the first high level programming language and eliminated the tedium of programming in machine language. The FORTRAN language was commercially released in 1957.

[110] Fitzgibbon (1957).

[111] A subgrid is sometimes called a harmonic and they are mathematically related (common factors) to the overall grid.

[112] Fitzgibbon letter to W. M. Wainwright (Geometrics Inc. 1957).

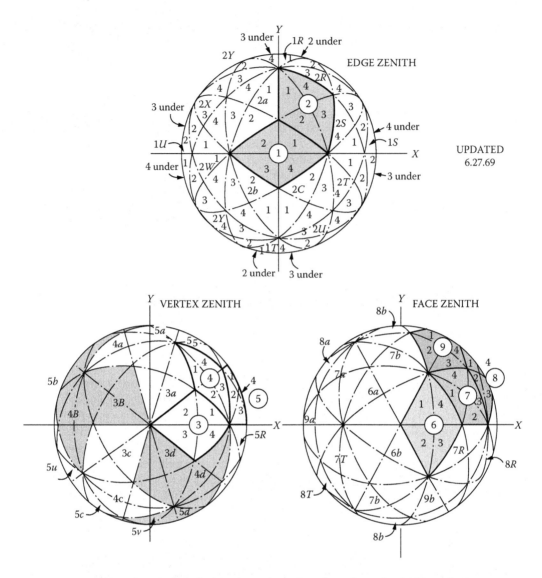

Figure 2.32. Triacon diamond and quadrant locations for spherical icosahedron.

The Cartesian coordinates for grid points presented another challenge. Their values depended on the sphere's radius and orientation in space and every vertex has different coordinates. This problem was solved by standardizing the orientations domes could take, based on which part of the spherical icosahedron reference polyhedron was uppermost.

Figure 2.32 shows edge-, vertex-, and face-zenith orientations for the icosahedron's basic triacon diamonds, as seen from above (positive z-axis). Due to the symmetry of the icosahedron, it turns out that all grid coordinates for a particular orientation can be derived from the coordinates of just the points within the view's shaded diamonds in Figure 2.32. For example, if the dome were edge-zenith, the topmost diagram in the figure, only the grid coordinates within diamonds 1 and 2 would be needed. All the other coordinates could be derived from these coordinates by applying a simple rotation equation or by manipulating the signs of their x-, y-, and/or z-component. Coordinates can be listed for a unit radius sphere and then multiplied by the radius of the actual sphere being built. This was a huge help in reducing calculations, and table books were prepared for just these diamonds. Today, PCs can calculate all coordinates in every diamond for these zenith orientations and any other in seconds.[113]

The most difficult spherical calculations involve truncations. These are cases where grid members are cut by the ground plane or a structural support. Where to cut the sphere was an important decision based on factors, such as required floor area, maximum height, or volume. Manufacturing and construction were also considerations. Most cuts would require a special band of panels or struts at the ground or supporting pylons. It is usually desirable to have triangular panels at the base so that their faces are always flat surfaces.

Minimizing the number of special connectors at the base was also important, since these directly influence construction cost and complexity. Today's computers and programmable scientific calculators can find optimal solutions very quickly, but this was not the case in the 1950s.

The author, in collaboration with Duncan Stuart, developed a comprehensive triacon subdivision program called SPHERES in 1968. This FORTRAN IV program produced triacon work tables and computer graphics for the five Platonic solids with triangular, diamond, or hexagonal grids in single or multiple layers for very dense grids. For a given frequency and Platonic solid, all central and surface angles, chord and arc factors, triangle areas, and dihedral angles were found. All grid coordinates for all grid locations on a full sphere were computed. Figure 2.33(a) shows the spherical LCD right triangle guide printout for the naming conventions used for

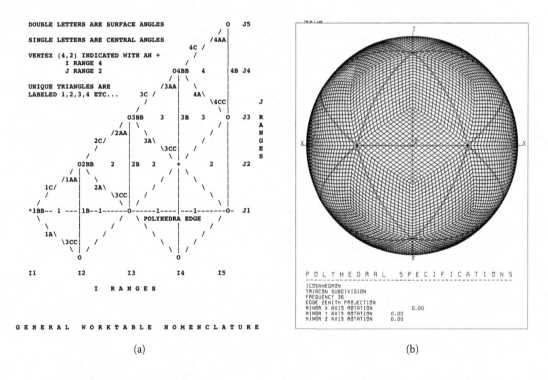

Figure 2.33. SPHERES computer program grid naming convention (a) and typical graphic display (b).

---

[113] (Kenner 1976).

angles, chords, and grid vertices. The program also outputs data for graphics plotters. Figure 2.33(b) shows a diamond pattern for a 36-frequency icosahedral grid. The icosahedron reference edges are shown, though they are not part of the final subdivision grid. Frequency is the number of segments into which the grid divides the polyhedron's edge.[114]

Computer subdivision tables were essential to all large projects in the late 1950s and much of the 1960s. Bernard Taylor of Synergetics, Inc. wrote a program in 1958, the year after Lewontin's program, to calculate the geometry of the American Society for Metals International Headquarters dome built in 1958. Other programs followed in the 1960s. Each new program would compute additional geometric information for unit spheres, such as *dihedral angles* (the angle at the edge of two planar triangles), triangle areas, part counts, and calculation checks. These made design and cost estimating easier. Program output formats had improved as well; tables were easier to read but they did not eliminate the need for hand calculations. Unit radius spherical geometry still had to be converted to the geometry of the dome's actual radius. Subdivision tables remained in use until the early 1970s, only to be replaced by programs running on time-share computers and later on minicomputers. By 1975, computing had become inexpensive enough that any spherical grid could be computed "on the fly" when a project needed it. What's more, the exact value of arc lengths, chords, and vertex coordinates were now computed, thus eliminating the hand calculations that were required when scaling unit radius spherical values to the actual radius of a project sphere. Furthermore, computers now offered graphic output![115] Drum and flatbed plotters, driven by the coordinate output of geodesic programs, could make basic *orthographic projection* engineering drawings or perspective renderings from any viewpoint. Generalized spherical table books were now obsolete!

Today, personal computers make quick work of geodesic calculations and graphics, and even modest-sized domes can be worked out with programmable scientific calculators, though large grids can be somewhat tedious.[116] The IBM 650 was the beginning of computerizing spherical subdivisions and provided a base for optimizing layouts that simply could not be done by manual techniques. The computer largely eliminated tedious manual calculations and allowed spherical work to be highly automated. For companies licensing Fuller's geodesic patent, standard calculation techniques were a tremendous help in quality control and project management efforts.

2.9   Summary

The mid-1940s through the 1950s were launching years for geodesics. Synergetic Geometry evolved from theory to practice with its first government and commercial successes. The Ford Rotunda, Union Tank Car, and Kaiser domes were winning architectural accolades. The Marine Corps demonstrations were successful. The Defense Department was to become a long-term patron of Fuller. The licensing success of the geodesic patent also demonstrated that other companies were willing to invest, and the

---

[114] SPHERES was initially developed for the IBM 1410 computer at the University of Detroit in 1968 and later run on an IBM 360 mainframe computer at Massachusetts Institute of Technology with CalComp drum plotter and Plot-10 graphics library support. SPHERES may have been the first geodesics computer graphics program. Note the use of I- and J-range references for each grid vertex in Figure 2.33(a), a holdover of the FORTRAN programming language where I and J were commonly used as integer DO loop indices (Popko 1968).

[115] Under contract to the National Aeronautics and Space Administration, Joseph Clinton prepared technical report "Advanced Structural Geometry Studies" detailing various spherical subdivision techniques. Each system was programmed in a computer; see Clinton (1971).

[116] Kenner (1976, Appendix 2, *Calculator Routines*, 166).

variety of projects further demonstrated the soundness of the geodesic concept. Projects such as Union Tank Car and Kaiser domes were widely respected by engineers and architects alike. Design practices and spherical methodologies improved steadily and CAD was born from necessity.

Thus far, we have only touched on Fuller's start-up years, keeping our focus on the birth of the geodesic dome and the first spherical subdivision techniques. What, we may ask, have we inherited from these early years?

Surprisingly, the single most important legacy has nothing to do with domes at all: our most important inheritance from Fuller is a way of thinking about design, problem solving, and "doing more with less," to use his expression. Fuller's Synergetic Geometry gives us a design framework rooted in both nature and science. Domes are just one by-product of Synergetics thinking – an important one for sure – but a by-product nonetheless. Thousands of students and professionals in pursuits other than domes have benefited from geodesics and Synergetic Geometry. The gift of Fuller and his associates is a creative process and a design imperative for conservation and service to mankind from which we continue to benefit.

In the context of *Divided Spheres,* our heritage is more specific: we inherit a framework for understanding the sphere, its symmetrical properties, and how to orderly subdivide it with great circles. The *alternate* and *triacon* subdivision methods we use today are direct descendants of Fuller's early work. The usefulness of these techniques can be seen in the diversity of geodesic applications today. In the next chapter, we sample some historic predecessors and modern applications.

Additional Resources

Buckminster Fuller Institute (BFI) is dedicated to accelerating the development and deployment of solutions that radically advance human wellbeing and the health of our planet's ecosystems by facilitating convergence across the disciplines of art, science, design and technology, www.bfi.org/.

Chu, Hsiao-yun and Roberto G. Trujillo. *New Views on R. Buckminster Fuller*. Stanford, CA: Stanford University Press, 2009.

Edmondson, Amy C. *A Fuller Explanation: The Synergetic Geometry of R. Buckminster Fuller*. Boston: Birkhäuser, 1987. Design Science Collection. Reprinted *Back-In-Action* book series (Pueblo, CO: Emergent World Press, 2007).

Fernández-Serrano, Martino Peña, and José Calvo López. 2019. "Projecting Stars, Triangles and Concrete – The Early History of Geodesic Domes, from Walter Bauersfeld to Richard Buckminster Fuller." *Architectura, Journal of the History of Architecture* (47): 92–114.

Fuller, R. Buckminster. *3 Way Geodesic Grid B Calculations on Present*. Department of Special Collections and University Archives, Stanford University Libraries, 1948. From Noah's Ark II notebook.

Fuller, R. Buckminster and E. J. Applewhite. *Synergetics: Explorations in the Geometry of Thinking*. New York: Macmillan Publishing Co., 1979.

Fuller, R. Buckminster and James Ward. *The Artifacts of R. Buckminster Fuller: A Comprehensive Collection of His Designs and Drawings*, Vol. 3: The Geodesic Revolution, Part 1, 1947–1959, New York: Garland, 1985.

Kenner, Hugh. *Bucky: A Guided Tour of Buckminster Fuller*. New York: Morrow, 1973.

López-Pérez, Daniel and R. Buckminster Fuller. *R. Buckminster Fuller: Pattern Thinking*. Zürich: Lars Müller Publishers GmbH, 2019.

Marks, Robert W. and R. Buckminster Fuller. *The Dymaxion World of Buckminster Fuller*. Garden City, NY: Anchor Books, 1973.

McAtee, Cammie. *Montreal's Geodesic Dreams: Jeffrey Lindsay and the Fuller Research Foundation Canadian Division*. Halifax, Nova Scotia, Canada: Dalhousie Architectural Press, 2017. Canadian Modern, Michelangelo Sabatino series editor.

Motro, René. *Tensegrity: Structural Systems for the Future*. Butterworth-Heinemann, 2006.

Pugh, Anthony. *An Introduction to Tensegrity*. University of California Press, 1976.

The Synergetics Collaborative (SNEC), a non-profit organization dedicated to bringing together a diverse group of people with an interest in Buckminster Fuller's to educate and support research and understanding of the many facets of Synergetics, its methods, and principles, www.synergeticscollaborative.org/.

Wigley, Mark. *Buckminster Fuller Inc. Architecture in the Age of RADIO*. Zürich: Lars Müller Publishers GmbH, 2015.

Zung, Thomas Tse Kwai. *Buckminster Fuller: Anthology for the New Millennium*. New York: St. Martin's Press, 2001.

③ Putting Spheres to Work

Buckminster Fuller's work in geodesic domes and spherical subdivisions was influenced by the work of others, just as his work is influencing the work of future generations. Today's geodesic designers benefit from the work of mathematicians and scientists in many diverse fields. We owe many analytic techniques to work carried out in descriptive geometry, biology, astronomy, climate, physics, and recreational equipment. This chapter touches on just a few of the many ways spherical design is applied and suggests how the use of Computer Aided Design (CAD) is further accelerating it.

3.1 The Tammes Problem

Pollen seems an unlikely place to look for examples of spherical subdivision. Nevertheless, we owe many useful concepts to early research in pollen grains. In 1930, the Dutch biologist P. M. L. Tammes published a landmark study *On the origin of number and arrangement of the places of exit on the surface of pollen-grains*.[1] This work bridges biology and mathematics, in particular, an applied area of discrete geometry called *circle packing*.

Tammes was interested in the number and locations of exit points in pollen grains and he compared many types. He found 4, 6, 8, and 12 common; 5 never occurs; and 7, 9, and 10

---

[1] (Tammes 1930, 1–84).

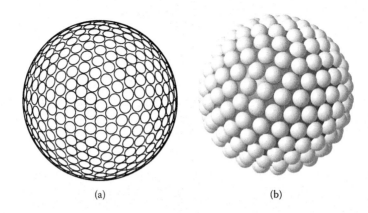

(a)   (b)

Figure 3.1. Circle and sphere packing.

occur less frequently. Very rarely does 11 occur.[2] He discovered that the distances between exit points were consistent within a type and their distribution on the grain was highly regular. These recurring numbers are no accident; they are optimal solutions to point distributions on the sphere. The similarity of pollen exits to circle packing lies in finding the optimal arrangement of a given set of points on the sphere. To measure how evenly distributed the points are, they are all surrounded by congruent circles large enough for two closest ones to touch but not overlap. Other circles will not be tangent, and the challenge is to make them as even as possible and minimize the total uncovered area outside the circles. Today, this optimization problem is called the *Tammes Problem* in his honor.

The best distribution of points occurs when the minimum distance between all points and their neighbors are maximized, or made as large as possible. A sphere surrounding each point offers a similar indicator—the closest spheres *"kiss,"* an expression borrowed from billiards describing two balls that just touch. Examples of sphere and circle packing are shown in the random point distributions in Figure 3.1. Although their *packing* is quite good, you can see enough open space around some spheres and circles and realize the point distributions could be improved further.

Circle packing has a large body of literature and includes work by Conway, Goldberg, Tóth, Tarnai, Gáspár, Fowler, Robinson, and many other talented mathematicians (see the bibliography for references). Two-dimensional (2D) and three-dimensional (3D) circle packing finds practical applications in planning the locations of cellular phone towers, geo-synchronous satellite orbits, sports equipment, pharmaceuticals, and the development of new materials.

Although we are not involved in pollen research in this book, we can make use of circle and sphere packing to evaluate our own subdivisions. For example, the circles on the surface of the sphere can be seen as small slices taken from the sphere that resemble contact lenses. They are called *spherical caps*. A *density factor*—the ratio of the area of all caps to the total area of the sphere—measures what fraction of the sphere's surface is covered by the caps.[3] If two subdivision techniques distribute the same number of points on the sphere, the subdivision with a higher density factor (larger radius caps) has the better distribution of the two. In later chapters, we show basic methods for subdividing spheres and use sphere and circle packing as density factors to compare our results. We will also see how density factors are applied to sports equipment and how golf ball designers optimize circular dimple patterns to improve the ball's aerodynamics.

---

[2] (Aste and Weaire 2000, 108).
[3] All caps on the sphere have the same and maximum radius without causing overlaps. See (Tarnai 2002, 289).

3.2   Spherical Viruses

In the mid-1950s, Francis Crick and James Watson (of genetic code fame)[4] developed a theory about spherical viruses that hypothesized they had cubic symmetry, symmetry similar to the axes of a 3D cube. They also believed that the only way for these viruses to build their outer protein shell (the capsid) was to use a common molecular building block over and over again. These protein subunits would necessarily have to be positioned and packed so each had a nearly identical environment. Whatever their geometry, it was highly regular and symmetrical.

Building on their theory, X-ray crystallographer Donald Caspar, pictured on the right, analyzed spherical viruses and found in addition to cubic symmetry, that they also exhibited five-fold rotational symmetry.[5] This meant the virus would likely have the symmetry of an icosahedron, which has both *cubic symmetry* as well as *five-fold symmetry*. The icosahedron has 12 vertices, 30 edges, and 20 faces. If you look straight at any one of its vertices toward the center of the solid, you can rotate the icosahedron around your line of sight in increments of one-fifth of a circle and each view looks the same. This is an example of five-fold symmetry (around the vertex). This finding was also consistent with Crick and Watson's theory. Their theory further suggested that viruses with cubic symmetry might be built from aggregations of smaller asymmetrical pieces in multiples of 12 and a maximum of 60 identical subunits. Figure 3.2 shows conceptual models of two spherical viruses with cubic symmetry. A model of the Human Immunodeficiency Virus (also called HIV or AIDS virus) shown in (a) shows the external appearance of the virus. The virus is coated (enveloped) in host cell membrane, which is drawn as a bluish green semitransparent layer in which various membrane proteins can be seen floating. The viral knobs (golden projections at the viral surface) insert into the matrix. These knobs allow the virus to attach to cells. The AIDS model (b) also has icosahedra symmetry (see Section 6.3.2 for a description of this type of symmetry).[6]

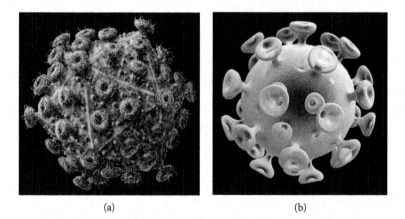

(a)   (b)

Figure 3.2. Spherical viruses with cubic symmetry.

---

[4] Francis H. C. Crick, James D. Watson, and Maurice Wilkins were jointly awarded the 1962 Nobel Prize for Physiology of Medicine *"for their discoveries concerning the molecular structure of nucleic acids and its significance for information transfer in living material"* (www.nobelprize.org).
[5] Image courtesy of Donald Caspar.
[6] Image (a) © Russell Kightley 2009; (b) image © TurboSquid 2009.

In the late 1950s, Caspar began collaborating with Aaron Klug,[7] who was involved in polio research. Robert Marks, an associate of Buckminster Fuller, noticed this work and introduced Klug to Fuller in 1959. Later, through Mark's writings,[8] Klug recognized similarities between his research data and the geometry of geodesic domes. Many domes use spherical icosahedra as their design base, and structural members are often arrayed in patterns of hexagons and pentagons.[9]

Together Caspar and Klug formulated a new concept about spherical viruses they called *quasi-equivalence*. They visualized common viral subunits as something similar to a dome made of quasi-equivalent (nearly equal) triangles. They also envisioned common viral sub-units clustered in an arrangement of pentamers and hexamers (pentagonal and hexagonal areas on the sphere defined by five or six triangles clustered around a central point). Subsequent research reaffirms how common icosahedral layouts and an affinity for great circle patterns.[10]

Most subdivisions of spheres try to create grids of nearly equal (quasi-equivalent) triangles. Groupings of triangles define larger shapes such as hexagons and pentagons, and the total number of triangles in a subdivided sphere (with polyhedral base) follow a specific rule dictated by a *triangulation number*. By using different parameters in the triangulation number formula, a wide variety of *triangular grids* and grid orientations can be made over the icosahedron (or other 3D solids). Some parameters produce grids that are parallel to the edges of the icosahedron's equilateral faces, while other parameters skew the grid at an angle or make them perpendicular to the edges. Today, we classify a subdivision type by the orientation of the grid to the faces of the 3D solid being subdivided. They are called Class I, II, and III.[11] Chapter 8 describes subdivision techniques that produce grids in each of these classes.

Figure 3.3 shows different geodesic grids with triangular, hexagonal, and pentagonal patterns at different orientations to an underlying icosahedron (shown with dashed lines). In Figure 3.3(a), the highlighted edges of the hexagons and pentagons cross perpendicular to the icosahedron's edges (shown dotted). This grid orientation is

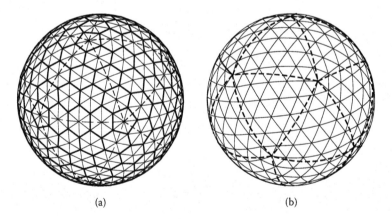

(a)   (b)

Figure 3.3. Icosahedral grid layouts.

[7] Aaron Klug won the 1982 Nobel Prize in Chemistry for his development of crystallographic electron microscopy and his structural definitions of nucleic acid-protein complexes (www.nobelprize.org).
[8] (Marks 1973).
[9] See (Šiber 2020) for geometric analysis of underlying symmetry.
[10] See (Wilson, 2016).
[11] Caspar and Klug named their classes P = 1, P = 2, and P = 3 and they roughly correspond to the modern names Class I, II, and III.

called Class II. The grid in (b) is skewed and not parallel or perpendicular to the edges of the icosahedron and is called a Class III grid. Caspar and Klug experimented with combinations such as these when they analyzed their X-ray data.

The Caspar and Klug quasi-equivalence theory was presented in a landmark conference paper in 1962.[12] Its major idea is to describe the ways large numbers of identical protein subunits can build closed containers, such as virus capsids by a self-assembly process. Their fundamental concepts about the geometry of certain viruses impacted virology for years to come. They discovered that icosahedral symmetry is a preferred structure of certain isometric virus particles and the triangulation number shows only certain numbers of hexamers and pentamers are possible. Their work developed the notion of "*skewed*" spherical grids, which today are referred to as Class III subdivisions. These grids were totally different than those developed by Fuller. For sure, Fuller's work inspired Caspar and Klug, but once their paper was published, it was incorporated into the literature in a way that did not link it to Buckminster Fuller.[13] Although our interests are in underlying geometric principles behind spherical subdivision, we can't help but marvel at how Caspar and Klug built up their theory of quasi-equivalence and made the connection between very big structures, such as geodesic domes, and the smallest of structures in viruses.

3.3   Celestial Catalogs

We still treat the sky as if it were one very large sphere surrounding the Earth. This viewpoint, although not true, is practical and serves astronomers today just as it did the ancients. Astronomers with ground-based and space-based telescopes are continually surveying the sky and capturing unimaginable quantities of data recording the positions, magnitudes (brightness), and other characteristics of celestial objects, not to mention all the photographic records that stream back constantly. More than two billion stars have been mapped with amazing 3D precision. And this does not count the constant stream of planetary images from space probes, orbiting satellites and rovers. Storing this huge quantity of data and making it readily accessible is an enormous archival challenge.

Researchers Peter Kunszt, Alexander Szalay, and Ani Thakar at Johns Hopkins University devised the Hierarchical Triangular Mesh (HTM) system[14] as one way to systematically subdivide the celestial sphere into small triangular reference areas, with each area acting as an efficient index to store and retrieve the astronomical data recorded within it.[15] It is used extensively for the Sloan Digital Sky Survey, which has created the most detailed 3D maps of the Universe ever made. It records details such as the positions and absolute brightness of hundreds of millions of celestial objects. It also measures the distances to more than a million galaxies and quasars.[16] Figure 3.4 shows how the sky can be divided into hierarchical sets of triangles, each set making a progressively finer and finer grid. All five spheres in Figure 3.4 are oriented the same way to make it easier to compare them. At the furthest left of the figure is the simplest subdivision, in which a single highlighted face is divided into four triangles by creating a triangle in the center whose vertices are formed by the midpoints of the surrounding

---

[12] (Caspar and Klug 1962, 1–24).

[13] (Urner 1991).

[14] (Szalay et al. 2005).

[15] Subdividing triangles by constructing a new one within it was a method used by Fuller in the 1950s. The technique was called the alternate method because subtriangles were created by adding chords between alternate sides of a triangle.

[16] Sloan Digital Sky Survey, Johns Hopkins University.

Figure 3.4. Hierarchical triangular mesh examples.

triangle's edge. Each of the four resulting triangles can be subdivided again using the same process. This recursive method, a subdivision of a subdivision of a subdivision, can make meshes as fine as needed, as demonstrated in the other spheres in the figure. If one area of the sky has more catalog entries, localized and finer meshes, just for that area, are also possible.

The spherical coordinate system, used by astronomers, records observational data (star information) with two coordinates: right ascension and declination. The system easily converts to latitude/longitude systems used by geographers or to an individual triangle in the grid called a *trixel*. A number identifies the trixel's exact location within a hierarchy and the portion of the sky it covers. An individual triangle is part of a cluster of four, and those four are one of four larger ones. The mesh can be continuously subdivided to meet the needs of growing catalogs. Forerunners of the HTM system include the Quaternary Subdivision System developed for cataloging and analyzing terrestrial data.[17] It is based on the same subdivision principle as HTM, but uses a different reference scheme for searching a particular triangle. In Chapter 8, we explain how to do this type of subdivision. We call the subdivision method *Mid-arcs*.

3.4   Sudbury Neutrino Observatory

The Sudbury Neutrino Observatory (SNO) is unlike any other observatory on Earth. Instead of sitting atop a dry mountaintop in Chile or Hawaii making observations through crystal clear skies, the observatory is located 1.3 miles (2070 meters) underground in an active Canadian nickel mine.[18] The two kilometers of rock above this ultraclean facility help block false signals from cosmic rays and other radiation from space. Only a minimum amount of cosmic radiation can reach SNO and interfere with its observations.

Particle physics is a basic science in which researchers look for the smallest or most fundamental components of matter and the forces that govern their interactions.

---

[17] (Dutton 1996, 8B.15–28).
[18] INCO's Creighton Nickel Mine in Sudbury, Ontario, Canada.

SNO, a collaborative research project between government agencies and universities, is designed to measure one of the smallest particles known: neutrinos. Neutrinos are subatomic particles so small and fast; they simply pass through the Earth unimpeded by matter. Our sun and other stars emit neutrinos as they generate and release energy. By detecting neutrinos from the sun, researchers learn about the physics of how the sun generates energy through nuclear fusion. SNO researchers have already discovered that the sun's neutrinos change from one type of neutrino to another on their journey from the sun to the Earth. This discovery solved a 30-year-old scientific problem and confirmed earlier theories about neutrinos' mass and energy.

SNO's outer structure, referred to as the Photomultiplier Tube Support Structure (PSUP), is a 60-feet-diameter (18 meter) geodesic sphere made of hexagonal and pentagonal stainless-steel modular sections. The surface of the sphere supports arrays of highly sensitive photo multiplier tubes (PMT), 9600 in all. Inside the PSUP is a second sphere, 39.4 feet in diameter (12 meters). This sphere is the world's largest acrylic vessel with 2-inch thick walls (5 cm); it holds 1000 tons of heavy water.[19] Figure 3.5(a) shows geodesic sphere structure with PMT array, photo taken from inside acrylic vessel. The PSUP's photo detectors are mounted between its structural sections. Every detector aims towards the center of the sphere, where they can record light traces from neutrinos passing through SNO at any angle. The completed PSUP is shown in Figure 3.5(b).[20] Highly purified water fills the space between the PSUP and the acrylic sphere as well as the cavern where the instrument is suspended.

When a neutrino passes through the observatory, it interacts with the heavy water contained in the central spherical vessel, causing a slight flash of cone-shaped light. The photo detectors on the PSUP record the time, location, cone shape, and number of photons emitted by the flash. This data tells researchers what the neutrino is and how it behaves. Figure 3.5(c) shows a computer graphic simulation of SNO detecting

(a)   (b)   (c)

Figure 3.5. SNO geodesic sphere.

---

[19] Heavy water is water in which deuterium, an isotope of hydrogen, replaces hydrogen atoms and H_2O becomes D_2O. The deuterium in heavy water enables the detection of all three types of neutrinos.
[20] PSUP photographs (a) and (b) courtesy of Lawrence Berkeley National Laboratories, Roy Kaltschmidt photographer.

neutrinos.[21] Some of the detectors are recording the passage of neutrinos (the small red cones in the image).

SNO's original mission is now complete. However, the facility continues to gather data, and new experiments are underway to advance solar physics. SNO is also capable of detecting a supernova, an exploding star, within our galaxy if one occurs while the detector is online. Supernovas release neutrinos earlier than photons (visible light); thus, it is possible to have some advance warning that a supernova will be visible with an optical or radio telescope.

3.5   Cartography

The essence of cartography is the transformation of information from a 3D spherical surface, the world, to a flat 2D surface, a map. This transformation is achieved through a spherical reference system and is often shown as a grid—tools and techniques also used by climate modelers and astronomers. In previous sections we have shown several cartographic examples, including Buckminster Fuller's Dymaxion maps, Cahill and Waterman butterfly maps, the HTM for celestial catalogs,[22] and Randall/Heikes' icosahedral climate model grids. Collectively, these systems (as well as many more not included in this book) are called Discrete Global Grids (DGGs). Figure 3.6 is a typical example.[23] Note that hexagons alone cannot form a complete, tessellation, a few cells will be pentagons.

Cartographers Kevin Sahr, Denis White, and A. Jon Kimerling describe DGGs as a set of spherical regions dividing up the Earth's (or a general sphere's) surface. Within each of these regions, there is a single point that can be used to assign data associated with the region. For spherical grids that define only regions, the *centroids* of the regions can be used as the single associated point for the region. Conversely, if a spherical system defines only points, the Voronoi regions of those points become the

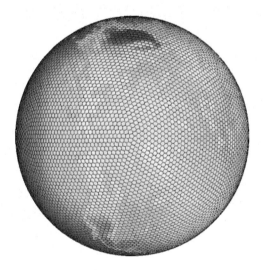

Figure 3.6. Geodesic global discrete global grid.

---

[21] Image © courtesy of MIT Laboratory for Nuclear Science.
[22] HTM is a variation of the Hierarchical Coordinate System; see (Dutton 1985).
[23] Geographic Information Systems DGG image created by Miguel Bartelsman.

regions.[24] Sahr, White, and Kimerling point out the following common characteristics of all DGGs:

- a base regular polyhedron (i.e., Fuller used the cuboctahedron for the Dymaxion map);

- a fixed orientation of the base regular polyhedron relative to the Earth (i.e., Cahill's vertex-zenith octahedron orientation);

- a hierarchical spatial partitioning method defined symmetrically on a face (or set of faces) of the base regular polyhedron (i.e., HTM's nested and reference system of triangular areas);

- a method for transforming that planar partition to the corresponding spherical/ellipsoidal surface (i.e., Randall/Heikes Voronoi region grid); and

- a method for assigning points to grid cells (i.e., centroids of regions).

Climate modelers would likely add one more DGG design characteristic to the list: DDGs provide a way to transform volume data to a spherical surface area or a summary statistic to a single point.

In Chapter 8 we will describe six schemas for subdividing spheres. Each result in a triangular mesh that can be grouped into squares, pentagons, hexagons, or diamond shapes, depending on the foundation polyhedron used, the intensity of the grid, and the subdivision technique used. Chapter 10 describes self-organizing hexagonal grids that may be used in cartography as well. Except for the two butterfly maps, shown in Figure 2.4, the cartographic examples above use one of these schemas.

3.6   Climate Models and Weather Prediction

It's impossible to ignore daily and seasonal weather, but more subtle and long-term climate changes are also becoming apparent. The National Aeronautics and Space Administration (NASA) says that over the past 40 years, the global temperature has increased 0.18°C per decade since 1981. Mountain glaciers all over the world have receded. The average thickness of the arctic sea ice has decreased 40%, and sea levels are rising around the world. Moreover, NASA reports that the growing season now starts seven days earlier and ends two to four days later than it did 40 years ago, and that ten of the warmest years on record have occurred since 2005.[25] Any one of these developments might not be a cause for alarm, but when seen together, their impact is considerable. The changes have already affected global agriculture, water supply, coastal habitation, energy generation, and energy demand. The effects are not the same for everyone. Some countries and societies are impacted more than others; some will adapt to change better than others.

Essentially, *climate modeling* is the use of computers to simulate the cause-and-effect relationships of the physical, chemical, and biological components of climate. Atmospheric general circulation models are simulation programs with dynamic cores that integrate equations for variables such as surface pressure, water vapor, solar gain,

---

[24] (Sahr, White and Kimerling 2003, 121–123).
[25] National Aeronautics and Space Administration, World Book at NASA—Global Warming, 2018, www.nasa.gov/worldbook/ global_warming_worldbook.html.

temperature, and fluid motion in spatial mediums, such as the atmosphere and oceans, land surface, and areas covered by ice. Although the techniques used can be applied to local weather forecasting, the emphasis is on large-scale global patterns where huge volumes of air and areas of landmass and oceans interact. Enormous amounts of spatial data, gathered over time or forecasted for future dates, make climate models computer-intense. They demand the highest processing speeds and memory possible and are typically run on the largest super computers.

Climate models need a systematic way to reference all areas of the planet, from the depths of oceans to the atmosphere above the oceans and land masses. The design and choice of a reference grid is one of the earliest decisions made by programmers of climate models because all data references and time-space calculations depend on it. The grid facilitates calculations and makes it easier to deal with 3D dynamics, data storage/retrieval, and computer memory. In essence, the grid is the way spatial data is aggregated and disaggregated and is the link between data, the climate model, and the computer they run on.

Initially, climate models used latitude/longitude grids, because they were familiar to geographers, and the coordinate system was primarily used to classify data and refer to locations on Earth. But as climate models became more sophisticated and simulations for small areas became possible, latitude/longitude grids presented drawbacks. The convergence of meridians at the North and South Poles creates more cells near the poles than at the equator. In addition, the poles had to be treated differently because the latitude terms 90° north and 90° south lose significance and any time-based simulations near the poles required smaller steps than at the equator to get stable results. Other grids were explored, since the polar regions greatly influence climate.

One solution to the "*pole problem*" has been to use grids resembling the same patterns as geodesic domes. Geodesic grids were first considered in the 1960s.[26] There are several ways to generate these grids. One way is to subdivide the icosahedron's equilateral faces into a series of smaller equilateral triangles and then project the resulting grid onto the surface of a circumscribing sphere. The projection process uniformly distributes the points and changes the shape of the triangles slightly, but the result is a grid of quasi-isotropic nearly equal triangles. The grid is also called a *tessellation*. Figure 3.7 shows the subdivided planar icosahedron (a) and its spherical projection (b). The scale and orientation of both figures are the same for comparison purposes.

Climate modelers David Randall and Ross Heikes at the Center for Atmospheric Sciences at Colorado State University use geodesic grids like the ones we have been

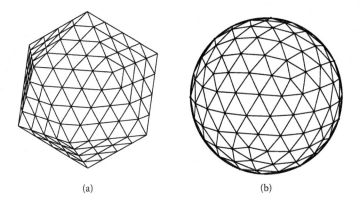

(a)   (b)

Figure 3.7. Icosahedral tessellation.

[26](Sadourny, Arakawa and Mintz 1968).

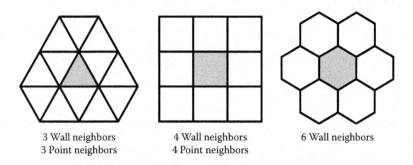

3 Wall neighbors   4 Wall neighbors   6 Wall neighbors
3 Point neighbors   4 Point neighbors

Figure 3.8. Tessellation wall-point neighbors.

discussing, but with an important enhancement. They use the triangular grid to derive a hexagonal grid, and it is the hexagonal grid on which they base their climate model. Why this extra step? In two dimensions, only three regular (all sides equal) tessellations—triangles, squares, and hexagons—cover all areas without gaps or overlaps.

Figure 3.8 shows all three grid systems surrounding a central cell shown shaded. For triangular and square grids, you'll notice that the cells neighboring the shaded one are either across one of its edges or across one of its points. Randall and Heikes call these "*wall-point neighbors*."[27] But notice that for hexagonal grids, all cells surrounding the shaded one are only across edges; there are no point neighbors. By eliminating point neighbors, you greatly simplify calculations when computing the differences between cells. All cells can be treated in the same way. We will see this grid system again when we look at Uber's Hexagonal Hierarchical Geospatial Indexing System.

Cells in a spherical tessellation have wall-point relationships, and it's possible to convert a triangular tessellation with wall-point neighbors into a hexagonal one with only wall neighbors simply by defining each point's Voronoi cell.[28] *Voronoi cells* are polygons that define an area (of points) that is closer to a designated point than to any other. If the original points are evenly spaced, or nearly so, the Voronoi cells will resemble a grid of hexagons. Randall and Heikes illustrate this in Figure 3.9(a). The Voronoi cells in green define the points (areas) closest to each of the points in the following triangular grid. The spherical versions of both grids are shown in (b).[29]

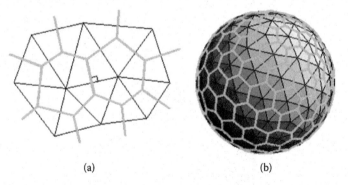

(a)   (b)

Figure 3.9. Hexagonal Voronoi cells from triangular tessellations.

---

[27] Illustration based on (Randall 2002, 35, fig. 5).
[28] Voronoi diagrams are named after Russian mathematician, Georgy Feodosevich Voronoy (1868–1908). However, they were recognized much earlier by René Descartes (1596–1650) and used by Lejeune Dirichlet (1805–1859). Voronoi tessellations are also called Dirichlet tessellations.
[29] Climate model color illustrations in this section courtesy of Ross Heikes.

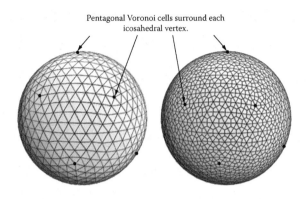

Pentagonal Voronoi cells surround each
icosahedral vertex.

Figure 3.10. Hexagonal Voronoi cells of a triangular tessellation.

When a Voronoi tessellation is made of a fully subdivided icosahedron, such as the one shown in Figure 3.10, the result is mostly hexagonal cells, but pentagonal cells also result. One pentagonal cell surrounds each of the icosahedron's 12 vertices. Some are pointed out in the figure. Each hexagonal and pentagonal cell is quasi-isotropic because their areas are nearly equal. Both spheres in Figure 3.10 are oriented and scaled the same for comparison.

When it comes to climate modeling, hexagonal geodesic grids offer many benefits over other grid systems. They are largely isotropic and do not suffer from oversampling near the poles like longitude/latitude grids. The poles are no different than any other area in the hexagonal grid. All cells in the grid are equal wall-neighbors. There are no point neighbors, as in other grid systems. Hexagonal geodesic grids can also accommodate different resolutions where the atmosphere might be represented by a coarse hexagonal grid, and land-surface areas represented by a finer grid as seen at the left. Randall says coupling the atmosphere and ocean models with multiple hexagonal grids simplifies modeling because the atmospheric state is *"interpolated down"* from large hexagonal cells to smaller ones below and fluxes are *"averaged up"* from small cells to larger ones as seen in the illustration to the left. Hexagonal geodesic grids can also be warped or morphed to meet local needs. Although hexagonal geodesic grids are more complicated to implement than rectangular longitude/latitude grids, their benefits far outweigh the additional work needed to set them up and use them in climate models. Figure 3.11 shows typical climate modeling output based on a geodesic grid.

Figure 3.11. Geodesic grid climate model.

Climate modeling is an intense field of research today. Most efforts involve cooperative agreements with government agencies, such as the US Department of Energy, and designs that take advantage of the latest super computers at laboratories across the country. Randall estimates that simulating a century's worth of climate with a full-coupled ocean-atmosphere model will require several weeks of round-the-clock supercomputer time.

3.7   H3 Uber's Hexagonal Hierarchical Geospatial Indexing System

Uber Technology Inc. offers another example of hexagonal geodesic grids, one closer to home. Uber's worldwide services include vehicles for hire, food and package delivery, couriers, and freight transportation. And at the neighborhood level, their partnership with Lime, includes electric bicycle and motorized scooter rentals. To manage such geographically dispersed assets and customer activities, Uber has developed **H3**, a grid system to visualize urban areas and optimize pricing and dispatch of ride-share cars and other supply-demand services.[30] It's a powerful example of how hierarchical hexagonal grids work hand – and – hand with geographic information systems.

Why did Uber use hexagons? To tile a plane surface, equilateral triangles, and squares can cover a surface without gaps or overlaps just as well. Hierarchical systems could use either of these geometries. However, as we pointed out in Figure 3.8 in the previous section, only the centers of hexagons are equally distant to the centers of the cells that surround them. Squares have two different distances; triangles have three. Hexagons, therefore, greatly simplify the mathematics of network and distance calculations.

Figure 3.12(a) shows **H3**'s global hexagonal areas. These cells are further disaggregated into finer and finer hexagons down to the regional, city, area (b) and even local neighborhood level (c) maintaining the link between location and real-time service information. **H3** enables Uber to analyze geographic markets, set dynamic pricing, manage local dispatching, and make decisions at the regional, city or neighborhood level.[31]

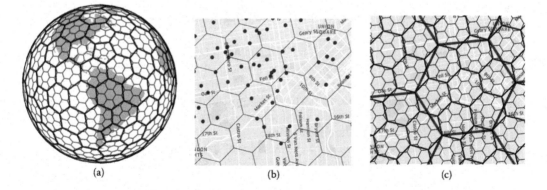

(a)	(b)	(c)

Figure 3.12. H3 examples of global to neighborhood indexing.

---

[30] Uber developed H3, as an open source grid system for optimizing ride pricing and dispatch, to make geospatial data visualization and exploration easier and more efficient (Brodsky 2018).
[31] H3 images courtesy of Uber Engineering Inc. (Uber 2018). See also "*Hexagonal Hierarchical Geospatial Indexing System*" at github.com/uber/h3.

3.8 Honeycombs for Supercomputers

Some of the largest supercomputers in the world are dedicated to global weather fore-casting and climate modeling. Supercomputers being used for these purposes are often tasked with the discretization of partial differential equations (PDEs). Solving billions of PDEs is a highly repetitive and computationally intense endeavor. With each new generation of supercomputers, the number of computational clusters or processors increases, along with their processing speed. Connecting these processors and main-taining efficient communication between them has become a formidable challenge. Honeycomb networks of processors can reduce bottlenecks between processors and improve PDE discretization.

There are two main ways to interconnect a computer's processors: switched and switchless. In switched systems, messages sent from one processor to another pass through a hierarchy of switches that enable the communications path between them. This approach has worked well in the past. But as processor speed increases, the num-ber of messages being passed from cluster to cluster also increases and soon the switch or hierarchy of switches is overwhelmed. The switch becomes a bottleneck, in effect, hindering further performance improvements.

The latest generation of supercomputers is switchless and employ many proces-sors working in parallel. In a switchless system, computational clusters communicate with one another through a set of paths and nodes much like a *network*. The network provides gateways for directing communications. Graphically speaking, a network might resemble a Cartesian grid with computational clusters located at grid points, and the grid itself as a set of communications pathways connecting every proces-sor with every other one. A lucky processor message might pass through only one or two links to get to its destination processor. Other messages might have to pass through many links to get to their destination. The challenge in processor design is to keep the number of internode links as few as possible, while making the links as short as possible. In most network layouts, there will be cluster nodes on the edge or ends of the network system, and they need a path that loops back and returns to the network. This usually means connecting this cluster back to the network with a "*long wire*," a path much longer than the usual ones. As the number of clusters increases, long wires are almost unavoidable. These long wires slow up any commu-nication they carry. The goal of system designers is to minimize or eliminate long wires altogether.

Inspired by weather forecasting and climate modeling, researchers Joseph Cessna and Thomas Bewley have developed a new logical and physical approach to solving processor communication problems in supercomputers. The approach is highly adapt-able to spherical applications, particularly those dedicated to PDE processing.[32] In ef-fect, Cessna and Bewley have applied spherical grids, like those already used by climate modelers, to supercomputer architecture.

In an earlier section, we showed examples of honeycomb (hexagonal) grids used in climate modeling. The Voronoi cells, derived from an initial triangular grid covering the Earth, produce a honeycomb grid. A significant feature of a Voronoi grid is that every cell (section of Earth, or volume of atmosphere) is separated from its neigh-boring cells only by its edges; there are neighboring cells adjacent to any of the cells' vertices (see Figure 3.8). Most processor activity in climate modeling and weather forecasting involve adjacent cells, so arranging computational clusters like a Voronoi's honeycomb network is appropriate.

---

[32] (Cessna and Bewley 2009).

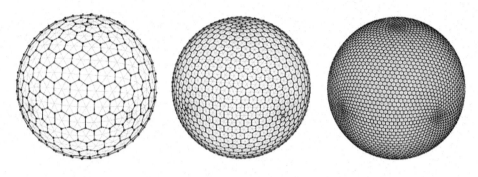

Figure 3.13. Honeycomb processor grids.

Cessna and Bewley found that computational clusters and the communication networks between them can be arranged in an analogous way. A honeycomb vertex is a computational cluster location. A honeycomb grid is a communications path. The number of networked clusters in any grid can be regulated according to the problem to which the supercomputer is applied.

Figure 3.13 shows a series of increasingly dense honeycomb grids, demonstrating that the number of processors can be varied according to the modeling requirements.[33] In other words, there isn't a one-to-one correspondence between a climate model honeycomb grid and the number of computational clusters and communication paths in the supercomputer. What is the same, however, is the hexagonal grid and the way the discretization of PDEs is performed. Cessna and Bewley found that honeycomb arrangements of computational clusters reduced the number of communication paths, all but eliminating the need for long wires between clusters, and optimizing the discretization of PDEs used for spherical applications.

Figure 3.14 shows typical honeycomb graphs over tessellations of equilateral triangles and hexagons, the two types of grids commonly used to cover a sphere (or Earth). The

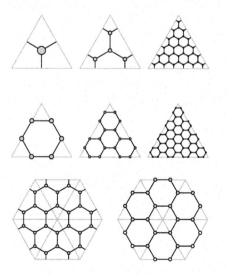

Figure 3.14. Typical processor and network arrangements.

---

[33] Images courtesy of Joseph Cessna and Thomas Bewley.

Figure 3.15. Examples of computation clusters and networks.

choice of how many computational clusters that could be used in a computer simulation can be regulated to match analysis requirements. Figure 3.15 shows sample clusters and their communications paths in layouts that can cover the surface of the sphere.

The honeycomb grid approach offers many advantages when arranging networks of computational clusters in supercomputers. Applications such as climate modeling and weather forecasting can optimize PDE discretization for local area simulation with this type of architecture.

3.9   Fish Farming

From the 1950s to 2020, global demand for fish products doubled as world population growth dramatically increased the number of people eating seafood. Skyrocketing demand combined with overfishing, climate change, and the adverse effects of pollution on water habitats are negatively impacting food supplies as well as economies that depend on fish products. Some of the highest valued species are now threatened, including cobia, grouper, amberjack, pompano, cod, halibut, and haddock.

According to the Food and Agriculture Organization, a specialized agency of the United Nations, more than 52% of today's fish for human consumptions comes from aquaculture or fish farms.[34] Aquaculture contributes to the global supply of fish, crustaceans, mollusks, and other aquatic animals, and production is substantial. With growth rates as high as 9% per year, aquaculture is outpacing all other animal food-producing sectors. Today, China, Viet Nam, and India export almost 23% of the worldwide total quantity of aquaculture products.[35] Asia dominates the aquaculture sector and will be responsible for more than 89% of the increase in production by 2030.[36]

Most aquaculture today is based on land or near shore. However, recent design innovations could extend aquaculture to offshore deep-water sites, increasing possibilities for would-be fish farmers lacking protected near-shore sites or near-shore aqua-environments that are suitable in terms of water temperature, quality, and/ or depth. These innovations would also be a boon to fish farmers facing shoreline zoning, protected shallow-water habitats, or other restrictions. The ability to farm offshore will greatly increase the number of locations that can host fish farms. Fish pens can be located wherever water oxygen levels, quality, and temperatures are most

---

[34] (Food and Agriculture Organization 2020, 7).

[35] FishStatJ – Fishery and Aquaculture Statistical Time Series Analysis Program, "*Global Aquaculture Production – Quantity and Value* (1950–2018 database)," 2020.

[36] (Food and Agriculture Organization 2020, 24).

(a)   (b)

(c)

Figure 3.16. Ocean Farm Technologies, Inc.'s Aquapod fish pen.

suitable. These conditions optimize fish growth and allow adequate separation between farms. Submersible pens protect fish and pens from storms and surface wave action, as well as potential collisions with ships. Fish pens must incorporate feeders and fish transfer/harvest mechanisms, and allow for remotely controlled monitoring and feeding, if they are to be economically viable. New high-tech fish farms now offer such possibilities.

Ocean Farm Technologies, Inc. has developed a fish pen using geodesic spherical design principles. Their patented design, called the Aquapod, is a 64-feet-diameter (19.7 meters) sphere with struts made of fiberglass-reinforced polyethylene (80% recycled content). Figure 3.16(a) shows a composite photo and an artist's drawing of an Aquapod viewed above and below water, while (b) shows an Aquapod completely submerged.

Wire-mesh netting covering the inside of the sphere makes the Aquapod both escape- and predator-proof, as shown in Figure 3.16(c).

The entire sphere, when assembled, is close to being neutrally buoyant in salt water. Fixed buoyancy can be set with closed cell foam, and buoyancy can be controlled with inflatable tubes mounted on the frame struts. Several inflatable tubes are visible above the water in (a). The sphere is anchored or moored in place but may occasionally be moved to other locations to maintain water quality or temperature.

Why a geodesic sphere? Geometrically speaking, a sphere encloses the most volume with the least amount of surface area. Submerged, the sphere is subjected to an almost uniform water pressure over its entire surface. A sphere is the best shape for ensuring that these pressure loads are distributed evenly when the pen is stationary or moves about. Geodesic layouts can be prefabricated from just a few standard components. In addition, they are easy to assemble on shore and repair at sea. In Chapter 8, we describe several layouts that minimize the number of different parts needed to build a sphere.

In 2008, Ocean Farm Technologies, Inc. worked with researchers at Massachusetts Institute of Technology Offshore Aquaculture Engineering Center to develop a

Figure 3.17. Prototype of a self-propelled Aquapod.

self-propelled design. Figure 3.17 shows a prototype configuration.[37] Pens can be repositioned to maintain water quality or temperature. A self-propelled pen could also eliminate the high cost of towboats that typically move cages from site to site today.

3.10 Virtual Reality

Virtual reality, it sounds like a contradiction in terms or, at best, pure escapism from the unpleasantness of daily life. In some ways, virtual reality is just that, though there are virtual realities that deliberately replicate stress and life-threatening situations. Aircraft flight simulators, for example, train pilots to handle emergencies.

Some real-life tasks are simply too difficult, unsafe, or costly to carry out without significant preparation or simulated practice. Examples of such tasks would include testing the first prototype of an advanced spacecraft, performing a first-ever difficult surgery, disarming a bomb, retrofitting new equipment in a nuclear reactor, or training for a deep mine rescue or an assault on a terrorist's camp. In such cases, virtual reality offers safe ways to solve problems, develop skills, and gather data regarding the resources and time needed to accomplish the task in the best way possible.

Technically speaking, virtual reality is a computer-generated 3D environment in which the user has complete freedom of motion and can view and interact with this environment in real time. In virtual-reality applications, a computer uses special devices or sensors to measure the user's movements and line-of-sight in order to send back visual, audio, and tactile feedback that give him the feeling he is actually in the world he is experiencing. Most systems today only offer limited movement. The user is often stationary and relies on devices like 3D mice, joysticks, or kinetic controllers (paddles, rackets, golf clubs, etc.) to move forward, backward, or change direction in the virtual world.

3D computer games are basic forms of virtual reality. Few players will admit how addictive they are. Home entertainment gaming computers like Nintendo's Switch, Sony's PlayStation, and Microsoft's Xbox offer photorealistic graphics and real-time animation of virtual worlds in which players engage in activities ranging from flying airplanes to playing golf, musical instruments, or first-person-shooter games. The global console games market was valued at about $26.81 billion in 2018 and is

---

[37] All Aquapod photos courtesy of Ocean Farm Technologies, Inc., Searsmont, Maine.

expected to grow to \$31.49 billion at a compound annual growth rate (CAGR) of 4.1% through 2022.[38]

These are impressive figures, but the future market for virtual reality is even brighter. The global market size is a little less than computer games, it is valued at USD 10.32 billion in 2019 but it's expected to grow at a CAGR of 21.6% from 2020 to 2027.[39] This growth is due to new applications that go beyond just home entertainment, finding new audiences particularly in military, law enforcement, SWAT training, construction, mining and other dangerous occupations. Users in other occupations are adopting it too—healthcare, virtual surgery, scientific visualization, product design, manufacturing, education, architectural design, real estate (virtual tours), and yes, even gambling.

What is new in virtual reality is the use of immersive systems that make you part of the virtual world—continuity of surroundings, conformance to human vision, freedom of movement and interaction, and instant physical feedback. Users can attach sensors to their bodies, use 3D computer mice, stand on tactile pads that measure their position and weight, and wear special cyber-gloves that follow any movement of the hand. Advanced head mount displays give users full 3D stereo views as they look around their virtual world. For sure, the combined effects of 3D graphics, tactile and motion sensors, and locomotion bring virtual reality much closer to real reality.

Until now, the missing sensation in virtual reality has been locomotion—letting people move freely over long distances within their virtual world in a natural way. Virtusphere, Inc. has solved this problem with a patented technology that literally immerses the user inside his or her virtual world.[40] The user is inside a *Virtusphere*, a ten-foot hollow sphere supported by a special platform of rotating wheels. The sphere, shown in Figure 3.18(a), is free to rotate in any direction but remains in place on its fixed support platform. The sphere is assembled from only a couple of different prefabricated spherical sections. The user enters the sphere through a circular hatch (removed in the figure to show the interior) which closes behind him. At this point, he is free to walk or run around inside the sphere in any direction or at any speed he wishes. The sphere is an ideal form for omnidirectional walking or running. Sensors mounted inside the structural frame's great circle arcs give complete coverage of the

(a)   (b)

Figure 3.18. Virtusphere and an assault training exercise.

[38] Research and Markets, "*Global Console Games Market Projections to 2030 – The Competition of Mobile Gaming is a Fierce Industry Hurdle*," Survey Report, 2019).

[39] (Next Generation Technologies – Virtual Reality Market Size, Share & Analysis Report, 2020–2027).

[40] (Latypov and Latypov 2003).

Figure 3.19. Virtusphere moon landing simulation.

user as the sphere rotates. The user wears a wireless head-mounted display with gyroscopes, which tracks his head movements while he walks or runs inside the sphere.[41] Special sensors track the combined rotation of the sphere (as the user walks or runs) and changes in the user's line of sight. The computer inputs this sensor data and outputs constant updates to the user's head-mounted stereo display. In (b), the soldier is preparing for an assault on a terrorist safe house.

What is unique about Virtusphere is the fact that the player can walk or run inside the sphere, and his movement will translate to movement within his virtual world. Imagine running down a street and looking in any direction, or standing on the edge of a steel beam high up a skyscraper under construction. Imagine landing on the moon as astronauts do while training, Figure 3.19. The combined effects of dynamic stereo vision, locomotion detection, environmental sounds and lighting, and tactile feedback (your feet moving you along, for example) add up to a truly startling experience.

Virtusphere's inventor, Nurakhmed Latypov, says that virtual reality gives a user a chance to feel, almost in real fashion, as if he or she is in some other world. In this world, users can see one another, talk, and interact in a 3D environment, and can even appreciate colors and sounds. Such immersion creates the impressive illusion of being in that other world.[42]

Our only question now is when can we get one for our home entertainment system?

3.11 Modeling Spheres

In the 1970s, everyone recognized Buckminster Fuller's geodesic domes. Nevertheless, few people understood the basic principles of spherical subdivision. Magnus Wenninger's classic book, *Spherical Models,* changed this and made the subject accessible, easy to understand, and fun by getting readers to make models.[43] He showed that the only tools you need to make complex, beautiful spherical models are a simple compass, a protractor, a straight edge, scissors, some paper and glue, a little patience, and a few back-of-the-envelope calculations.[44] He literally put the principles of spherical subdivision at the reader's fingertips.

---

[41] All Virtusphere photos courtesy of Virtusphere, Inc.
[42] (Latypov 2003, 1).
[43] (Wenninger 1979); republished by Dover with a new appendix in 1999.
[44] See Geodesic Domes by Euclidean Construction (Wenninger 1978).

Wenninger started modelers off with models of the basic Platonic solids (the tetrahedron, octahedron, cube, dodecahedron, and icosahedron). From there, he progressively built the reader's projective geometry skills, so he or she can tackle increasingly complex models. Using various Platonic frameworks, Wenninger went on to show how to make models of all three types, or classes, of spherical grids. These same classes are used today in a broad range of applications, from arranging dimples on a golf ball, to developing vaccines against viruses, to studying how climate change affects the Earth.

Spherical polyhedra, like their planar versions, exhibit many forms of symmetry that can be used by modelers. Small spherical sections can be prefabricated in paper, wood, or sheet metal, and then duplicated and assembled to complete a full sphere. Today's CAD modelers employ the same techniques, only in a digital form. 3D sketches of spherical models use projective geometry methods. Digital computer models of geodesic arcs are analogous to circular paper bands, and mathematical constraints join 3D digital geometry, just as glue bonds paper strips together. Building paper models develops the same skills employed by CAD modelers.

Wenninger built simple models as well as very complex spherical subdivisions that resemble basket weaves. In these designs, spherical patterns interlock and pass under and over one another, much like woven fabric. He relied on the same projective geometry techniques he wrote about in *Spherical Models* (with the Dover mathematics appendix), but he also used modern spherical calculator programs and personal computer polyhedral modelers like *Stella*[45] to make templates for more complex models. Figure 3.20 shows him completing a basket weave spherical model.[46] Several more are under construction on his modest workbench and his bookshelves display a small fraction of the thousands of polyhedron and spherical models he has made over the years.

Wenninger[47] passed away in 2017; his indelible influence on polyhedral modeling remains and will continue to inspire the curious for generations to come.[48] For those who knew him personally, he will always be remembered for his signature spherical design he called "*Order in Chaos.*"[49]

Order in Chaos **by Magnus Wenninger**

Figure 3.20. Magnus Wenninger, polyhedral modeler.

---

[45] Stella is a personal computer polyhedral modeler from Software3D developed by Robert Webb.

[46] Photos by Edward Popko.

[47] Father Magnus Wenninger is a monk of Saint John's Abbey, a community of Benedictine monks at Collegeville, Minnesota, USA.

[48] See Bibliography for complete listing of his publications.

[49] Computer model and image courtesy of TaffGoch.

3.12   Computer Aided Design

No other technology has put spheres to work more than computers. Earlier in Section 2.8, the first application of CAD in spherical subdivision was noted, a specialty FORTRAN computer program written in the mid-1950s to produce reference tables of chords, arcs, face angles, and x, y, and z coordinates of spherical grids (see Section 2.8 Covering Every Angle). By the early-1980s, CAD programs had started to automate another design task, 2D drafting. Manual drafting, a labor intense and often error prone process, was particularly difficult because designs constantly change and geodesic shapes and spatial arrangements are usually complex. Most CAD systems, at that time, required main-frame computers and specialty graphics workstations. Only large aerospace, automotive and mechanical design companies could afford them.

Today's CAD systems, however, are far more accessible and can easily deal with the complex geometry of spherical subdivisions. They tend to be PC based and many offer very precise 3D geometric modeling and rich displays. Users build up their designs by defining individual parts. They are combinations of geometric primitives such as plates, rods or surfaces and detail features such as holes, fillets, or chamfers. Completed parts are grouped together into assemblies where their relative positions and motions are mathematically defined (constrained) with respect to each other. Assemblies can be combined thus making larger and larger functioning systems.

CAD is a huge advantage in spherical work. Users can easily define points, lines, arcs, and planes on a sphere; change them instantly and visually inspect layouts from all viewpoints. They can virtually walk through their designs, detect interferences and simulate motion and forces. Many CAD systems compute part and assembly mass properties, such as weight, center of gravity, displacement, buoyancy, or surface area (a useful number if you are ordering paint). Completed product designs can be virtually disassembled and used to determine packaging, shipping, and even storage.

For many users, CAD is simply the start of a process leading to the ultimate goal—manufacturing. Beyond very precise geometry, CADs business value often lies in the analysis, attributes, and logical connections it creates as designs are built-up and that this data flows, down-stream, to coordinate suppliers and drive automation and manufacturing. Attributes might include logical-physical connections, materials, coatings, design standards, revision logs, inventory control, annotations for technical documentation or sourcing information. Numeric control devices such as 3D printers, lathes, spot welders, millers, or robots can input geometric data and make tooling choices to produce finished goods. Some attributes might be sent to report generators for bidding, quality control, inventory, or scheduling.

CAD has not abandoned 2D drafting. Drawings are still required in most commercial applications for bid-packages, field construction documents, bills-of-material, or illustrations in manuals. What's different today is 2D drawings are no longer the object of the design process; they are *by-products* of 3D design and are generated automatically from different views or sections of the model. Labor savings aside, the huge advantage is that drawings are accurate and kept in sync with the design as it evolves. Today, 3D modeling is the center of the design process; everything else—documents, analysis, visualization, simulation, and manufacturing—is driven from it.

David Price (aka TaffGoch) is one of the most diverse and skilled geodesic CAD modelers. He has been modeling for many years. The figures below are just a sampling of his hundreds of 3D geodesic layouts, visualizations, and dome recreations. Taff's models are geometrically complex and rich in detail. He uses personal computer tools available to anyone and, what's more, his 3D models are widely available on the

Figure 3.21. Geodesic reciprocal frame sphere.

internet for anyone to use. Taff fully exposes his modeling techniques; they are vast warehouses of practical how-to techniques for anyone wanting to learn CAD. The following examples are just a sampling of Taff's work.[50]

Figure 3.21 shows a geodesic *reciprocal frame sphere*. The modern term for reciprocal frames is *nexorade*. It is an icosahedral tessellation he labels as a Class I, Frequency 2, modeled in the CAD program *SketchUp*™ and rendered in the program *Kerkythea*.

Figure 3.22 depicts the first five *icosahedral rotegrity spheres*, in their order of increasing triangle density (frequency). The sphere on the left is the simplest icosahedral rotegrity. The term "*rotegrity*" and its layout method were first introduced in 1970 by Dick Boyt.[51] Reciprocal-frames, nexorades, and rotegrity layouts are discussed in Chapter 10.

Figure 3.22. Geodesic rotegrity spheres.

---

[50] All images courtesy of TaffGoch.

[51] (Dick Boyt invented the term "*rotegrity*." In an interview with Joseph Clinton, Boyt said it came from two terms; rotation and tensegrity. [1971]).

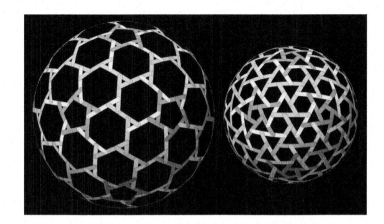

Figure 3.23. Nexorade (or rotegrity) spheres.

Figure 3.23 shows that rotegrity spheres can be assembled from the same four arc "*straps*." The difference being, to which midpoint an endpoint is attached. One mid-point produces large pentagons & hexagons, with small triangles. The other midpoint produces small pentagons & hexagons, with large triangles. Each sphere is composed of 210 straps.

Figure 3.24 demonstrates the amazing flexibility of CAD to model intricate, almost lace-like patterns on a sphere.

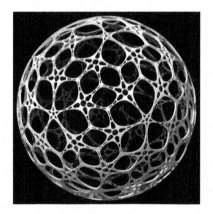

Figure 3.24. Model of Amazon's HQ campus biosphere.

3.13   Octet Truss Connector

Putting spheres to work oftentimes means using them as a design references; the final design or product may not show the spherical influence at all. The *Octet Truss Connector* is one such example.

The *Octet Truss* was a revolutionary construction system for pre-fabricated, lightweight, long-span structures (without intermediate columns) for aircraft hangers, shopping centers, sports arenas or exhibition halls. It is based on juxtaposition octahedral tetrahedral units made of pipes joined by aluminum connectors. A typical truss section is shown in Figure 3.25. The design detailing and fabrication were

Figure 3.25. Typical section of the Octet Truss.

developed at the Synergetics Inc. office in Raleigh, North Carolina by architect T. C. Howard.[52] It was the refinement of an idea patented by Buckminster Fuller in 1961.[53]

A demonstration structure was built for an exhibition at the Museum of Modern Art (MoMA) in New York City.[54] It spanned 100-feet long, 35-feet wide, and 60-feet high and weighed approximately 8000 pounds, see Figure 3.26.[55] The tremendous strength of the structure was due to the triangulation of members, plus a connector design which kept all forces directed along the centers of the pipe struts. It was reported that a typical *Octet Truss* section, weighing only 65 pounds, could support a load of 12,000 pounds.

Figure 3.26. MoMA exhibition Octet Truss.

[52] (Chu 2009, 139).

[53] Fuller 1961, "*Synergetic Building Construction,*" US Patent 2,986,241.

[54] Installation view of the exhibition, "*Three Structures by Buckminster Fuller,*" September 22, 1959 throughout Winter 1960. The Museum of modern Art, New York Photographic Archive, Exhibition Albums, 6.

[55] Photo courtesy of Museum of Modern Art Photo Archives.

(a)   (b)

Figure 3.27. Single Octet Truss connector and six-way ring.

To support the trussing geometry, an innovative cast aluminum connector was de-veloped. A CAD model of the connector and assembly was created to analyze its design, one is shown in Figure 3.27(a).[56] Each connector has four "*ears*" and a hub. The hub accepts a slipover pipe that is held in place by using a special knurling tool to crimp the pipe-end into the hub's retaining grooves. The four connector ears are cast at angles that allow multiple pipes to join in an octahedral-tetrahedral configuration. Once connectors are fitted to each end of the pipe, they are assembled into octahedral-tetrahedral units by riveting pipe connectors together. All Octet Truss struts use the same pipe length and relative connector orientation. This standardization lowers man-ufacturing costs and units can be prefabricated on or off the construction site. These are significant advantages of this truss system.

Figure 3.27(b) shows six hubs assembled to form a ring. Notice in (b) that surfaces of ears around the ring are tangent to an imaginary plane (colored blue). This plane is one of the four great circle planes of the reference spherical cuboctahedron.

Figure 3.28(a) shows a common truss nine-way cluster; up to 12 pipes can be joined to a single hub. The dihedral angles between a connector's ears are the key to this

(a)   (b)

Figure 3.28. Nine-way Octet Truss connector and design references planes of the spherical cuboctahedron.

---

[56] SOLIDWORKS 3D models by Edward Popko.

design. They are derived from the dihedral angles between the cuboctahedron's great circle planes. Figure 3.28(b) superimposes these planes on the connector cluster to show the relationship between the two.

The *Octet Truss Connector* is an elegant design. It clearly demonstrates how subdivided spheres or spherical polyhedra can serve as design references and yet not become a physical component or feature in a final design.

3.14   Dividing Golf Balls

Despite images of players strolling along lush fairways and enjoying leisurely cart rides from hole to hole, golf is an intense and competitive game. Professional golfers can swing a club at 130 miles per hour with enough force to send a ball more than one-sixth of a mile at speeds in excess of 160 miles per hour. Golf's record-setting tournaments are a testament to the game's well-prepared athletes who devote countless hours of practice to their sport. But players at all levels also benefit from specialized high-tech equipment, including clubs and the golf ball itself.

Almost all golf balls have a surface with many depressions, called dimples. All else being equal, the flying distance of a ball with dimples can be almost twice that of a smooth ball. Dimples deliver distance by generating small-scale turbulence around the ball's surface. This turbulence, somewhat counter intuitively, reduces drag on the ball because it makes the wake (drag) behind the ball much smaller. A ball with a rough surface simply flies better than a smooth one. Golf ball dimples represent one of the most creative and diverse applications of spherical subdivision principles.

The United States Golf Association (USGA) defines the maximum weight (1.62 ounce), minimum diameter (1.68 inch), and spherical symmetry requirements of competition-level golf balls.[57] Spherical symmetry requires that a ball be designed and manufactured to behave as though it were symmetrical to enhance its control and predictability of flight; both are safety considerations. In other words, the ball must perform the same in flight, no matter how it is oriented when placed on the tee. Designers meet those requirements in an amazing variety of ways, covering the ball's surface (8.87 inch²) with between 200 and 1000 dimples arranged in everything from simple bilateral patterns to complex polyhedral symmetries. Golf balls are wonderful examples of optimizing design tradeoffs where manufacturing techniques, aerodynamics, and design overlap. Amazing creativity is displayed in these 8.87 inch², an area somewhat larger than a business card. If a golfer does not recognize the difference when he plays, he will when he pays; high-performance, professional-quality balls can cost four times as much as a knock-about ball.

In 1931, John Vernon Pugh disclosed in British Patent Provisional Specification Serial No. 377,354 that by using an icosahedral lattice for defining dimple patterns on golf balls, it is possible to make a geometrically symmetrical golf ball.[58] The lattice he developed divided the sphere into 20 equilateral spherical triangles, thus defining a spherical icosahedron. Pugh detailed how to plot the icosahedron onto a sphere accurately.

---

[57] United States Golf Association and the Royal & Ancient Golf Club, Equipment Testing, Appendix III—The Ball, www.usga.org/equipment/guide/book/appendix3ball.html.
[58] (Nesbit and Stiefel 1991).

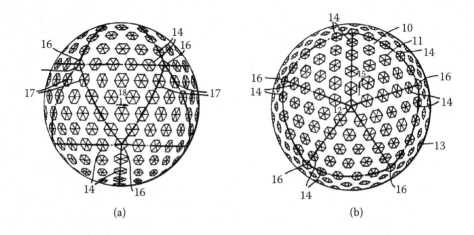

(a)   (b)

Figure 3.29. Early icosahedral dimple arrangement.

In the United States, most balls had a relatively simple dimple layout that placed *"parallel"* rows of dimples in an octahedral layout. Things changed quickly when Frank Martin and Thaddeus Pietraszek patented the first icosahedral layout with modern dimples. Figure 3.29(a) shows their 1978 hex-pent dimple arrangement.[59] This was a pivotal patent; it focused attention on the icosahedron, and it showed the limit-less design possibilities of dimples. Within a year, Frank Martin had teamed up with Terence Melvin and others to secure yet another icosahedral patent. Although the essence of their patent claim focuses on ball coverings, the skin of the ball, their patents show circular dimple arrangements based on a spherical icosahedron.

The goal of ball designers is to design a pattern of dimples that improves the ball's flight characteristics (mostly flight distance) and directional control. By analyzing wind tunnel and robotic hitting machine data, designers concluded that the best way to achieve these goals was to increase the number of symmetry axes on the ball, and spherical icosahedra offered the best polyhedron base to work from. Despite many designs that use other bases, the icosahedron, or a small variation of it, remains by far the most common today.

In the mid-1950s, the icosahedral framework revolutionized dimple layouts. The icosahedron has six axes of vertex rotation, in contrast to other approaches that offer fewer. As a result, the additional axes increased flight stability. The ball's initial orientation on the ground, when launched, became less important. In 1985, Steven Aoyama was awarded a landmark patent that established the icosahedron as the framework of choice. Figure 3.30 demonstrates the flexibility of Aoyama's schema in a series of different dimple layouts.[60] In his patent[61] he subdivides icosahedral faces into four smaller triangles: a central equilateral triangle.

surrounded by three congruent *isosceles triangles* (see Mid-Arcs, Section 8.5.8). In effect, this subdivision creates a variation of the icosahedron called the *icosidodecahedron*. These 3D forms are covered in detail in Chapter 6. The Mid-arcs schema adds six great circles to the sphere and divides each of the icosahedron's 20 faces into four smaller triangles. The additional great circles and smaller triangular areas define a totally new layout schema.

[59] (Martin and Pietraszek 1978, sheet 1/2, figs. 2–3).
[60] (Aoyama 1985, sheets 4, 5, 7, and 9 or 11, figs. 10A/B, 11A/B, 12A/B, and 14A/B).
[61] (Aoyama, op.cit.1985).

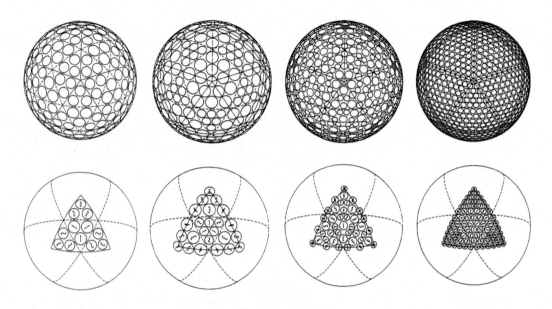

Figure 3.30. Golf ball patent 4,560,168.

Aoyama's patent drawings show how the density of dimples can be regulated by simply changing the size and mix of dimples within the four working areas. The upper row of vertex-zenith views shows entire balls covered, with the dimple design shown below them. From left to right, three different-size dimples cover the first ball with 240 dimples, whereas 6, 10, and 13 sizes create 312, 692, and 912 dimples on the others. With 16 different sizes (not shown in the figure), the ball is covered with 1212 dimples. Three of the four configurations in the figure show dimples along the icosahedron's face edges, but no dimples cross any of the six faces that subdivide great circles. These are kept dimple-free for manufacturing flexibility. Unlike our earlier examples of spherical caps, there may be more than one radius dimple.

Dimple research and development is as competitive as ever, often pitting hobbyists against international sporting equipment companies. More than 250 US patents have been awarded for dimple designs since Aoyama's work. The dimple wars are intense, and the battlegrounds shift from time to time, sometimes focusing on the overall layout, other times on the design of individual dimples and how they pack. As of this writing, more than 40 patents are pending based on application submissions dating back to 2017. Some are incremental refinements of older designs; others develop new designs like zero-ground balls (balls that minimize their contact with the ground), dimples within dimples, noncircular dimples, or random dimple patterns. Figure 3.31 is a small sampling of patented dimple designs.[62] We will explore golf ball dimples further in the next chapter when we discuss their polyhedra symmetry. Golf balls may be small, but you will surely appreciate how creative their designers subdivide spherical polyhedra to solving difficult geometric and manufacturing problems, keeping improved flight, control, and player scores in mind.

---

[62] Patent illustrations from (a) (Kashima 2008, sheet 2/3, fig, 4); (b) (Simonds 2006, sheet 1/12, fig. 1); (c) (Maehara 2001, sheet 6/6, fig. 10); (d) (Machin 1995, sheet 5/14, fig. 5); (e) (Gobush 1991, sheet 12/13, fig. 4); and (f) (McGuire 2002, sheet 3/8, fig. 2).

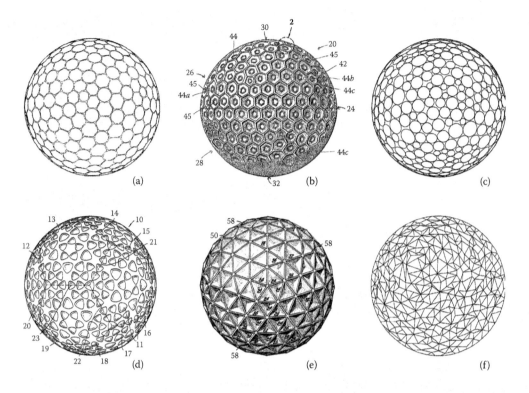

Figure 3.31. Samples of golf ball spherical subdivisions.

3.15   Spherical, Throwable Panono™ Panoramic Camera

If you can toss a ball, you might be the next Ansel Adams. The throwable panoramic camera Panono is an amazing invention by Jonas Pfeil.[63] Throw this ball-shaped camera into the air or support it on a monopod and you will have a spherical 360° panoramic picture centered on the ball when it comes down or is stationed. It is an ingenious application of spherical design.

The camera is about the size of a grapefruit, 4 inch diameter, and weighs a pound. Inside the ball are an accelerometer to compute the time the toss reaches its apex and a computer to trigger the 36 cameras at that precise moment taking a high-resolution set of images. Software combines the images into a seamless 108 megapixel stitched 16,384 × 8192 pixel panorama. Unlike wide-angle cameras and projection systems, such as IMAX, the throwable's images are truly spherical; there are no blind spots and they can be shown in a planetarium-like setting or simply viewed on a PC or pad display with pan/zoom in any direction. Figure 3.32 (a) shows a typical stitched 360° Panono image; all the separate images taken by the camera have been combined into one seamless panoramic. Software such as *krpano Panorama Viewer,* can view this image, look in any direction, pan,

---

[63] Jonas Pfeil first conceived of the camera as a Diploma thesis at the Technical University of Berlin entitled "*Throwable Camera Array for Capturing Spherical Panoramas*" in 2010.

(a)   (b)

Figure 3.32. Typical Panono 360° panoramic and normalized detail from it.

zoom and extract any detailed view such as the normalized section shown in (b); its portion of the panoramic is outlined in (a).[64]

Camera inventor Jonas Pfeil's highlights the importance of the number and spherical location of the cameras:

> *The cameras are positioned on the ball's surface by numerically optimizing their orientations and ensure full panoramic coverage. Also, the smallest amount of overlap between neighboring cameras is maximized to facilitate blending between the images.*

The Panono takes pictures several ways. It can be stationary, mounted on a monopod or hand held with a selfie-stick. But there is another way, just toss it up. An accelerometer measures the launch acceleration and the onboard program integrates it to determine the launch velocity and the rise time to the throwable's highest point. When the rise time is reached (apex of the throw), camera motion is zero and the program triggers the cameras producing a blur-less set of panoramic images. The camera can also be mounted on a mono-pod and set off the ground or hand-held and triggered remotely via smartphone.

Each toss produces 36 JPEG format images which are combined after downloading to a PC or tablet by way of standard USB connection. The camera uses rechargeable batteries, much like a cell phone. Editing and viewing software allows complete 360 pan/zoom and printing of panoramas.

3.16 Termespheres

Artist Dick Termes is one of the most unique painters in the world. He renders intricate 3D freeform and geometric wrap-around panoramic perspectives called *Termespheres*—his spherical canvas. To do so, he has invented a new method system of perspective—*Six-point Perspective.*[65]

Termes' canvases range in diameter from 12 inches to 48 inches and some as large as 90 inches. Artwork can be free-form designs, interlocking geometric patterns and illusions much like Escher etchings or layouts with polyhedral symmetry such as Japanese Temari Balls

Artist Dick Termes in studio

---

[64] Images courtesy of Stig Nygaard.
[65] (Termes 2020).

Figure 3.33. Notre Dame of Paris, 1995.

Media: Polyethylene sphere, acrylic, 24″ diameter, displayed rotating from a ceiling motor

(see Section 3.23). Other paintings can be 360°, total wrap-around panoramic views of real or imaginary spaces or architectural settings.

Termes unique projections and his use of light and color are as beautiful as they are engaging. What's more, they are sculptures; see Figure 3.33.[66] In order to fully experience them, you must participate. They can rotate from ceiling motors and appear to float in front of your eyes. You must walk around them or rotate them in your hands. In architectural spaces, your viewpoint changes as you move about. You can see objects such as furniture, stairways, windows, and doors change in perspective. It is like you are rotating in a circle but you are just looking straight ahead. The scene constantly changes and appears to reverse its direction of rotation and flip its sense of curvature from convex to concave at the same time.[67] The visual effect is striking; the viewer's sense of the space, motion and perhaps even time constantly changes.

Termes has superb drawing and painting skills and he has had to develop a perspective system to deal with his adopted form. He calls it *Six-point Perspective*; it's his reference framework that allows him to map or project the world he sees (or imagines) onto this spherical canvas. To best understand his system, one must see how it expanded the traditional one, two and three-point systems.

First introduced in the 15th century to make drawings or paintings more realistic on 2D surfaces (paper or canvas), perspective systems employed one or two vanishing points to project geometric objects from the viewers eye point and to account for the visual effect of objects getting smaller and smaller as they recede into the distance

---

[66] All paintings, images, and drawing courtesy of and by Dick Termes.

[67] For centuries, astronomers have used globes to map the stars and constellations. Some globes project the stars from the viewpoint of one standing on Earth, centered within the globe, looking up at the sky. Others map the star's from *"outside looking in"*—perhaps as the Creator looks back on his work.

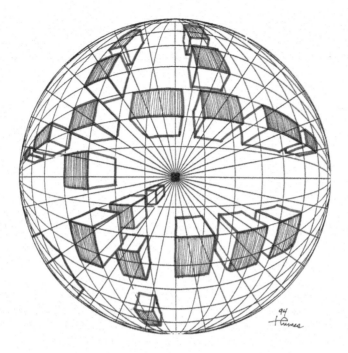

Figure 3.34. Five and six-point perspective reference grid.

away from the viewer.[68] The two vanishing points are located on the horizon, typically chosen to be the imaginary eye-level of the viewer. One is located to the left, the other to the right of the viewer's direct center view of the paper or canvas. A third vanishing point, occasionally used by architects, places a point high above or below the center of the reference horizon. It's for very tall objects or objects that drop way below or above the horizon and is used to diminish their size as they rise above or fall below the viewer. It has the same effect in drawing or painting as one sees in a tall building tapering and becoming smaller and smaller as it rises away from the viewer into the sky or a well appearing smaller and smaller in diameter the deeper it goes into the ground.

For spherical work, Termes has added three more points to the traditional system. Figure 3.34 is a schematic of his *Six-point Perspective*. Six vanishing points allow him to map – 3D spaces onto a spherical surface. The advantage of his system is the artist can accurately project any panoramic real or imagined scene up, down, and all-around 3D space onto any size sphere and the viewer can see those spaces from any angle simply by moving around the completed work or rotating it in his hand. What the viewer sees constantly changes with his motion as if he was inside the space looking around or outside looking in.

Six-point Perspective essentially maps the spheres surface into one continuous image, a series of perspective areas each vanishing to or projecting out from the center of the sphere. Figure 3.34 shows the spherical reference grid and its middle vanishing point at the center of the sphere. If you think of yourself inside the sphere, the

---

[68] Single point perspectives existed for at least a hundred years, but the first known picture to make use of multipoint linear perspective was created by the Florentine architect Fillipo Brunelleshi (1377–1446) and written about by architect Leon Battista Alberti in 1435, *De Pictura* (On Painting). In 1415, Brunelleschi employed two-point perspective in painting the Baptistery in Florence from the front gate of the unfinished cathedral. It wasn't until 1583 that the first treatise on perspective by a professional artist appeared, *Casso di Risparmio di Vignola's Le due regole della Prospettiva Pratica*.

vanishing points radiate out rather than into the sphere. A cube is rendered to show how a simple object is projected. In most of Termes' art work, the six-point grid is not visible, it's only a temporary reference while painting or drawing. Note too, the vanishing points in the *Six-point Perspective* system resemble the vertices of the spherical octahedron.

Termespheres are wonderful examples of subdivided spheres and how reference systems can enable an illusion so convincing one thinks they are part of scene's space and time.

3.17   Space Chips™

Space Chips™ are a wonderful way for kids to explore the Platonic and Archimedean solids and go further into spherical constructions.[69]

Designed by kid's toy inventor Dick Esterle, you can think of chips as spherical Legos. They intuitively illustrate the connection between 2D straight-edged shapes and the spherical aspiration of three-dimensions.

Space Chips™ are a simple system of snap-together, flat, flexible, plastic triangles, pentagons, and square rhombi with equal length edge-slots around their perimeter that connect with other *Chips*. Holes in the centers allow for added struts to expand into other configurations. The slots allow wide variations in joinery and accommodate any angle naturally occurring between the *Space Chips™*. It's easy to build a simple polyhedron and challenging to build higher-order spherical forms. It doesn't take long to realize that an infinite number of combinations are possible. There are no *"wrong"* ways to use *Space Chips™*.

Figure 3.35 shows a small sampling of what's possible: (a) Platonic solids, (b) spherical icosidodecahedron (20 triangles + 12 pentagons), and (c) *Buckyball* with 60 dodecahedral shapes using 540 space chips in 6 colors.[70]

Space Chips™ teach many spatial principles. Colors, for instance, are more than decoration—they can be color-coded to show dual-polyhedra and various edge/face/vertex symmetries. Even a quick glance at the "v" shape interlocked edges suggest great circles and relationships and each chip's mid-face hole suggest the vertices of yet another 3D figure somehow magically defined and trapped inside. Although young children won't know their technical names, they will sense that the corners (vertices) of their platonic sculptures touch (are tangent to) an invisible ball (circumsphere), see Section 6.2.3.

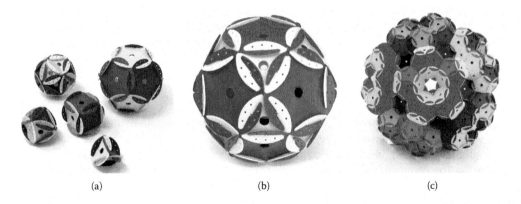

(a)   (b)   (c)

Figure 3.35. Space chip platonics (a), icosidodecahedron (b), and Buckyball (c).

---

[69] Space Chips is a registered trademark of Monkey Business Sports.
[70] Space Chips name, design, and images © Dick Esterle.

3.18   Hoberman's MiniSphere™

Toys and games embody many spherical principles, and they are a playful way to see spherical subdivisions and 3D geometry in action. Charles Hoberman's expandable *MiniSphere* and Mark White's *Code World* are two examples. Both are embodiments of patented technologies for totally different applications—one for expandable roof construction systems, and the other for genetic coding.

The *MiniSphere* is a series of interconnected plastic trusses that can expand and contract, causing its spherical form to greatly increase or decrease in size without changing its basic geometry. One version is shown at the right in its folded-up configuration. It's a fascinating, dynamic toy that one can watch for hours, wondering how it works without locking itself up or distorting its shape as it changes size. In some large exhibition versions, a motor facilitates expansions and contractions.

Hoberman, who was interested in expandable truss systems, developed several patented designs that could expand and collapse. These designs offer high levels of prefabrication, lightweight elements, ease of transportation, and the ability to span large spaces with minimum support.[71] The MiniSphere is an embodiment of this patented truss design.

The core of this invention is a set of trusses, each made of a great circle "*loop-assembly*." A loop is made of many smaller scissor-pairs of lightweight tubular elements jointed to one another or to small hubs. A loop can expand or fold up, changing its size but not its basic geometry. In one version, three mutually orthogonal closed loops are arranged, each forming a great circle on their respective Cartesian *xy*-, *yz*-, and *zx*-planes. The three loops intersect each other at six places and are joined via small hub connectors. In this arrangement, even though one loop-assembly passes through (intersects) another, they do not inhibit each other's synchronous unfolding or folding, as they are expanded or collapsed together. Their intersections always maintaining their orthogonal alignment. Smaller scissor-pair trusses fill the spaces between the three main closed loop-assemblies. These trusses are very flexible, and many combinations are possible. Figure 3.36, from

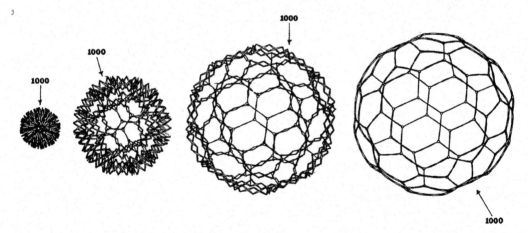

Figure 3.36. Radial expansion/retraction truss structures.

---

[71] See Hoberman's reversibly expandable doubly-curved truss structure (USPTO Patent 4,942,700) and radial expansion/retraction truss structures (USPTO Patent 5,024,031) for examples of unfolding structures and toy transformations.

Hoberman's patent, shows a typical expansion sequence.[72] An arrow follows one of the loop-assemblies made of scissor-pairs through its successive expansion. At its most expanded position, the scissor-pairs are so extended they look like a double-thick strut. Folding is simply the reverse sequence.

3.19   Rafiki's Code World

Rafiki's Code World is a very different spherical experience that proves you can learn subtle geometrical principles while having fun. This copyrighted logic game challenges players to define a sequence of spherical moves to match color combinations drawn from a deck of color-coded challenge cards. This innovative game is not played on a flat board; rather, it is played on a sphere made up of three different spherical polyhedra. The main parts of this game are a brightly colored globe defined by two polyhedra surrounded by a glider, a third polyhedron (see figure at left).[73] The globe is a spherical icosahedron with its 20 color-coded triangular faces and circular insets. Across its faces is a continuous track or groove that defines the edges of another spherical polyhedron, the dodecahedron. These two polyhedra, the icosahedron and dodecahedron, are *duals*, which mean they have a symmetrical relationship. The 12 vertices of the icosahedron are centered midface to each of the dodecahedron's 12 faces and the dodecahedron's 20 vertices are centered within the icosahedron's 20 faces. Thus, one polyhedron has as many faces as the other has vertices and vice versa. You can easily verify these relationships in the globe. Note, also, that the 30 edges of the dodecahedron and icosahedron cross perpendicular to each other. Both polyhedra, therefore, have the same number of edges. We will take a closer look at polyhedra and their dual relationships in Chapter 6. We use them extensively when we subdivide spheres.

A glider, shaped like the edges of a spherical tetrahedron, surrounds the globe. The tetrahedron is one of the five Platonic solids. It has four large equilateral faces. Its six edges are represented in the model by wide green arcs spanning four vertex hubs colored red, blue, green, and yellow (not visible). Although the glider is free to move about the globe, pins under the hubs maintain the glider's orientation within the dodecahedral groove. Game inventor Mark White calls the entire unit, glider and globe, a *"stone,"* short for the sorcerer's stone (a legendary tool of alchemists) that figures prominently in the game. Each time the glider is repositioned, a new configuration of four faces and four circle colors are mapped (decoded) by the hubs. There are 120 possible color configurations. The challenge of *Rafiki's Code World* is to solve various *Code World* puzzles with a sequence of glider moves. Puzzles are drawn from a deck of cards, which shows combination pairs of colored triangle-circles or just circles. White says that puzzles have an infinite number of valid solutions; some sequences are shorter than others and herein lies the spatial challenge.

From a purely polyhedral point of view, *Rafiki's Code World* game is a wonderful kinematic introduction to dual polyhedral relationships and color mapping problems. But, like Hoberman's MiniSphere, Mark White's *Code World* is more than a game; it's an embodiment of another totally different invention, one based on genetic mapping. White sees a relationship between symbol-sets (analogous to the colored faces in *Code World*) and sequences of the amino acids in a protein (the way the glider's position

---

[72] (Hoberman 1990, sheet 7/12, figs. 13–16).
[73] Image courtesy of Mark White.

spans particular colored faces). His patent application describes the model this way: it is really a spherical slide rule for biologists.[74]

3.20   V-Sphere™

V-Sphere™ from Verdes Innovations S.A. is a fun spherical puzzle.[75] It's a spherical octahedron (see Section 6.3.3) with eight fixed-colored triangles each surrounded by twin circle tracks. Each track has 12 colored moveable rectangular pieces that slide about. One track has only 11 which leaves an empty space that allows pieces from one track to pass through or to be rerouted around to another position on a different track. The object of shuttling pieces around is to surround each triangle with it's like-colored pieces. Sounds simple? Until you try it.

3.21   Gear Ball—Meffert's Rotation Brain Teaser

Created by award winning toy makers Oskar van Deventer and Uwe Meffert, *Gear Ball* is a brain-teaser spherical puzzle.[76] Think of it as a spherical Rubik's Cube with gears. Both employ twisting motions to rotate and restore like-colored pieces to the faces of a cube. In *Gear Ball*, colored plastic pieces are interlocked and moved around by gear-sets that shuttle parts over the surface of an inner ball. The object of all the twisting is to reposition pieces so that like-colored ones move onto and restore each of the six single-colored cube faces. It's extremely challenging to anticipate how a twisting rotation will move pieces to a desired position yet minimally disturb ones you have already grouped like the ones shown on the ball to the right.

3.22   Rhombic Tuttminx 66

VeryPuzzle's *Rhombic Tuttminx 66* has out-Rubiked Rubik's Cube.[77] This 5.7-inch diameter twist puzzle is based on the spherical icosidodecahedron[78] and shuttles 266 small plastic parts around 42 wheels. Like the Rubik's Cube, the challenge is to rotate various wheels in such a way that colored tiles can be shuttled about the surface of the sphere, from wheel-to-wheel, to group them together with other like-colored pieces and re-assemble the original 32 solid colored face patterns. The challenge, of course, is not to disturb groupings already joined together while moving pieces around. Each rotating wheel of parts is centered on a face or vertex of the spherical icosidodecahedron. Parts for the same face have the same color and directional herringbone texture. VeryPuzzle designers say that the striking color patterns are based on a visually pleasing geometric relationship called Voronoi Cells (see Figures 3.9 and 3.10).

---

[74] (Mark White 2003).

[75] Image courtesy of Verdes Innovations S.A., www.v-spheres.com.

[76] Gear Ball image courtesy of distributor Recent Toys International, Amsterdam. Gear Ball is made by Meffert's.

[77] Image courtesy of VeryPuzzle, Zhongshan, China, www.verypuzzle.com.

[78] The icosidodecahedron is an Archimedean solid with icosahedral symmetry. It has 30 vertices, 60 edges, and 32 faces (twenty are triangular, 12 are pentagonal). See Table 6.5.

3.23   Japanese Temari Balls

Sanuki kagari (embroidered) *temari* ("hand balls") are one of the most beautiful examples of spherical subdivision and traditional folk-art. They are a highly valued Japanese craft practiced by cultural preservation guilds and hobbyists alike. Figure 3.37 is just a small sample of the huge variety of layouts, colors, embroidery, and needlework. Craftsmanship and design complexity are indicators of rank in guilds and hobbyist self-improvement.[79]

Temari Balls are a form of folk art, originating in China, and introduced to Japan in the 7th century. Typically, they are 4 inches to 6 inches in diameter and have cyclic or regular polyhedral symmetry. Traditional balls were carefully hand-embroidered, often made from old kimono threads (a sign of remembrance), and given to children by parents or grandparents as gifts on special occasions. Balls might contain secret wishes for the child or contain a noise-making object like a bell. Temari Balls are precious heirlooms; antiques are highly prized by collectors.

The extraordinary variety and beauty of these designs speak for themselves. What is less visible and appreciated, however, is the spherical subdivision techniques artisans use to define the reference points, great circles and polygonal areas used for their subsequent threadwork. What's more, traditional techniques do not require any calculations.

Measurements use simple paper strips like tape measures. The basic unit of measure is the circumference of the ball. It's marked on the tape and by folding it in various ways, crease marks show half- or quarter-circumference distances that are then used to position pins at the poles and equator. The beauty of basing a layout on the ball's circumference is that the same reference methodology applies to any size Temari; measurements are always relative to the ball's circumference. If one is learning how to make Temari Balls, these layout techniques are excellent ways to learn spherical subdivision.

Figure 3.37. Embroidered Temari Balls.

---

[79] Image by permission © Steven Bowes.

3.23.1   Basic Ball and Design Layouts

The cores of Temari Balls vary a great deal. Traditional ones might start with cloth scraps wadded up to make a ball shape and held in place by tightly wound threads in a random great circle pattern. Modern ones might start with a Styrofoam ball covered with batting and bound tightly by yarn and threads.

All design layouts begin by dividing the surface of the *mari* (ball) into spherical polygons by temporarily placed pins on the ball and winding reference threads, mostly great circles, between them. If the reference threads become part of the final design, they are tied off where they intersect pins and the pins are removed before artistic stitching begin.

There are three traditional layouts but many variations exist. The three basic ones are shown in Figure 3.38.[80] The combination numbers indicate how many times a great circle crosses the ball's equator:

(a)   Simple division (*tanjun toubun*) bi-polar resembling the equator and lines of longitude, often with equatorial stitching;

(b)   8-combination division (*hachitoubun no kumiawase*) spherical octahedron, cube and tetrahedron references;

(c)   10-combination division (*juutoubun no kumiawase*) which is used to define dodecahedral and icosahedral designs.

3.23.2   Platonic Layouts

Many Temari designs are based on spherical platonics and most begin by defining the spherical octahedron (eight-combination division *hachitoubun no kumiawase*), Figure 3.38 (b). An octahedral layout can be used as-is or further subdivided to create its dual, the cube, or refined further to define a tetrahedron from the cube's edges and vertices. Arcs from spherical cubes are used to define icosahedral layouts.

The octahedral layout starts with two polar pins set half-diameter distance apart. Points on the ball, midway between the poles, define the equator. Four equally-spaced pins along the equator and great circle threads are wrapped through each antipodal pin-set mark the octahedron's thee principal great circles. Three more great circles are wrapped through pins of adjacent faces adding three more to a total of six great circles total. Figure 3.38(b) shows all six. Notice in (b) that sets of three great circles intersect

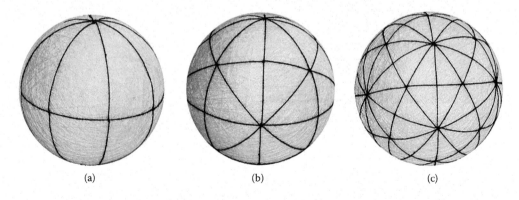

(a)   (b)   (c)

Figure 3.38. Traditional Temari Ball reference designs.

[80] Images courtesy of Ginny Thompson at TemariKai.com.

at the midpoint of each octahedral face. These points are the vertices of the spherical cube, the octahedron's dual. The dual relationship can be seen in Figure 3.38(b).

Although the tetrahedron is not a dual of the cube or octahedron, it's a self-dual and has a symmetry relationship to the cube. By selecting opposing pairs of cubic face vertices and great circle segments between them, the tetrahedron is defined.

In summary, the octahedron is defined first, various combinations of great circles define face midpoints, these midpoints define its dual the cube and selective sets of vertices/edges of the cube define the tetrahedron.

Temari reference layouts can become quite complex and subtle as shown in Figure 3.38(c). But they all have their roots in the basic layout of poles and equator—both based on the relative circumference measure. The physical layout sequence in Temari Balls mirrors many of the mathematical procedures described later in this book.

3.24 Art and Expression

Subdivided spheres appear to defy gravity; they seem to want to be in space like stars, or under water like bubbles. And when they are held up, they feel as if columns of light alone could support them. Most of our spherical subdivisions are highly symmetrical—thus, balance is maintained in what's kept and what's left out. It's like your hand. Your fingers, along with the spaces between them, make up something larger.

Artists often equate mass and monumentality. However, subdivided spheres point out contradictions in this equation. Subdivided spheres can be massive, yet light, solid, airy, space-filling, or full of holes.

Here we sample the work of George Hart, Duncan Stuart, Vladimir Bulatov, Bathsheba Grossman, Rinus Roelofs, Francesco de Comité, Dick Esterle, and Nick Sayers.[81] No explanation is needed—they are here to delight and to educate. For the spheriest, subdivision art is where surfaces, edges, and points meet light, texture, material, and you, the viewer.

Spaghetti Code
Aluminum, 2 meters,
2004 © George W. Hart

Five-dimensional Hypertetrahedron
Pencil and illustration board, 20 inch × 30 inch,
undated, estate of Duncan Stuart

---

[81] Five-dimensional hypertetrahedron image used by permission of the Estate of Duncan Stuart; Spaghetti Code image courtesy of George Hart; Dodecahedron XII image used by permission of Vladimir Bulatov, Abstract Creations, Corvallis, Oregon, USA; Quintron image courtesy of Bathsheba Sculpture LLC; Interwoven Sphere image courtesy of Rinus Roelofs; Hamiltonian Circuits image courtesy of Francesco De Comité; Train Ticket Sphere image courtesy of Nick Sayers.

Nobbly Wobbly™
*Child and pet's rubber ball toy based on
the 6 great circles of the icosidodecahedron.
Image courtesy of © Dick Esterle*

Interwoven Sphere
*Metal tubing, 60 elements,
2008 © Rinus Roelofs*

Dodecahedron XII
*Stainless steel and bronze, diameter 2 inch,
2007 © Vladimir Bulatov*

Hamiltonian Circuits on a Dodecahedron
*Computer 3D rendering,
2009 © Francesco De Comité*

Quintron
*Direct-metal painting in steel, diameter 4 inch,
2009 © Bathsheba Grossman*

Train Ticket Sphere
*Train tickets slotted together,
diameter 48 cm, 2010 © Nick Sayers*

Additional Resources

Science and Nature

Caspar, Donald L. D. and A. Klug. "Physical Principles in the Construction of Regular Viruses." *Cold Spring Harbor Symposia on Quantitative Biology* 27 (1962): 1–24.

Conway, John H., Heidi Burgiel, and Chaim Goodman-Strauss. *The Symmetries of Things*. Wellesley, MA: A K Peters, Ltd., 2008.

Mathematics, Geodesics, and Cartography

Brodsky, Isaac, "Geospatial Analytics with H3+kepler.gl," *Spatial Data Science Conference 2019*. New York City: Columbia University, 2019.

Kitrick, Christopher J. *"Dymaxion Map Transformations – Technical White Paper." ResearchGate*, DOI: 10.13140/RG.2.2.27550.69444, July 2018.

Ringler, Todd D., Ross P. Heikes, and David A. Randall. "Climate Modeling with Spherical Geodesic Grids." *Computing in Science and Engineering IEEE Computational Science and Engineering* 4.5 (2002): 32–41.

Szalay, Alexander S. et al. "Indexing the Sphere with the Hierarchical Triangular Mesh," Microsoft Corp. 2005. Technical report MSR-TR-2005-123, 2005.

White, Denis, A. Jon Kimerling, and W. Scott Overton, "*Cartographic and Geometric Components of a Global Sampling Design for Environmental Monitoring." Cartography and Geographic Information Systems* 10 (1992): 5–12.

Polyhedra and Spherical Art

Cromwell, Peter R. *Polyhedra*. Cambridge [Eng.], Cambridge: Cambridge University Press, 1997.

Cundy, H. M. and A. P. Rollett. *Mathematical Models*. Oxford: Oxford University Press, 1961.

Hart, George W. and Henri Picciotto. *Zome Geometry: Hands on Learning with Zome Models*. Emeryville, CA: Key Curriculum Press, 2001.

Holden, Alan. *Shapes, Space and Symmetry*. New York: Columbia University Press, 1971.

Mizuta, Ai (Mayumi), Ginny Thompson English collaborator. "Temari – The Geometric Shapes & Needlework." Format: Kindle Edition, Reference: Amazon.jp amzn.to/2TkkkOS, 2020.

Termes, Dick. *"Parts of the Whole: 1-6 Point Perspective on a Sphere."* Spearfish, SD: Termesphere. Termespheres.com, 2020.

Web, Robert, *"Stella4D."* PC Software2D, www.software3d.com, 2020.

Wenninger, Magnus J. *Polyhedron Models*. Cambridge [Eng.]; New York: Cambridge University Press, 1971.

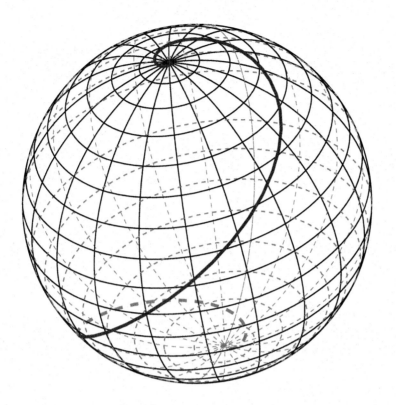

④ Circular Reasoning

We live on a small part of a large sphere. Most of the time our senses tell us that it is flat. We perceive our world as Euclidean, with straight lines and distances measured in feet or meters. We see parallel lines in roads and railroad tracks. Objects come in various sizes and we can enlarge or reduce any image, preserving all the angles. Our daily life reinforces these perceptions.

But on a sphere, none of these impressions are true. The shortest path joining two points on a sphere is called a *geodesic* and it is an arc of a great circle defined by the intersection of the sphere and a plane through its center and distances are measured in arc length. Geodesic arcs are the "*straight lines*" of spheres. On a sphere, every pair of distinct great circles intersect, not once but twice, at points opposite one another called *antipodal points*. There isn't even a unique shortest path between these opposite points, either. We see this on a globe where there are an infinite number of geodesics (longitudinal lines are arcs of great circles) that pass through the North and South Poles; *all* of them are the "*shortest path*" between these two antipodal points.

There are other differences between Euclidean geometry and spherical geometry. There are no "*rectangles*" with straight sides and four right angles, either. And somewhat surprisingly, there are polygons with just two "*straight*" sides. Triangles are different too. When the three corresponding angles of one triangle are equal and in the same order as another triangle, the two triangles are congruent. But you cannot *"enlarge"* or *"shrink"* the size of a spherical triangle and preserve its angles as you can with similar triangles in Euclidean geometry. There are no two ways about it: spherical geometry is simply non-Euclidean.

No geometry on a sphere can satisfy all of Euclid's five postulates (see sidebar to the left) for plane geometry, so any geometry we define is *"non-Euclidean"* by nature. Non-Euclidean geometry is one of many geometric systems where some of Euclid's first four postulates hold but his fifth one, the so-called parallel postulate, does not. Euclid's fifth postulate, restated in a simpler way, says that through a point not on a line, there is no more than one line parallel to the first line. In Euclidean geometry, parallel lines—lines in the same plane that do not intersect—are fundamental. However, on a sphere, the parallel postulate does not hold, either because there are no lines parallel to the given line through the point, or there will be more than one line through the point that does not meet the original line.

Mathematicians since Euclid realize many self-consistent geometries are possible where one or more of Euclid's postulates do not hold. Spherical geometry is one of them.

In this chapter, we explore the properties of spheres and see where we can employ Euclidean principles and where we can't. As the chapter title says, we use *"circular reasoning."* We start with arcs of great circles joining non-antipodal pairs of points and proceed to construct spherical polygons on the surface of the sphere. Since spherical triangles are the easiest to use when we want to subdivide a sphere, we concentrate on properties of triangles, such as shape, symmetry, and area.

Spheres are homogeneous; that is, a region about every point looks like a region about every other point, in every direction we look. To compensate for this sameness, we introduce coordinate systems so we can locate specific points and distinguish one region from another. The earliest coordinate systems go back to astronomers studying the celestial sphere, or geographers mapping the surface of the Earth, many of whom were also astronomers. As in plane geometry, the choice of coordinate systems is very important for the sphere. Often a problem can be stated and solved much more easily in one coordinate system than another. A basic task is to determine which system to use in which situation and how to convert results found in one system so they can be used in another.

We begin with the fundamental notions on which everything else is based; namely, points and circles.

4.1   Lesser and Great Circles

Any plane cutting through a sphere intersects the sphere's surface in a circle. If the plane goes through the center of the sphere, the circle is called a *great circle*; and if not, the circle is a small *circle*, also called a *lesser circle*. A family of parallel planes in space will cut the sphere in one great circle and infinitely many lesser circles and some of these circles are particularly useful in geodesic applications. In Figure 4.1(a), every circle is a lesser circle defined by a plane that intersects the sphere but does not pass through the sphere's center. In the figure, the plane defining each lesser circle is perpendicular to one of three Cartesian axes. Some circles have such small radii that they appear as points, but the intersection of the sphere and a plane is a point only when the plane is tangent to the sphere at the point so it does not cut through the sphere.

A slice of a sphere also creates something else besides the great or lesser circle, namely two spherical caps. The spherical caps of a great circle are hemispheres or two half-spheres. For a lesser circle, one spherical cap is smaller than a hemisphere and shaped somewhat like a contact lens. A spherical cap has a center on the sphere and that is different from the center of the disc in the plane that cuts out the circle on the sphere. If we choose a collection of points on the sphere and consider a spherical cap centered on each one, and if all the caps have the same size and the largest possible radius without causing any caps to overlap, we can compare the total area of the caps and the area of the sphere to get a measure of how well distributed the points are on the sphere (see Figure 4.1(b)). The dashed caps are on the other side of the sphere. The radii of the cap's lesser circles are all the same. Some are tangent to others, but none overlap. Thus, for this number and distribution of points on a sphere, the total area of the caps as a percentage of the total area of the sphere tells us how uniformly the points are distributed. In Figure 4.1(b), the caps cover about 71% of the surface of the sphere. This important use of lesser circles and spherical caps will be explored again later in the chapter.

Great circles are the geodesics of the sphere; they are the essential tool of spherical subdivision. Unlike lesser circles and spherical caps, a great circle is a circle on the surface of a sphere having the same diameter as the sphere; they are the largest possible circle on a sphere. A great circle shares its center with the center of the sphere. Any two points on a sphere, plus the sphere's origin, uniquely define the plane of a great circle—so long as the two points on the surface of the sphere are not coincident or on exact opposite sides of the sphere. A sphere's great circles are all equal in diameter, curvature, and planar area.

An infinite number of great circles can be drawn on any sphere (a few of them are shown in Figure 4.2) by randomly selected planes, each passing through the

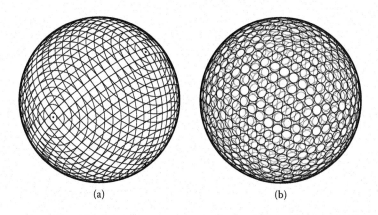

(a)   (b)

Figure 4.1. Lesser circles and spherical caps.

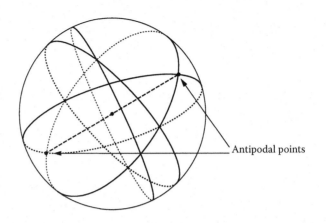

Figure 4.2. Random great circles.

sphere's center. Notice that every one of the great circles intersects every other great circle two times. Notice the line of intersection of two great circles intersect the sphere's surface at two points opposite each other (antipodes).[1] One pair of great circles, their line of intersection and intersecting antipodes are indicated in the figure.

In Figure 4.2, it is easy to see each great circle divides the sphere into two symmetric parts called *hemispheres* and that great circles are the largest circle possible. The amount of *curvature* of a great or lesser circle can also be measured. It is equal to the reciprocal of the circle's radius (the smaller the reciprocal, the "*flatter*" the circle's curve). Great circles have the least curvature of any circle on a sphere while small circles have higher curvature on the same sphere because they bend more sharply. For example, if the radius of the sphere in Figure 4.1(b) is equal to one, the curvature of the great circles will also be equal to one whereas the curvature of the cap's lesser circles with radius of 0.063 (each cap has the same radius in the figure) is 15.9. It is clear, from this example, that arcs of great circles are the only possible choice for locally shortest curves in the plane of the circle and curves with this property are the straight lines of spheres and the shortest distance between any two non-antipodal points. This distance is often called the *great circle distance*.

Great circles have another relationship with their spheres. The area of a sphere is given by the classic formula:

$$\text{area of sphere} = 4\pi r^2.$$

Notice in the equation that the sphere's area is four times the planar area of one of its great circles. There is a handy way to remember this: you need twice as much paint for the cathedral dome (half a sphere) as you do for the floor under it.

4.2   Geodesic Subdivision

Great circles and their geodesic arcs are essential in spherical subdivision. Great circles are used to subdivide a sphere, and great circles are the result of spherical subdivision. We need to take a closer look at their properties to understand this dual role.

We have said that geodesics are the straightest lines on a sphere and they are arcs of great circles. In common usage, a geodesic is the shortest distance between two points

---

[1] Antipodal points have the same Cartesian coordinates, but opposite signs for each component. For example, on a unit sphere, point (–0.5, 0.5, 0.707) is opposite (0.5, –0.5, –0.707).

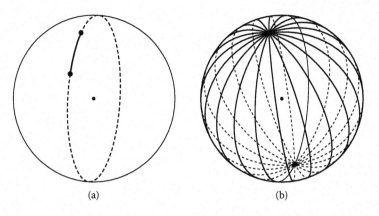

(a)   (b)

Figure 4.3. Short- and long-path geodesics.

on a sphere. But from a mathematical perspective, there are two geodesics between a pair of points that are not antipodal: a short path and a long path. Figure 4.3(a) shows both—the geodesic short path is solid, while the long path is dashed—between two points. In spherical subdivision, we almost always use the short path geodesic.

The situation is a little different when the points are antipodal. In this case, an infinite number of geodesics pass through them. Both long and short paths have the same arc distance and none are unique in length. Figure 4.3(b) shows a few of them and illustrates two facts again: every great circle bisects every other great circle, and every great circle intersects every other great circle at antipodal points. In (b), every great circle intersects at the same antipodes, but this is certainly not the case in Figure 4.2, where every pair of intersecting great circles creates different antipodes.

In spherical subdivision, we regularly intersect geodesic arcs to derive points that define a polygon or grid on the sphere's surface. We will develop these techniques in detail later, but here, we can get a feeling for the general subdivision process. In Figure 4.4(a), points a and f, c and h, and the sphere's origin define two great circles that intersect at antipodes 1 and 2 (Appendix C explains the simple procedure for doing this). We can say that points 1 and 2 are derived from, or are the result of, intersecting two great circles.

We can derive points to make very complex grids and spherical subdivisions by intersecting different combinations of great circle arcs. Figure 4.4(b) outlines how this is done. The process starts by defining a series of reference points around the edges of a spherical triangle. In this example, there are 24 edge reference points including the three triangle vertices. The triangle's edges themselves are geodesic arcs between the triangle's vertices. The reference points are represented by small dots around the edge of the large spherical triangle in (b), and some are labeled a through h. We will discuss how these reference points are defined in a later chapter. For now, we will focus on using them to derive new points defined by intersecting geodesics.

By selectively picking pairs of edge reference points, for example, points a and f with points c and h, or b and e with d and g, we define four geodesic arcs that intersect at antipodal points 1 and 2, 3 and 4. Points 1 and 3 fall within the spherical triangle and define useful vertices on a subdivision grid. Points 2 and 4 fall on the opposite side of the sphere and are ignored. We can repeat this process—select pairs of edge reference points, define their geodesic arc, and select one of their intersection antipodes— over and over until all the subdivision vertex points within the triangle are defined.

To review geodesic subdivision again from a slightly different perspective, in the previous example, we started with 24 reference points around the edge of a large spherical triangle, as shown in Figure 4.4(b). Three of those points are the vertices of

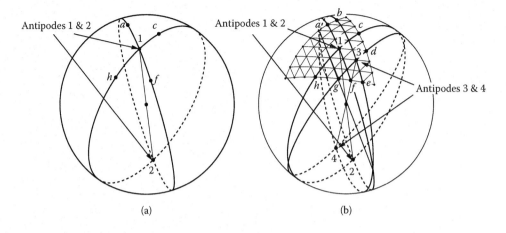

Figure 4.4. Defining points on a sphere by intersecting great circles.

the large spherical triangle we want to subdivide. Notice that these 24 reference points define sets of eight great circles; each great circle in a set runs somewhat *"parallel"* to the edges of the large spherical triangle. By intersecting combinations of one great circle from each of the three sets, we create 42 antipodal points, 21 of those points falling within the large spherical triangle we are subdividing; the other 21 fall on the opposite side of the sphere and are ignored. The 45 total vertices shown for this sub-divided triangle consist of 24 edge reference points plus 21 points derived from the intersection of great circles.

Once the vertices of the grid are defined, the grid members themselves can be either short geodesic arcs, such as the ones shown in Figure 4.4(b), or straight-line chords between grid points. The former results in a grid of small spherical triangles; the latter results in planar triangles. If we want to subdivide the whole sphere, we simply make copies of the subdivided spherical triangle and then position them to cover the rest of the sphere without gaps or overlaps.

We have just given the briefest of introductions to subdividing spheres. This method demonstrates a basic technique for using great circles to generate points that make up a subdivision grid. These techniques have several variations, and we will devote two chapters to them. Some variations start by subdividing a different shaped triangle, an equilateral triangle, or perhaps a right triangle. The number and spacing of the edge reference points around it may also vary. Other variations change the number of great circle intersections needed to derive a single subdivision point.

Regardless of these variations, the technique of deriving points by intersecting great circles or geodesic arcs remains the same. In Chapter 8, we use this important technique in several ways, and in Chapter 9, we compare their effects in subdivision. Depending on your application, one variation may result in a more suitable distribution of points than another and we will examine some ways of telling which are best for a given situation. Chapter 10 introduces another very different method of subdivision where grids are self-organizing and not based on great circle intersections.

4.3   Circle Poles

Every great and lesser circle has two poles, each equidistant from every point on the circle's circumference. The poles are points on the sphere and define an axis that is perpendicular to the plane of the circle, passing through the circle's center and the

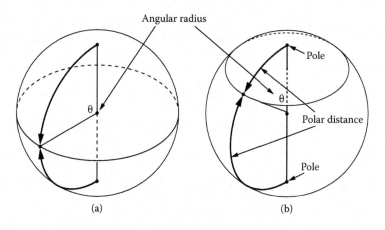

Figure 4.5. Circle poles and polar distances.

sphere's origin. Figure 4.5 shows a great (a) and lesser (b) circle, their poles, axes, and points on the circumference.

Polar distance is the arc distance on the sphere between the point and either the North or South Pole. All points on the equator are the same distance from the North or South Pole, measured in degrees or *radians* (90° or π/2 radians) as shown in Figure 4.5(a). For the lesser circle shown (b), the polar distances are somewhere between 0° and 180°.

The *angular radius* of a circle (lesser or greater) is the interior angle between the circle's closest *pole* and any point on its circumference. This angle is labeled θ in Figure 4.5(a) and (b). Note that the angular radius of any great circle is always 90°. The angle between the two points in the sphere measured from the origin is called the *central angle*.

One last relationship between great and lesser circles is worth mentioning. Pairs of points on a sphere that are not coincident or antipodal, together with the origin of the sphere, define two circles. In geodesics, we are usually interested in the shortest path between the pair, but the same two points also define the circumference of a lesser circle, as shown in Figure 4.6. The center of this circle is inside the sphere and at the middle of the chord between the two points.[2] The plane of the lesser circle is

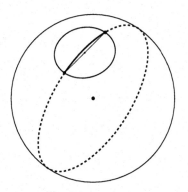

Figure 4.6. Points define great and lesser circles.

---

[2] The line between these two points is a chord. The perpendicular distance from the midpoint of the chord (indicated with a dot) to the great circle is called the *sagitta,* and to the center of the great circle is called the *apothem.* Both are somewhat antiquated terms but useful in spherical work.

perpendicular to the plane of the great circle, defined by two points and the origin of the sphere. If those two points are antipodal, the lesser circle becomes a great circle, and an infinite number of them pass through these opposite points, as shown earlier in Figure 4.3(b).

4.4   Arc and Chord Factors

Distances between points on the surface of the sphere are measured in one of two ways: either along the surface following the great circle arc between them, or a straight line that cuts "*through*" the sphere as a chord.

The *chord distance* is the direct straight-line distance between any two points on a sphere. The chord's length depends upon the radius of the sphere and is directly proportional to it. *Chord factors* express this length relative to both the radius of the sphere and the central angle between the two points measured at the origin. Chord factors are very useful in geodesics. If the sphere's radius is one, a unit sphere, the chord factor can be scaled up or down for any real-world application. That is, a chord's factor can be multiplied by the true radius of the intended sphere to arrive at the exact dimension of a required implementation. For any sphere with radius r, we can express the general relationship of a chord's length to its central angle θ as

$$\text{chord factor} = 2r\sin(\theta/2).$$

Arc factors are conceptually related to chord factors. Instead of measuring the direct straight-line distance between points, arc factors measure the arc length (geodesic distance) between the points in radians. The arc factor is simply the radian measure of the central angle of the arc. One radian is the same length as the sphere's radius "*wrapped*" around the circumference of the sphere or one of its great circles. The number of radians or degrees in a great circle is

$$1 \text{ radian} = 360°/2\pi \ \text{ or } \ 180°/\pi = 57.2958°.$$

Thus, the number of radians in a degree is

$$1° = \pi/180° \text{ radians}.$$

An arc factor provides two pieces of information around the arc at the same time: it defines the size of the angle and the length of the arc it makes on the surface of the sphere. The arc factor is simply the radian measure of the central angle θ between two points on the unit sphere. The formula is

$$\text{arc factor} = \pi/180° \theta r,$$

where θ is the central angle between two points.

4.5   Where Are We?

The sphere is an omnidirectional object; every view is the same. There is no top, bottom, left, or right side to a sphere unless we impose one. To answer the question, "*Where are we?*" we need a reference system, and several important ones are used in geodesics. The ones we consider here are all based on an origin and axis system.

We start with the oldest and most intuitive system, the altitude-azimuth system, introduced by ancient astronomers. From there, we look earthbound and see what the familiar latitude-longitude system tells us about the sphere. Neither of these systems considers the radius of the sphere. The stars are so far away, the Earth's radius is simply ignored. We will also look at two other systems, spherical and Cartesian systems, where distance and radius do play a role.

Why do we need so many systems? For a unit sphere, anyone these systems will do, if we make certain assumptions. But in practice, it often is easier to solve certain geometry problems by using one system rather than another. It is common practice to solve part of a problem in one system, and then convert the result to another system for the final resolution. Reference systems help us to find where we are, just as the ancient astronomers figured out their location by looking at the sky.

4.6 Altitude-Azimuth Coordinates

The altitude-azimuth coordinate system, also called the horizontal coordinate system, is a two-dimensional (2D) reference system that astronomers and celestial navigators have used for centuries to locate bodies in the sky. The sky is treated as if it were a huge spherical shell surrounding the Earth, with the observer inside, at the center. Altitude-azimuth coordinates are easy to understand and are widely used because they are based completely on references relative to the observer or a reference point at the center of the sphere.

The references in this system are the observer's horizon plane (the earth or sea), *zenith* and *nadir* (points directly overhead and beneath the observer), and the *meridian plane* passing through the observer's location north and south and the zenith-nadir axis. Figure 4.7 shows the general arrangement of these references relative to an observer at the center. Point P on the sphere (i.e., a star, planet) is located by two angles: its altitude (Alt) and *azimuth* (Az).

Referring to Figure 4.7, the altitude is the angle from the observer's horizon to point P and can be any value from $0°$ (the body is on the horizon) to $90°$ (the body is at the observer's zenith or directly overhead). It is measured along the arc of a great circle

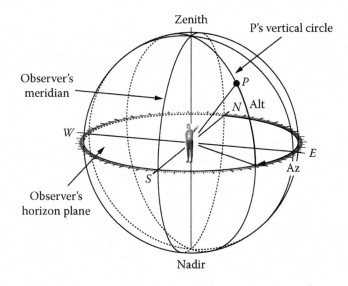

Figure 4.7. Altitude-azimuth coordinate system.

called a *vertical circle*. The observer's elevation is not relevant.[3] He or she can stand on the deck of a ship or atop a mountain and will still achieve the same result.

The azimuth angle is measured from the observer's meridian. Two conventions are in common use. In North America, azimuths are measured from the north, increasing in a clockwise direction. East is 90°, south is 180°, and so on. In Europe and in celestial navigation, azimuth is measured from the south, with west being taken as positive. The azimuth of the eastern point on the horizon is –90° and the azimuth of the west is 90°.

Altitude-azimuth coordinates do not include the distance to the point. Distances to the stars are so great that these small discrepancies are not significant.[4]

Although widely used, the altitude-azimuth coordinate system does have some drawbacks. When applied to astronomy, the main drawback is the fact that its reference frame moves with the Earth. It is difficult to track celestial bodies without having accurate time and location information, because their coordinates are constantly changing, a result of the Earth's second-by-second rotation around its axis, and its yearly revolution around the sun. Altitude-azimuth coordinates also are specific to a particular location on Earth, but a measurement can be made, the time noted, and the result easily converted to any other coordinate system. For the same date, time, and point in the sky, coordinates will be different for observers at other locations, if the body is even observable from all of them. Oftentimes, other systems are used and then converted to altitude-azimuth coordinates for a particular time and observer location.

4.7   Latitude and Longitude Coordinates

Undoubtedly, the most familiar coordinate system is the one using latitude and longitude. All points on the sphere can be located with latitude-longitude coordinates. Unlike the altitude-azimuth coordinate system, this system combines lesser circle references with greater circles. Here we describe the general features of latitude-longitude coordinates and conversions to the Cartesian system.

Before we start, we should point out that terrestrial latitude-longitude coordinates describe positions on a sphere. The Earth is not a perfect sphere, as it is slightly flat at the poles and bulges a bit at the equator. Technically speaking, the Earth is a spheroid or an ellipsoid. For discussions of coordinates, however, we will treat the Earth as if it were a perfect sphere.

The concept of latitude has been known for centuries, as astronomers recorded everyday observations of the sun's movement throughout the year and the altitude angle above the horizon of Polaris, the North Star. Even without corrections for atmospheric distortion and for Polaris offset from the true north celestial pole,[5] latitude can be determined to within a few miles just by measuring the North Star's altitude above the horizon.

---

[3] In celestial navigation, a sextant measures the altitude of celestial bodies above the horizon, which might be the sea or Earth's land horizon. In these cases, the user's height must be factored in. The result is a small angle called dip of the horizon, which must be subtracted from the observed angle above the horizon. Otherwise, the sextant angle reading is too large, which will have the effect, if not corrected, of placing the observer further from his actual line of position.

[4] For objects closer to the Earth, such as the moon, *parallax* must be considered.

[5] The star Polaris is not exactly on the Earth's north polar axis. Instead, it is offset by a small amount and circles the true point by an arc distance of ¾°, thus tracing a small circle around the north celestial pole of about 1½° diameter. In celestial navigation, a small correction angle must be applied to a sextant's measured angle when Polaris sights are taken. The amount of this correction depends on the time the sighting is taken. Twice each day, there is no correction. At two other times, a maximum positive and negative correction is required.

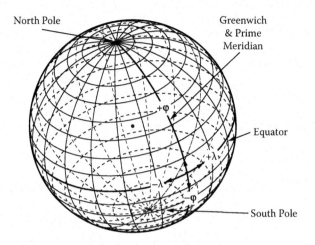

Figure 4.8. Latitude and longitude system.

Referring to Figure 4.8, the latitude of a point, usually denoted by the Greek letter *phi* (φ), is the angle from the equator to the point. Latitude circles are lesser circles defined by points of equal latitudes. The equator, latitude 0°, is the only one that is a great circle.

Longitude references, denoted by the Greek letter *lambda* (λ), are great circles through the North and South Poles and are also called meridians of longitude. The line connecting the two poles is called the axis. Longitude measures angles relative to the Greenwich Meridian[6] and increases 0–180° east or west. There are no common sign usages, except for time zone differences where east longitudes are positive.

4.8 Spherical Trips

Because latitude-longitude coordinates are so familiar, we can take some mental trips around the world to review what we have learned about spheres and see other characteristics of spherical triangles.

We take a companion on our first trip and begin at the North Pole. The first thing we notice is that every direction is south. Remember, there are an infinite number of great circles (longitude meridians) through two antipodal points (North and South Poles). We decide to walk south, but each of us sets off in a different direction. After a long walk, we both meet up again at the South Pole. We arrive at about the same time, because we traveled at approximately the same rate and our overall great circle distances were the same.

The only difference between our two trips is how far apart we were at any time. This distance is called the *separation angle*—the angle, measured from the origin of the sphere, between two meridians at a given latitude. If we each took the same great circle but walked in opposite directions from the North Pole, the most we could be separated is 180°. The fact that our separate courses started and ended at common points also means that our paths defined the sides of a spherical polygon—a two-sided one. A two-sided polygon, called a *lune* or *digon*, is only possible on a sphere; they are always something less than a hemisphere. If we had made our trip by traveling the same great circle, but had walked in opposite directions, our paths would have defined a hemisphere.

---

[6] The Greenwich Meridian has not always been the prime meridian. Throughout history, it has changed many times, favoring the dominant maritime nation at the time.

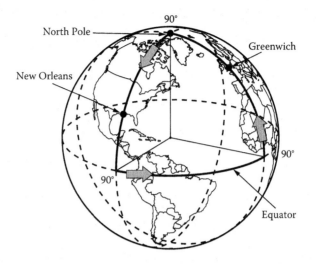

Figure 4.9. Spherical triangle with three right angles.

Our second trip involves three courses. Figure 4.9 shows our trip plan. Starting again at the North Pole, we decide to walk south to the equator by way of longitude 90° west. Along the way, we pass through New Orleans, Louisiana. When we reach the equator, we turn left 90° and proceed to walk one quarter of the way around the equator (or 90°) until we get to the prime meridian, longitude 0°. We then make another 90° left turn and walk due north, past Greenwich (near London) until we arrive again at the North Pole. We have just walked a triangular course and arrived back where we started. We have also demonstrated that spherical triangles can have three 90° angles. The North Pole is the only place on Earth where you could travel south, then east, and then north and return to your starting point. From the South Pole, you could accomplish the same thing by traveling north, then east or west one quarter of the way around the equator, and then south again to make a similar trip.

Our last trip to the equator tells us something else about spherical triangles—any two triangles where two angles and a side are equal are not congruent on a sphere even though they are in plane Euclidean geometry.

We can see this with a slight variation in a third walking tour. Once again, we leave from the North Pole and walk to the equator. But instead of walking along the equator exactly one quarter of the way around the sphere, we decide to walk a little less (or more) and then make the same left 90° turn to get back to the North Pole. This route still forms a triangle, but this one only has two 90° angles instead of three. The distances from the North Pole to the equator on the two polar legs of the trip remain the same as before, but the equator side between the two 90° angles is different. Unlike in Euclidean geometry, this new triangle is not congruent to one we made in the second trip, even though they both have an equal side (North Pole to equator) and two equal angles (90° angles at the equator).

Our last two trips each had three courses and formed the edges of a large spherical triangle. But what if we take a trip with four courses and try to form a square or rectangle? Let's say we start at mid-latitude 45° north longitude. This trip might shake your faith in the utility of the compass. We start by walking 500 miles due east, then turn left 90° and walk north for another 500 miles. We make yet another 90° left turn and walk west another 500 miles, then make another turn left and walk due south another 500 miles. We have just walked a long, "*squarish*" route, where each course was 500 miles and each turn was 90°. But we did not end up where we started from!

How did this happen? The reason is that although the distances were equal on the north-south legs along arcs of longitude (great circles), the two east-west legs did not cover the same number of degrees of latitude. We traveled more degrees at the higher latitude than at the lower one.

4.9   Loxodromes

Latitude and longitude are so familiar we can be fooled when navigating. Consider the use of a compass. A compass performs a very specific function; it always points in the same direction—north.[7] Think of it this way: when you change directions, the compass still points north. In effect, you are turning under the compass; the compass continues pointing in the same direction regardless of which way you point. The value of a compass is that it allows you to maintain a course over a long distance by ensuring that you cross every meridian (longitude) at the same angle. This course is called a *rhumb line* or a *loxodrome*, from the Greek words *loxos* (oblique) and *dromos* (running, from *dramein:* to run).

A loxodrome is a line on a sphere of constant bearing that cuts across all meridians at any constant angle except a right one. A loxodrome is also called a *spherical spiral* and it displays as a logarithmic spiral on a polar projection. If the course is not due north-south or east-west, a loxodromic spiral will eventually spiral to either the North or South Pole and then endlessly loop that pole in smaller and smaller circles. The title graphic of this chapter shows a loxodrome.

A loxodrome displays as a straight line on a Mercator map. When plotted on any other map projection, it is a curved line. A loxodrome is a great circle, only if it is along the equator or a meridian. A rhumb line course between two points that is not along the equator or a meridian is always longer than the *great circle route* between the same two points. And there are an infinite number of rhumb lines between any two nonpolar points that are not on the same latitude or longitude. They only differ in the number of times they encircle the Earth.

Figure 4.10 shows rhumb lines or loxodromes between two sample points, 1 and 2, and how they compare to a direct great circle route, shown with a bold line.

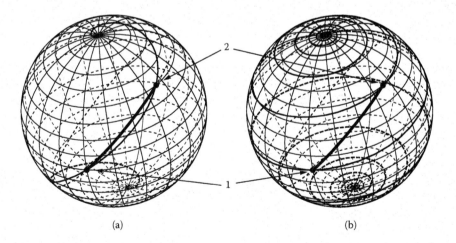

(a)   (b)

Figure 4.10. Loxodrome courses 56° and 84.2° with a great circle route.

---

[7] A compass is influenced by many factors. The magnetic North Pole is not exactly located at the geographic North Pole. Electrical currents (nearby wires), and large ferrous metal sources near the compass (engines, pipes, etc.) influence which way the compass needle points.

In Figure 4.10(a), the loxodrome takes the most direct route from point 1 to point 2 by following compass course 56°. In (b), the loxodrome follows a different course of 84.2°. It still leads to point 2, but must circle the Earth one time to do so. In both illustrations, the great circle route between the two points is the shortest route possible. Notice, too, if you miss your destination on any rhumb course other than along the equator, you will eventually end up at the North or South Pole.

Can we ever use a compass to follow a great circle route, if using a compass to navigate any direction other than due north-south or east-west puts us on a rhumb line or loxodrome? The answer is a qualified yes. You can approximate the great circle route by taking a series of short rhumb line courses. You would travel a particular compass course for a while before changing to another course, and then another, until you arrived at your destination. To do this, the navigator divides the overall track between his starting and ending points into a series of small segments. Between each segment is a waypoint, a location where he or she will change course slightly. The length of a course segment is about the distance the navigator expects to travel in a day, if by ship, or in an hour or so, if by plane, depending on the plane's speed.

On a ship, the course change might take place at the same time every day. For instance, navigators might change at local noon, the time when navigators traditionally used their sextants to determine their latitude by the sun's noon-time meridian passage. Today, airplanes and ships are equipped with autopilots, and course corrections are made automatically from time to time in order to stay close to a theoretic great circle course. The more rhumb lines or legs the course is divided into, the closer individual rhumb lines will approximate the great circle route.

When a great circle and rhumb line course are shown together on a Mercator map, the great circle route appears as a curved line, while the rhumb line is straight. On a sphere, they appear just the opposite, the great circle being the shortest *"straight line"* course, while the rhumb line is the longer curved course between points unless the points are directly north or south of each other or along the equator.

To give an example of just how dramatic the difference between navigating a rhumb line and great circle course can be, consider a trip between London, England, and Anchorage, Alaska. The latitude-longitude of London's Heathrow International Airport is 51° 28′ 15.72″N, 0° 27′ 30.98″W. Anchorage International Airport is located at 61° 10′ 14.69″N, 149° 59′ 33.66″W. The great circle distance between them, the most direct route, passes over Greenland and near the North Pole, and is 4475 miles (7201 km). The rhumb line bearing to Anchorage from London is about 276° (almost due west by compass) and the distance is 5731 miles (9223 km). Clearly, the rhumb line distance is longer, almost 28% longer. These differences cannot be ignored by airlines or shipping companies. The time and fuel-cost implications are considerable.

4.10   Separation Angle

We have described the geodesic between any two points on a sphere as an arc of a great circle expressed in degrees. This same angle has different names, depending on the context of its use. In spherical geometry, it is an interior angle or arc angle. In astronomy, it is called a *separation angle* and it is the angle measured between celestial objects. On Earth, it is sometimes used synonymously with the great circle distance between pairs of latitude-longitude points. For example, in our first spherical trip previously, two explorers walked in opposite directions from the North Pole on their way to the South Pole. Their distance apart at any given moment was a function of their latitude-longitude positions. Using φ, λ, and the sign conventions already discussed to represent latitude-longitude of two points (indicated by subscripts), the general formula for separation angle is

$$\cos(\text{separation angle}) = \sin(\varphi_2)\sin(\varphi_1) + \cos(\varphi_2)\cos(\varphi_1)\cos(\lambda_2 - \lambda_1).$$

A degree of the great circle arc and the distance traveled on the Earth's surface are related. One degree equals 60 minutes, and each minute equals one nautical mile, or 1676 feet. This is one reason that ocean and air navigation use nautical miles, not statute miles, to measure distance. Along the same great-circle meridian (changing only latitude), one degree equals a nautical mile. But a minute of longitude along a line of latitude is a nautical mile only at the equator.

The above equation is quite useful, but it assumes the Earth is a perfect sphere. Some caution must therefore be exercised when finding the separation angle for very short arcs or for arcs very near the poles because, in trigonometry, the sin function is sensitive to rounding errors for values close to 0°.

4.11   Latitude Sailing

Approximating great circle routes with rhumb line segments is a modern-day technique. It is only possible when you know your longitude, your distances, and when to change course.

Before the world's land masses and relative directions and distances were accurately established, explorers would combine coastal and offshore sailing, known as latitude sailing, to reach their destination. Instead of going directly to their destination, as we would today, earlier explorers had to sail north-south or along familiar coasts until they reached the latitude they desired. Then they would sail due east or west using a compass or sun and star sightings to maintain the direction until they arrived at their desired port. Latitude sailing was therefore also called "*easting*" or "*westing*." Navigators would check their latitude by taking regular sightings of Polaris, the North Star, or by measuring the elevation of the sun at their local noon, when the sun was due north or south of their ship. The angle, with corrections, told them their latitude and was measured with a wide variety of instruments, including the *astrolabe*, *quadrant*, *cross staff*, and *sextant*.

A cross staff is shown in Figure 4.11.[8] For general sailing, only minor corrections in the angle were needed when sighting Polaris. Sights of the sun and moon, however,

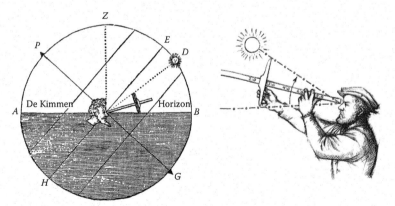

Figure 4.11. A cross staff measures the horizon angle to the sun.

---

[8] Image caption in Robert McGhee, Canada Rediscovered (Libre Expression and Canadian Museum of Civilization, 1991): "Pilot calculating the height of the sun above the horizon using a Jacob's staff (cross-staff), 16th century."

required a seasonal and daily correction. Nautical almanacs produced by observatories provide these correction factors.

4.12 Longitude

Longitude and time are related. The Earth rotates through 360° of longitude in 24 hours, or 15° per hour. A celestial object will cross two meridians at different times, and the time difference converted to degrees is the difference in longitude between the two places.

In practice, a clock on board a ship would be set to the local time of a reference location, such as the Old Royal Observatory in Greenwich, England.[9] The Royal Observatory at Greenwich would prepare celestial tables, giving the exact time that the stars, planets, moon, and especially the sun would pass the Greenwich Meridian[10] (longitude 0°) for every second of every day of the year.

At sea, a mariner would use a sextant or other device to observe one of these bodies and record the exact time it crossed the meridian (due north or south) of his ship. This was the moment at which the body was highest in the sky. The exact moment is hard to know, because the body appears to dwell at its high point for a short period. But averaging sites before and after meridian passage led to accurate timing. Consulting his almanac, the mariner would note the time difference between the body crossing his meridian and Greenwich. This is the time it took the Earth to rotate under the body the distance between the two places. Since the Earth rotates 360° every 24 hours or 15° per hour, it is a simple matter to compute longitude. But again, accurate timing and an almanac with the times of meridian passage for celestial bodies were required.

Here's an example. If the sun crosses the meridian at Greenwich, England, at noon Coordinated Universal Time (UTC), but doesn't cross the ship's meridian until 16:19 UTC, it took 4 hours 19 minutes for the Earth to rotate that much. The Earth rotates 15° in one hour (360° in 24 hours); thus, the ship is at 64° 48.0′ west longitude, the same as Bermuda.

4.13 Spherical Coordinates

Spherical coordinates, also called spherical polar coordinates, are naturally suited to geodesic work and share some characteristics with latitude-longitude coordinates. Unlike the previously discussed systems, spherical coordinates include the sphere's radius. A point can be accurately plotted in three dimensions and located on the surface of a sphere, if you know the sphere's radius (a length) as well as two polar coordinate angles.

The radius, usually represented by lowercase Greek letter *rho* (ρ), is the distance from the origin to the point. This book uses unit radius spheres, so ρ is always equal to one. The first polar angle, represented by *phi* (φ), is the geodesic arc along the great circle from the positive *z*-axis to the point. This arc is also called the zenith distance or colatitude and the angle's value range is $0 \leq \varphi \leq 180°$. This is the same polar angle described earlier in this chapter in Section 4.3. The azimuth, the other polar angle, is represented by *theta* (θ). The azimuth is the angle between the *x*-axis and the point's great

---

[9] In 1736, John Harrison, an English clockmaker, built the first clock capable of keeping accurate time at sea in conditions of extreme motion, and humidity and temperature change. For the next 50 years, he developed a series of sea clocks (marine chronometers) each more accurate than its predecessor. By the mid-19th century, an industry had developed and chronometers based on Harrison's designs had become affordable enough that most ships had one aboard set to Greenwich time. Prior to sea clocks, Greenwich time was found by astronomical observations such as Lunar-Distance methods.

[10] Today, astronomic tables are prepared and published jointly by Her Majesty's Nautical Almanac Office and the US Naval Observatory for use by the navies of both countries.

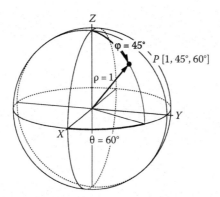

Figure 4.12. Spherical coordinate system.

circle intersection with the xy-plane. This is the same angle in the azimuth-altitude system, except it is measured from the x-axis instead of the north reference pole. Theta's value range is 180°, and the positive direction is counterclockwise around the z-axis.

Figure 4.12 shows the spherical coordinates of a point P, where (ρ, φ, θ) coordinates are $(1, 45°, 60°)$. Notice that the azimuth angle (θ) will be undefined when (φ) is 0° or 180°. The effect is to define an infinite number of great circles passing through the z-axis, as illustrated earlier in Figure 4.3(b).

Some caution is needed when using the spherical coordinate system. The (ρ, φ, θ) convention used in this book follows the standard notation of mathematicians. However, in fields like physics, angles (φ) and (θ) are reversed. To add to the confusion, polar angle (φ) is sometimes measured from the xy-plane rather than the $+z$-axis, and azimuth angle (θ) is sometimes measured from the x-axis 0 to $\pm 90°$ rather than counterclockwise 0 to 180°.[11] Scientific calculators with spherical coordinate functions have the same drawbacks. This book and the algorithms shown in Appendix C uses the mathematics convention (ρ, φ, θ), with the angular specifications described above.

4.14 Cartesian Coordinates

Without a doubt, the Cartesian system is the most familiar coordinate system. A point's position in three-dimensional (3D) space, relative to the origin of an orthogonal (perpendicular) axis system, is specified by three distances (positive, zero, or negative) in the x, y, and z directions. The Cartesian origin and some sample points, along with their coordinates, are shown in Figure 4.13.

Some very useful relationships exist between these points. The distance of a point from the origin is simply

$$\text{distance} = \sqrt{x^2 + y^2 + z^2}.$$

The distance between any two points is

$$\text{distance} = \sqrt{(x_1 - x_2)^2 + (y_1 - y_2)^2 + (z_1 - z_2)^2}.$$

Many transformations can be performed on Cartesian coordinates. Key ones described in Appendix C and include scaling, translation, and rotation. *Scaling* simply

---

[11] For standards and conventions, see report ISO 31-11 Quantities and units—Part 11: Mathematical signs and symbols for use in the physical sciences and technology, International Standards Organization, 1992.

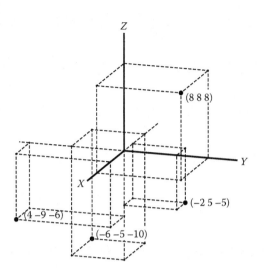

Figure 4.13. Cartesian coordinate system.

moves points closer to or farther from the origin by multiplying all coordinates by a constant value. The directions of points relative to one another remain the same. *Translation* moves one or more points to a new location. If more than one point is being translated, the move does not change the angular relationships of any of these points to one another. In a sense, the points are like passengers seated on a bus—they all move together, but their positions with respect to each other remain the same.

Rotation is very important in geodesics. In a rotation, one or more points spin around an axis, such as children rotating on a carousel. Their distances to the rotation axis remain the same; only their positions around the axis change. Just like in translation, rotations do not change the relationship of the rotated points to one another. See Appendix B for a complete explanation of rotations used in geodesic subdivisions.

Although Cartesian coordinates can define any point in 3D space, they are not independent when the points lie on the surface of a sphere. They must be related by the equation $r^2 = x^2 + y^2 + z^2$, where r is the radius of the sphere. A simple relationship between the Cartesian and spherical coordinate systems is illustrated in Figure 4.14. It is a simple matter to convert from one system to another, and converting back

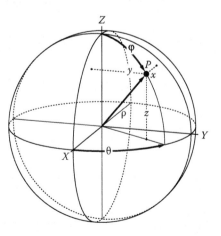

Figure 4.14. Cartesian and spherical coordinates.

and forth is common practice in subdivision work. Referring to the earlier spherical coordinate example, where P's coordinates were (1, 45°, 60°); the Cartesian value is approximately (0.3536, 0.6124, 0.7071).

Converting from spherical to Cartesian coordinate systems is quite easy. Scientific calculators can do the conversion just by changing the coordinate specification mode. The equations are

$$x = \rho \sin \varphi \cos \theta \qquad \rho = \sqrt{x^2 + y^2 + z^2} ,$$

$$y = \rho \sin \varphi \sin \theta \qquad \tan \varphi = \left(\frac{\sqrt{x^2 + y^2}}{z} \right) , \quad \text{and}$$

$$z = \rho \cos \varphi \qquad \tan \theta = \frac{y}{x} .$$

4.15 ρ, φ, γ Coordinates

The (ρ, φ, λ) coordinate system is not a new coordinate system; it is simply a variation on the latitude-longitude system discussed earlier. Converting latitude-longitude coordinates to Cartesian coordinates involves intersecting great circles and lesser circles. The equator is usually treated as the Cartesian xy-reference plane. In the (ρ, φ, λ) system, the xz-plane acts as the equatorial reference plane. This simple change in orientation makes it easy to find the Cartesian coordinates of *triacon* grid points, a spherical subdivision technique we explain later in Chapter 8. The (ρ, φ, λ) coordinate system is handy in other geometric situations as well.

The three-part coordinate (ρ, φ, λ) specifies a radius (ρ) and two angles (φ and λ). Since we work with a unit sphere, radius $\rho = 1$. The "*latitude*" angle φ is measured from the "*equator*," which is the zy-plane and is in the range of 0 to ±90°. In a similar fashion, the "*longitude*" angle is measured relative to the zy-plane from the $+z$-axis. Its range is also 0 to ±90°. Figure 4.15 shows the basic setup. A point P at (1, 0°, 0°) is at the zenith or $+z$-axis. In the figure, the great and lesser circle references are shown every 5°. Small dots show points at great and small circle intersections every 5°, and the point P at coordinates (1, 45°, 45°) is emphasized. It is interesting to note the loxodrome appearance of the intersection points, where both α and β increments

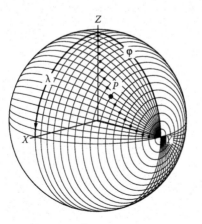

Figure 4.15. (ρ, φ, λ) Coordinate system.

were equal. We will make use of the (ρ, φ, λ) coordinate system in Chapter 8, when we develop the triacon subdivision schema.

To derive the Cartesian coordinates for a sphere of radius ρ at the intersection of a great and lesser circle with z-axis polar angles φ and λ:

$$x = \rho \sin \lambda \cos \varphi,$$

$$x = \rho \sin \varphi,$$

$$z = \rho \cos \lambda \cos \varphi.$$

Now that we know how to define points in space and on a sphere, let's look at how we can define spherical polygons by creating edges between points.

4.16 Spherical Polygons

Spherical polygons have two or more sides, always defined by the arcs of great circles. In this section, we look at the most common ones: lunes, quadrilaterals, and regular polygons since any spherical polygon can be broken down into two or more back-to-back spherical triangles. We will look at spherical and planar triangles in detail, since the subdivision schemas we develop later also define triangles.

4.16.1 Lunes

Lunes, also called digons or bigons, are two-sided polygons defined by the intersection of two great circles. Lunes are unique to spheres and do not exist in plane geometry. Anyone who has seen a beach ball is familiar with lunes, even if they do not realize it. Technically, a beach ball is a *hosohedron*, a spherical polyhedron made completely of lunes.

A lune is less than a hemisphere, and the great circles that define one lune define another congruent one on the opposite side of the sphere. Figure 4.16(a) shows both of them, one is shaded, and its *lunar angle*, *theta* (θ), is indicated. This same angle is also called the dihedral angle or *angular separation*, depending on the context of where it is used. A key point to remember is that the angle where the great circles intersect is the dihedral angle between them; it is measured on a plane perpendicular to the line of intersection of the two great circles defining the lune. If another great circle, such as the one shown in (b), divides a lune into two congruent spherical triangles, the two triangles are said to be *colunar*.

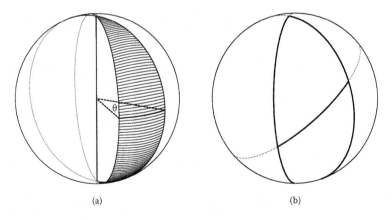

(a)   (b)

Figure 4.16. Lune, lunar angle, and colunar triangles.

A lune demonstrates another useful spherical concept. For a dihedral angle of θ, the area of a lune is the proportion of the sphere covered by the lune. The surface area of a sphere is $4\pi r^2$, so the area of a lune is $\theta/2\pi \times 4\pi r^2$ or $2r^2\theta$, where θ is expressed in radians. For a unit sphere, a lune's area is twice the lunar angle. If the lunar angle is measured in radians, the area is expressed in steradians or square radians. If the lunar angle is measured in degrees, the resulting area is spherical degrees. These units are explained shortly.

4.16.2 Quadrilaterals

Quadrilaterals are four-sided polygons and can be rectangles, squares, trapezoids, or rhombuses. A complex spherical subdivision problem can often be made easier by using quadrilaterals or by dividing the quadrilateral into two back-to-back triangles by adding a diagonal arc or chord.

Like other spherical polygons, the sides of quadrilaterals are always arcs of great circles. None of a quadrilateral's sides are parallel. Although an infinite number of quadrilaterals are possible, only a few of them with special angles can cover an entire sphere without gaps or overlaps.

Figure 4.17(a) shows quadrilateral faces of a spherical cube. Each face angle is 120° (not 90°), while the sides are 70.5288°. Figure 4.17(b) is a spherical icosahedron subdivided by rhombic quadrilaterals. The long axis of each rhombus along the sphere's surface corresponds to an edge of the spherical icosahedron (dashed). For the icosahedron, 30 large rhombus quadrilaterals (accented in the figure) cover the sphere. The resulting polyhedron is called a *rhombic triacontahedron*. Each accented quadrilateral is congruent and covers the entire sphere without overlaps or gaps. Its angles are 120° and 72°, and all four sides are 37.377°. The quadrilaterals within the large accented ones are not all congruent, and none of them individually could cover the entire sphere. The guidelines for spherical quadrilaterals are the following:

- *Rectangle.* Four equal angles, two opposite sides are equal;

- *Square.* Four equal angles, four equal sides;

- *Rhombus.* Two opposite pairs of equal angles, four equal sides;

- *Trapezoid.* Two pairs of equal and adjacent angles; only two opposite sides are equal (share equal angles).

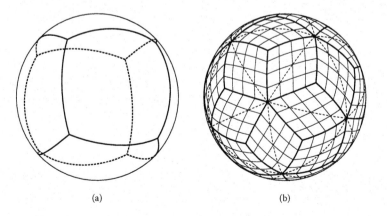

(a)   (b)

Figure 4.17. Spherical square and rhombic quadrilaterals.

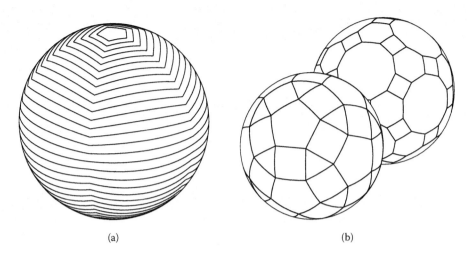

(a)   (b)

Figure 4.18. Spherical polygons.

Some caution must be exercised when working with quadrilaterals. Geographers often refer to areas bounded by two different latitudes and longitudes as spherical quadrilaterals. Strictly speaking, these areas are not spherical quadrilaterals because latitude references, except for the equator, are lesser circles. The sides of all spherical polygons, quadrilateral or otherwise, are always arcs of great circles.

4.16.3   Other Polygons

In addition to lunes, triangles, and quadrilaterals, many other regular convex spherical polygons exist. A *convex polygon* is one where any line or arc between its vertices defines a side or lies totally inside the polygon. The vertices of regular spherical polygons lie in a plane in 3D space. They are also the vertices of a regular Euclidean polygon inscribed in a sphere.

We can demonstrate this by showing the infinite variety of sizes that even a single polygon can take. Figure 4.18(a) shows a set of spherical pentagons. Within a single pentagon, its angles are equal and its sides are as well. But the angles of any two pentagons are not the same. The smaller a spherical *pentagon* becomes, the more its angles approach (but never equal) a regular planar pentagon. The larger the pentagon is, the more it's perimeter approaches a great circle.

Figure 4.18(b) shows two spherical Archimedean polyhedra with vertices centrally projected onto the surface of a surrounding sphere. They provide additional examples of regular spherical polygons such as hexagons and dodecagons.

No matter how many sides a spherical polygon has, each side and each angle is less than 180°, and the sum of the sides is always less than 360°. Later, we will discuss the concept of spherical excess and the area of triangles. *Excess* is essentially the sum of the interior angles of the spherical polygon less the sum of the interior angles of its planar version. The difference is its area. We will see that excess applies to spherical triangles and any spherical polygon because spherical polygons can be subdivided into two or more spherical triangles.

4.16.4   Caps and Zones

A plane passing through a sphere defines either a lesser or greater circle. The plane also slices off a spherical cap. Lesser circles define small caps that resemble contact lenses, while a great circle defines a hemisphere, the largest possible cap.

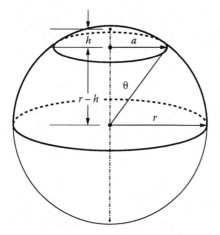

Figure 4.19. Spherical caps.

Figure 4.19 shows a lesser and great circle cap superimposed on each other, with both centered over the same point (the larger dot) on the surface of the sphere. Both caps have the same polar axis. The two radii for the lesser circle, along with its height, are indicated along with the formulas to calculate them.

Caps play many roles in spherical work. One of the most useful is analyzing the distribution of points on a sphere. When caps are centered over every point on the sphere and their angular radii are all the same and as large as possible without causing any caps to overlap, the area of all the packed caps can be compared to the sphere's total area, indicating their packing density. If two spheres distribute the same number of points, the one with the higher density distributes points more evenly. If we call this *density factor D,* then the relationship is

$$\text{density} = \frac{n}{2}(1 - \cos\theta),$$

$$\text{surface area}_{\text{cap}} = 2\pi rh = \pi(a^2 + h^2),$$

where n is the number of points and θ is the angular radius of the cap.[12] The more even n points are distributed on the surface of the sphere, the larger the common radius of all caps will be, and thus the greater the final density.

In order to find the density, we must know the area of a spherical cap cut out by a plane at distance h from the tangent plane to the sphere at the center of the cap. By a special property of 2D spheres, already known in some form to Archimedes, the most brilliant mathematician-scientist of antiquity, this area is given by $2\pi rh$, where r is the radius of the sphere.

Figure 4.20 shows the largest possible caps for each of the spherical Platonic solids. In clockwise order: (a) tetrahedron, (b) dodecahedron, (c) icosahedron, (d) octahedron, and (e) cube. We can show how to compute the density and measure the distribution of points by using the six spherical caps centered at the endpoints of the three Cartesian coordinate axes of the spherical octahedron shown in Figure 4.20(d). By looking at the slice of this sphere by a coordinate plane, we can see that there are six nonoverlapping spherical caps with radius $r - r/\sqrt{2}$ so h equals $r - r/\sqrt{2}$ and the area of each cap is

---

[12] (Tarnai 2002, 289).

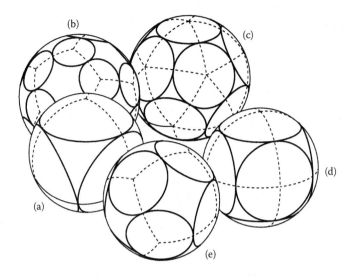

Figure 4.20. Spherical platonic caps.

Spherical Platonic Caps					
	Angular	Circle	Cap	Total	Total
Polyhedron	Radius	Radius	Area	Caps	Area	Density
Tetrahedron	54.7356	0.8165	2.6556	4	10.6223	0.8453
Octahedron	45.0000	0.7071	1.8403	6	11.0418	0.8787
Cube	35.2644	0.5774	1.1530	8	9.2239	0.7340
Dodecahedron	20.9052	0.3568	0.4136	20	8.2721	0.6583
Icosahedron	31.7175	0.5257	0.9384	12	11.2607	0.8961

Table 4.1. Spherical platonic caps.

$2\pi r(r - r/\sqrt{2}) = 2\pi r^2 (1 - 1/\sqrt{2})$. The total area of the six caps is then $12\pi r^2 \left(1 - 1/\sqrt{2}\right)$ while the area of the whole sphere is $4\pi r^2$. The ratio of the area of the caps to the area of the sphere is then $3\left(1 - 1/\sqrt{2}\right) = 3\left(\sqrt{2} - 1\right)/\sqrt{2} = (3/2)\left(\sqrt{2} - 1\right)\left(\sqrt{2}\right) = (3/2)\left(2 - \sqrt{2}\right)$, about $(3/2)(0.5858) = 0.8787$ so a little more than 7/8 of the sphere.

The highest possible density is 1.0, a case where caps are centered over two antipodal points and each cap covers an entire hemisphere. As expected, these two hemispherical caps cover the entire sphere with no spaces or gaps. Table 4.1 summarizes their angular radius, circle radius, and surface area on a unit radius sphere. The total number of caps and the resulting density are also tabulated. Note that the caps covering the icosahedron have the highest density; they cover the highest percentage of the sphere's overall surface area. We will revisit spherical caps in the next chapter when we look at various point distribution strategies.

4.16.5   Gores

Gores are a series of 2D polygons, each polygon defined on its own plane, that when tiled together form a continuous strip approximating a 3D surface. The points

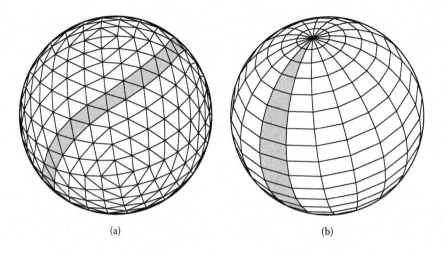

(a)   (b)

Figure 4.21. Triangular and quadrilateral gores.

distributed on a sphere can be connected with chords to define a covering mesh of triangles. A sequence of adjacent triangles creates a gore. Some series can wrap all the way around a sphere, such as the one shown in Figure 4.21(a). It is easy to imagine cutting a gore out of paper, creasing it where two triangles share an edge, and folding it to make a faceted model of the sphere.

Although triangles are the most flexible way to approximate a spherical surface, gores are not limited to triangles; any planar polygon with every vertex on the surface of the sphere will suffice. Figure 4.21(b) shows a gore made of quadrilaterals and a triangle at the pole.

4.16.6 Spherical Triangles

Triangles play a dual role in spherical subdivision: they are the result of subdivision, and they are used to subdivide. You have already noticed that points on a sphere can often be connected to form triangles, and these points can define either spherical or planar versions. Here we look at the properties of spherical triangles and what relationships they have to planar ones defined by the same points on a sphere.

We have already seen that a spherical triangle is formed by the intersections of three great circles. Figure 4.22 shows the details. Unlike plane triangles, the sides of a spherical triangle are measured in angles, and they are always less than a semicircle. The sum of the three angles of a spherical triangle always exceeds 180°, but is always less than 540°. A spherical triangle can also have two or even three 90° angles. We saw several examples before in Section 4.8.

Surface angles are the "*corners*" of a spherical triangle. They occur where two great circles intersect. All spherical triangles have three surface angles and, although there is no hard and fast rule, we will adopt the convention of using uppercase letters like A, B, and C to refer to them. The sides opposite these angles are traditionally given lowercase letters like a, b, and c. Figure 4.22 labels a spherical triangle this way. Surface angles are measured on a plane tangent to the sphere at the point of intersection.

The sides of a spherical triangle are segments of great circle arcs. In reality, their "*length*" is arc-length and they are measured by a central angle, also called an *arc angle* or *interior angle*, and usually denoted in lowercase. These terms are often used

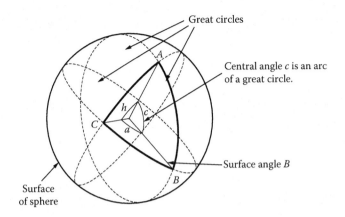

Figure 4.22. Spherical triangle ABC.

interchangeably in spherical trigonometry. Figure 4.22 locates central angles *a*, *b*, and *c*. It is easy to see how this angles measure the portions of the great circle arcs that forms a sides of spherical triangles.

We have already encountered dihedral angles when we covered lunes. The dihedral angle is the angle between the planes of two intersecting great circles, and it's the same angle as the surface angle of the three vertices *A*, *B*, and *C* in our spherical triangle (Figure 4.23).

Spherical triangles have many of the same characteristics as their planar cousins, but there are some important differences. Table 4.2 compares them. Like plane triangles, if two sides of a spherical triangle are equal, the angles opposite them are equal. If two sides of a spherical triangle are unequal, the angles opposite these sides are unequal, and the greater side lies opposite the greater angle.

But there is another important difference. The sides of a spherical triangle are entirely determined by its three surface angles. Two spherical triangles are congruent only if their corresponding surface angles are equal. This is not true in planar triangles, where specifying the triangle's three apex angles defines an infinite number of similar triangles. And unlike planar triangles, the sum of the three sides (central angles) of a spherical triangle is less than 360°, while the sum of the surface angles is always greater than 180°, but less than 540°.

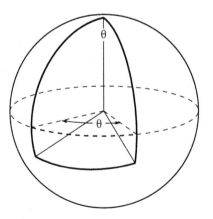

Figure 4.23. Dihedral angle.

Spherical and Planar Triangles Compared	
Spherical Triangles	**Planar Triangles**
Triangles lie on the surface of a sphere	Triangles lie on a plane
Angles are formed by the intersection of two great circles	Angles are formed by the intersection of two lines
Any angle is less than 180°	Same
Sides are arcs and measured in degrees or radians	Sides are straight and measured in length
Each side is less than the sum of the other two sides	Same
The sum of the sides is less than 360°	N/A
Any side is less than 180°	N/A
If two sides are equal, the angles opposite them are equal	Same
If two sides are unequal, the angles opposite these sides are unequal, and the greater side lies opposite the greater angle	Same
Sum of the surface angles is always more than 180° but less than 540° The sum greater than 180° is called spherical excess	Sum of the angles is always 180°
Can have one, two or even three 90° surface angles	Can have, at most, one 90° angle
If two surface angles are equal, then the sides opposite those angles are also equal.	Same
Area depends on the arc-length of the sides and the radius of the sphere	Area depends on the length of the sides
There are no similar triangles on a sphere that are not congruent	Similar triangles have the same angles but different length sides
Two triangles are congruent, if they have the same angles and sides in relationships: ASA, SSS, or SAS	Same
Angle-Angle-Angle, AAA, specifies a unique triangle	Angle-Angle-Angle specifies an infinite number of similar triangles

Table 4.2. Spherical and planar triangles compared.

4.16.7   Congruent and Symmetrical Triangles

In Euclidean geometry, triangles are congruent if the combinations of their angle (A) and side (S) parts are the same: *ASA, SSS, SAS*. These relationships are the same for spherical triangles plus one more: *AAA*. *Congruent spherical triangles* have all their angles and sides equal and both triangles perfectly coincide. If two non-equilateral triangles have equal angles and sides, but in reverse order, they are said to be *symmetric triangles*. Symmetric triangles are equal, part for part, but they cannot be made to coincide because their respective sets of parts are arranged in opposite orders. Each triangle is the opposite hand of the other. Let's look at some examples of congruent and symmetrical triangles.

The same three great circles that form one spherical triangle form eight triangles. How do we get eight for one? This is quite a bargain! It is not hard to see how this happens. The three great circles that form any spherical triangle intersect each other at six

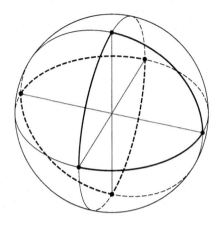

Figure 4.24. Three orthogonal great circles define eight equilateral triangles.

points, three antipodal pairs. Every time one spherical triangle is formed, the planes of the great circles that define it have also defined seven other triangles. The simplest example is when three great circles are orthogonal or 90° to each other. Notice in Figure 4.24, all eight triangles that result are congruent equilateral triangles and every surface angle is 90°, something that is only possible with spherical triangles (this is also the geometry of the spherical octahedron). If the great circles are not orthogonal, four pairs of symmetrical triangles result. Figure 4.25 illustrates this point. Notice that each of the four pair of triangles have antipodal vertices and are symmetrical but not congruent to one another.

4.16.8   Nothing Similar

A surprising difference between planar and spherical triangles is there are no similar triangles on a sphere. Triangles can be acute, obtuse, right, or equilateral. They can be congruent. But they cannot be similar; that is, they cannot have the same surface

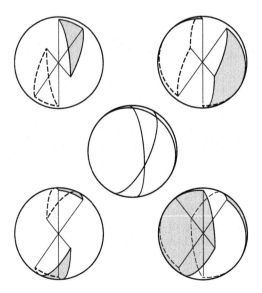

Figure 4.25. Three great circles define eight spherical triangles.

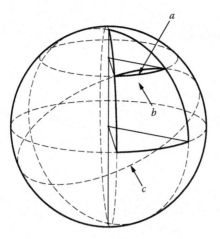

Figure 4.26. No similar triangles.

angles but be proportionately larger or smaller sizes. At first, this seems impossible because similar triangles are so fundamental to plane geometry. But on a sphere, three angles always specify a unique triangle. An angle-angle-angle (*AAA*) triangle specification defines a unique spherical triangle, whereas on a plane, it defines an infinite number of similar triangles, each with the same angles but sides of different lengths. The following three examples demonstrate this essential concept.

Figure 4.26 shows an isosceles triangle formed by two meridians, each perpendicular to the equator. A smaller triangle is inscribed in the larger one. Side *a* is the arc of a great circle between the two medians, a requirement for spherical triangles. Arc *b* is from a lesser circle and does not qualify as a component of a spherical triangle, even though it looks like it is parallel to side *c*. Although both triangles have a common apex angle, the other two face angles are clearly not the same.

Let's look at similar triangles from another perspective. In plane geometry, we can cluster four equilateral triangles together to make a larger one. We can also subdivide a large equilateral triangle, making four smaller, congruent ones. In this example, and the next one, we show that neither of these two constructions are possible on a sphere.

In Figure 4.27, an equilateral triangle (a) is replicated three times and the copies appended to each side of the original triangle in (b) and (c). The resulting configuration of four triangles (c) is not, however, a larger equilateral triangle with the same angles. Instead, a new triangle with still larger angles is created, as indicated by the outermost sides in (d).

In our third example, it is easy to make a triangle within a triangle by creating sides between pairs of midedge or medial points. The new triangle is sometimes called a *medial triangle*. In plane geometry, the medial triangle of an equilateral triangle results in four smaller, congruent equilateral triangles, and the sides of the medial are parallel to the sides of the original triangle. But none of this is true on a sphere, and Figure 4.28 shows why.

Two views of the same equilateral spherical triangle and its medial triangle are illustrated. The result is a single equilateral triangle in the middle, surrounded by three isosceles triangles. Figure 4.28(a) clearly shows that the central triangle is equilateral, while (b) views a typical apex isosceles triangle that surrounds it. The original equilateral angles are all 90°. The newly created medial triangle in the center is equilateral and its vertex angles are each 70.5°. The three surrounding isosceles triangles are congruent and have vertex angles of 90°, 54.7°, and 54.7°. The edges of the original

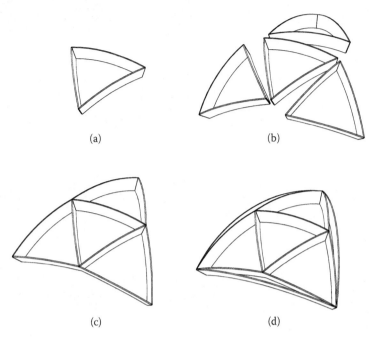

(a)   (b)

(c)   (d)

Figure 4.27. Appending equilateral triangles.

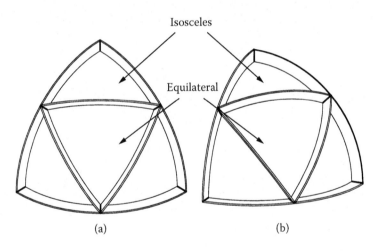

Figure 4.28. Subdividing a spherical equilateral triangle.

equilateral triangle are each 90°, while the interior equilateral triangle has 60° edges. None of the edges are parallel to one another.

These three examples clearly demonstrate that there are no similar triangles on a sphere that are not congruent. Every spherical triangle is either unique or congruent, and three vertex angles are all that is needed to specify one. Next, we will look at spherical triangles with special properties.

4.16.9   Schwarz Triangles

Subdividing a sphere often involves subdividing just a small subarea and then replicating the geometry to cover the rest of the sphere. This subarea must be a special

spherical polygon if it is to cover, or tile, the rest of the sphere's surface without causing overlaps or leaving gaps.

Hermann Amandus Schwarz's (1843–1921) work includes differential geometry
and spherical surfaces. He discovered special spherical-right-triangles that could be
replicated to cover a sphere through *reflections* and rotations without leaving any gaps
or causing overlaps. Today, we call these triangles *Schwarz triangles*, and they are very
useful in spherical subdivision because we can subdivide just this one triangle and then
replicate the subdivision layout to cover the rest of the sphere. There is an additional
benefit with Schwarz triangles: many are right triangles, and right spherical or planar triangles are the easiest to work with in trigonometry. Buckminster Fuller called
Schwarz triangles "*least common denominator*" (LCD) triangles because he thought of
them as the smallest triangle that can represent the whole sphere.

Schwarz triangles are different for each of the spherical Platonic solids. We list their
angles here and show them in Figure 4.29 (clockwise from the top in the same order).
The faces of each spherical Platonic solid are accented.

- 60°, 36°, 90° Icosahedron;

- 60°, 45°, 90° Octahedron;

- 45°, 60°, 90° Cube;

- 60°, 60°, 90° Tetrahedron;

- 36°, 60°, 90° Dodecahedron;

The faces of each Platonic solid shown in Figure 4.29 have 6, 8, or 10 right Schwarz
triangles. Half of the triangles are left-handed and half are right-handed. The tetrahedron has 12 left-handed and 12 right-handed triangles, while the cube, octahedron,
dodecahedron, and icosahedron have 24, 24, 60, and 60 left-handed and right-handed
Schwarz triangles, respectively. In a later chapter, we describe the triacon subdivision
method, which is based entirely on right Schwarz triangles. Because of this, the triacon
method can subdivide all the spherical Platonic solids shown in Figure 4.29.

60°, 36°, 90° Icosahedron
60°, 45°, 90° Octahedron
45°, 60°, 90° Cube
60°, 60°, 90° Tetrahedron
36°, 60°, 90° Dodecahedron

Figure 4.29. Platonic Schwarz triangles.

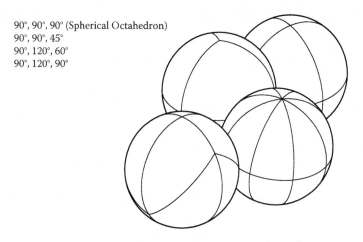

90°, 90°, 90° (Spherical Octahedron)
90°, 90°, 45°
90°, 120°, 60°
90°, 120°, 90°

Figure 4.30. Other right triangle tiles.

Although there are an infinite number of different right spherical triangles, only a few of them can cover the surface without gaps or overlaps. There are two ways to know if a triangle can do this: determine its area or calculate its vertex sums.

For a triangle to cover the entire sphere, its area must divide evenly into the total area of the sphere. Another way to know if a spherical triangle can tile a sphere is to examine its three vertex angles. For a triangle to tile an even multiple of times, each of its angles must add up to 360°, as they are arranged surrounding a common vertex point.[13] For example, in the icosahedral Schwarz triangle, the number of times any of their three angles share a common vertex on the sphere are $6 \times 60°$, $10 \times 36°$, and $4 \times 90°$ and each adds up to 360°.

Figure 4.30 shows some additional spherical right triangles that pass these two tests and can perfectly cover the whole sphere. But note that several of these Schwarz triangles have more than one right angle. From back to front in clockwise order, they are

- 90°, 90°, 90° (Spherical Octahedron);

- 90°, 90°, 45°;

- 90°, 120°, 60°; and

- 90°, 120°, 90°.

Within this new group, only the spherical octahedron is suitable for spherical subdivision. Although all of them can cover the sphere, their *tiling* arrangements are not suited for subdivision because they do not have the necessary edge-to-edge or vertex-to-vertex symmetry we need. For the subdivisions demonstrated in this book, we use the smaller Schwarz triangles that subdivide the five spherical Platonic solids shown in Figure 4.29. We will use Schwarz triangles later in Chapter 8 to help generate small-area grids when we use subdivided Schwarz triangles to assemble 3D spheres with Computer Aided Design.

4.16.10 Area and Excess

Thus far, we have seen how great-circle arcs define the angles and sides of spherical polygons. Here we see how those same angles define the polygon's area as well. Later,

---

[13] (Dawson 2008).

we will use an amazing property called spherical excess, which relates the sum of a spherical triangle's angles to its area. But first, we need to see how area is measured.

4.16.11   Steradians

Steradians are used in many applications of spheres.[14] While the area of any spherical polygon can be measured in square units such as inches, feet, meters, and so forth, a more general way to express spherical area is to use units relative to the sphere's radius. While radians measure angles relative to the radius of a circle, steradians measure area relative to the radius of a sphere. Steradians can be thought of as "*square radians*" and they are abbreviated (*sr*). Steradians are part of the International System of Units (SI) and many high-end scientific calculators include this unit.

The number of steradians in a sphere is equal to the surface area of the unit sphere (or any other sphere—the size of the sphere does not matter). An entire unit sphere, therefore, measures 4π steradians, or about 12.57 steradians in all. One steradian is about 8% of the sphere's total surface area. To get a visual idea of a single steradian, look at Figure 4.31 which shows a spherical cap whose surface area (not the cap's lesser circle area) is 1 steradian. If this is a unit sphere (radius = 1.0), then the area of the cap is r^2. The choice of a spherical cap is completely arbitrary; the surface area could be any shape.

4.16.12   Solid Angles

The *solid angle*, abbreviated (ω or Ω), is the name of the area projected onto the surface of a unit sphere and is measured in steradians. The solid angle of the spherical cap in Figure 4.31 is 1 steradian. Spherical polygons that have the same area have the same solid angle regardless of their shape. Solid angle is simply the ratio of the polygon's area to the square of the sphere's radius. Again, the area's shape does not matter. The maximum solid angle possible on a sphere is the total area of a unit sphere itself, which is 4π or 12.5664 steradian. The general formula for a solid angle is

$$\Omega = \text{area/radius}^2.$$

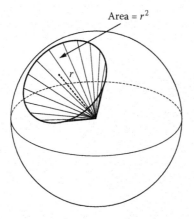

Area = r^2

Figure 4.31. Steradians and solid angles.

---

[14] Steradians are used in many areas of science and physics. For example, in the calculations of radiant intensity, the intensity of electromagnetic radiation (how brightly a light bulb shines) is measured in watts per steradian. Astronomers using radio telescopes use steradians to measure the density of radiant sky sources (stars or other celestial objects).

4.16.13 Spherical Degrees and Square Degrees

Steradians are one way to express area, but it can be expressed in *spherical degrees* or *square degrees* as well. Neither of these are SI units, but they are frequently encountered in spherical work and are easy to work with.

Just as degrees are used to measure parts of a circle, square degrees are used to measure parts of a sphere. A spherical degree (*sd*) is 1/720th of the area of the whole sphere. Each degree can be visualized as a spherical triangle with angles 90°, 90°, and 1° or one-half of a 1° lune. The sphere in Figure 4.32(a) is divided into 720 areas, each equal to one spherical degree, 360 in each hemisphere. One triangle of one spherical degree is accented. One spherical degree equals $(4\pi r^2)/720$ or $(\pi r^2)/180$ steradian, which is about 0.0175 steradian.

Square degrees, on the other hand, are units based on a small spherical square with one degree of arc on a side. Astronomers often use square degrees to refer to areas on the celestial sphere. For example, a NASA image of the sky might show a field of view of six square degrees. Square degrees are abbreviated $\deg^2$ to distinguish them from other units.

You can think of a square degree as a small postage stamp on the surface of a large sphere; one is shown in Figure 4.32(b). Remembering that there are $180/\pi$ degrees in a radian, you can figure out that the number of square degrees in a full sphere is $4\pi(180/\pi)^2$, or about 41,252.96 $\deg^2$. Therefore, one square degree equals $(\pi/180)^2$ steradians, or about 0.0003046 steradian. To review:

1 radian	$180°/\pi$ degrees or about 57.295780°
1 steradian	$(180°/\pi)^2$ square degrees or 3,282.8064 square degrees
Area of the whole sphere	4π steradians or about 41,252.9612 $\deg^2$
Solid angle (in steradians)	$\Omega = $ area/radius^2
Solid angle of entire unit sphere	4π steradians

Armed with ways of expressing spherical area, let us look at how we find the area of spherical polygons. Area is related to excess, one of the most useful properties of spherical polygons.

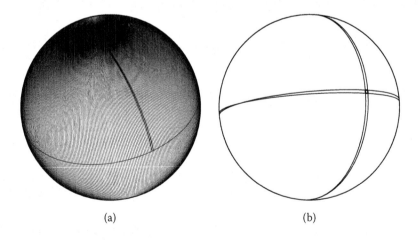

(a)   (b)

Figure 4.32. Spherical degrees and degrees squared.

4.17   Excess and Defect

A major difference between planar and spherical triangles is the fact that the sum of their angles is different. The interior angles of planar triangles always add up to 180°, but on a sphere the sum is always greater than 180° but less than 540° (or between π and 3π radians). There is a direct relationship between the area of a triangle and the sum of its interior angles. The amount the sum exceeds 180° or π is called the *spherical excess* and is denoted E (or delta or Δ).

Spherical excess is very useful because it is also the area of the triangle on a unit sphere. If a spherical triangle (or any convex polygon for that matter) is very small relative to the size of the sphere, its excess will be small and the polygon begins to behave like a regular planar polygon. On the other hand, if the spherical triangle approaches the largest size possible, like an equilateral where each vertex angle is just under 180°, then the excess approaches half the area of the total sphere. We will illustrate these two extremes next.

Spherical excess was first described by Thomas Harriot (c. 1560–1621), but Harriot did not publish his work. Excess was independently rediscovered by French-born Dutch mathematician Albert Girard (1595–1632) and excess is sometimes referred to as the Harriot-Girard theorem. The theorem states that excess, E, is the difference between the sum of the angles of a spherical triangle and the sum of the angles of a plane triangle π (radians) or 180° and is measured in radians or degrees accordingly. The formula is

$$E = (A + B + C) - \pi.$$

where A, B, and C are measured in radians. The area of a triangle is simply the relationship area = Er^2, expressed in steradians or spherical degrees depending on the units used. For a unit sphere, excess is the area of the triangle.

Figure 4.33 illustrates the difference in spherical excess between large and small triangles. The large triangle (*a*) has angles 90°, 90°, and 90°. Its excess is 90°. In steradians, this excess is 1.5708 or $\pi/2$. Since this particular triangle is 1/8th the area of a unit sphere and eight of them equal the whole area of the sphere, this confirms that excess (E) is the area of a sphere: 8(1.5708 steradian) = 12.566 steradian, the total area of a sphere.

The two small triangles in Figure 4.33(b) both have angles 45.05°, 45.05°, and 90°. Their excess is 0.1°, which is quite small. When triangles are small relative to the size of the sphere, the sum of their angles approach 180° and they behave similarly to planar triangles.

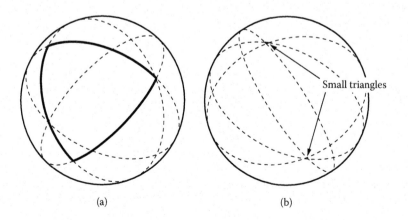

(a)   (b)

Figure 4.33. Large and small triangle spherical excess.

Spherical Defect							
	Surface Angles			Interior Angles			
Triangle	A°	B°	C°	a°	b°	c°	Defect°
Large	90.0000	90.0000	90.0000	90.0000	90.0000	90.0000	90.0000
Small	45.0500	45.0500	90.0000	3.3842	3.3842	4.7845	348.4472

Table 4.3. Spherical defect.

Spherical defect is a concept similar to excess. The defect of a spherical triangle is the difference between 2π radians and the sum of the triangle's side arc lengths. It is usually denoted D. The formula is $D = 2\pi - (a+b+c)$ in radians or $360° - (a+b+c)$ in degrees. The triangles shown in Figure 4.33 and summarized in Table 4.3 illustrate that the larger a triangle becomes, the closer its spherical defect begins to approach zero. On the other hand, the defect of a smaller triangle, relative to the size of the sphere, approaches $360°$.

Although triangles are the most common polygon formed when subdividing spheres, squares, pentagons, hexagons, and others shapes can result, and their area is also based on spherical excess. The area, S, of a spherical polygon with n sides and sum θ of its interior angles on a sphere of radius r is

$$S = \left[\theta - (n-2)180°\right]r^2.$$

A simple example shows the relationship of excess; the sides and sum of its angles follow the same rules we have been discussing. The four angles of a spherical cube's face, shown earlier in Figure 4.17(a), are each 120° and total 480°. Thus, the area of the face on a unit radius sphere is 480° – (4 – 2)180° = 120 spherical degree or 2.09 steradian. We can easily check this. The cube has six equal spherical faces; thus 6(2.09 steradian) = 12.566 steradian, the area of a full sphere.

4.17.1 Visualizing Excess

The preceding equation for spherical excess is extremely useful in spherical subdivision work. Albert Girard used the area of lunes and their proportional relationships to the sphere's area to derive the relationship and, in so doing, he proved that a triangle's area *is* proportional to its spherical excess. His logic sharpens our circular reasoning by synthesizing the relationship of lunes, triangles, and their areas, giving us an extremely useful formula for geodesic work.

We said earlier that a lune is a spherical polygon with just two angles (both equal), and two antipodal vertices (with both sides formed by 180° arcs of great circles). A lune's area is $2r^2\theta$, thus it is directly proportional to its lunar angle (θ). For a unit sphere, a lune's area is simply twice its lunar angle expressed in radians.

In this example, we define a spherical triangle with three orthogonal great circles. In other words, the great circles are positioned at 90° angles to each other and arranged similarly to the Cartesian coordinate system reference planes. Figure 4.34(a) shows this arrangement.[15] As expected, eight spherical triangles are defined (four pairs in which each pair is symmetric).

---

[15] This arrangement of great circles also defines a spherical octahedron..

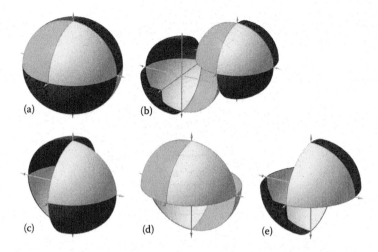

Figure 4.34. Lunes and spherical excess.

To illustrate Girard's theorem, we will concentrate on the surface of just two triangles, the yellow pair. Figure 4.34(b) pulls the sphere apart to make it easier to see these two triangles, which are on opposite sides of the sphere. Notice that every vertex in one yellow triangle is antipodal to every vertex of its symmetrical counterpart. Arrows show their relationship. Next, notice that the surface of each yellow triangle is defined by a set of three overlapping lunes colored red, green, and blue.

When great circles define a lune, they define two identical lunes that share the same two vertices (points where the great circles intersect). Figure 4.34(c)–(e) show these pairs of identical lunes (i.e., those of the same color) and how they define the surface and vertices of their respective yellow triangles. It is easy to see that for each lune, one lunar vertex defines one of the triangle's vertices, while the other lunar vertex defines a vertex in the symmetric triangle. Notice, too, that half of every lune's surface covers one of the two yellow triangles' surfaces.

We can now summarize the relationships between the areas of the six lunes, the overall sphere, and both triangles; they lead directly to the excess formula and a spherical triangle's area. Simply stated: the sum of all lunar areas (there are six) is equal to the area of the sphere, plus four times the area of the two triangles. The *"four"* in the equation is explained by the fact that there are two extra overlaps in two symmetric triangles. Since pairs of lunes (red ($_r$), green ($_g$), and blue ($_b$)) have the same area and we know the area of the whole sphere is $4\pi r^2$, we can express their area relationships and the area of a single triangle (T), such as

$$2r^2\theta_r + 2r^2\theta_r + 2r^2\theta_g + 2r^2\theta_g + 2r^2\theta_b + 2r^2\theta_b = 4\pi r_2 + 4_{\text{area}}(T).$$

Girard's theorem for excess is a simplification of the preceding equation.[16] While excess is expressed in degrees or radians, area is expressed in steradians, square degrees, or degrees squared. Since the areas of the lunes are proportional to their lunar angle, which is also an angle of the triangle, the excess of a spherical triangle is directly related to the lunes that define it. Girard's theorem on spherical excess and the area of a triangle beautifully summarizes the relationship between spherical triangles and lunes.

[16] For complete derivation, see Eric W. Weisstein, "Girard's Spherical Excess Formula," MathWorld—A Wolfram Web Resource, mathworld.wolfram.com/GirardsSphericalExcessFormula.html.

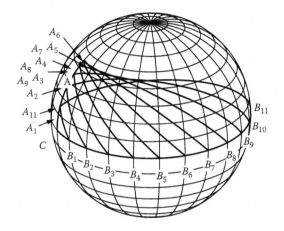

Figure 4.35. Excess in successive right triangles.

We have established the relationship between a triangle's vertex angles, excess, and its area. We have also seen that when spherical triangles (or any polygon) are relatively small compared to the size of the sphere, they behave almost like regular 2D triangles. But large spherical triangles behave very differently than plane triangles.

Examining a diagram based on one by James Wertz and Wiley Larson,[17] we can see how excess changes as triangles get larger. Figure 4.35 shows a series of 11 right spherical triangles with a latitude-longitude reference grid of 10° and 15°, respectively. The 11 triangles become successively larger from left to right. They all share the same right-angle vertex on the left, where $C = 90°$ and each triangle has the same angle $B = 45°$. All triangles partially overlap and their base sides are all coincident to the equator. The smallest triangle's base side is 15°, and each successive triangle increases its base side by 15° (indicated by the reference longitudes). The largest triangle's base somewhat wraps around the sphere and is 165°.

Our example shows 11 right triangles with angle-side-angle (*ASA*) specifications. They each have the same vertex angle C, a side that is a multiple of 15°, and the same vertex angle B. Although B is the same angle, we have subscripted them 1–11 to make their correspondence with their A angles 1–11 more evident. In two dimensions, these 11 triangles would all be similar triangles but, on a sphere, each triangle is quite different because their vertex angles $A_{1\text{-}11}$ are different; this also affects the two non-base side lengths of each triangle.

Table 4.4 lists the characteristics of these 11 triangles on a unit sphere. Keep in mind that every triangle has the same C and B angles, only angle A changes from triangle to triangle (a consequence of the different length side between C and B). What we notice from the table is the three vertex angles, A, B, and C, for the smallest triangle (first row entry) are almost like a regular 2D isosceles triangle. This is confirmed by the fact that there is almost no excess at all because their sum is near 180°. Figure 4.35 confirms the smallest triangle on the left. It's also the least "*distorted*" one.

In contrast, vertex A for the largest triangle (last row) is almost three times larger than vertex A for the smallest triangle. The triangle's excess is quite large too; as a result, it looks quite distorted. This series of triangles demonstrates that as the triangles get larger, their excess increases, their area increases, and they behave less like planar

---

[17] (Wertz 1999, 172, fig. 508).

| | | | | | | | | | | Area | |
| | | | | | | | | Area | | (spherical | |
A°	a°	B°	b°	C°	c°	A°+B°+C°	E°	(steradian)	degree)	Area (d²)
46.920	15.0	45.0	14.511	90.0	20.754	181.920	1.92048	0.03352	1.920	110.036
52.239	30.0	45.0	26.565	90.0	39.232	187.239	7.23876	0.12634	7.239	414.750
60.000	45.0	45.0	35.264	90.0	54.736	195.000	15.00000	0.26180	15.000	859.437
69.295	60.0	45.0	40.893	90.0	67.792	204.295	24.29519	0.42403	24.295	1392.012
79.455	75.0	45.0	44.007	90.0	79.271	214.455	34.45471	0.60135	34.455	1974.109
90.000	90.0	45.0	45.000	90.0	90.000	225.000	45.00000	0.78540	45.000	2578.310
100.545	105.0	45.0	44.007	90.0	100.7299	235.545	55.54529	0.96945	55.545	3182.511
110.705	120.0	45.0	40.893	90.0	112.208	245.705	65.70481	1.14677	65.705	3764.608
120.000	135.0	45.0	35.264	90.0	125.264	255.000	75.00000	1.30900	75.000	4297.183
127.761	150.0	45.0	26.565	90.0	140.768	262.761	82.76124	1.44446	82.761	4741.870
133.080	165.0	45.0	14.511	90.0	159.246	268.080	88.07952	1.53728	88.080	5046.585

Characteristics of Spherical Right Triangles

Table 4.4. Characteristics of spherical right triangles.

triangles. This series of 11 triangles demonstrates in yet another way that there are no similar triangles on a sphere.

4.17.2   Centering in on Triangles

Spherical subdivision creates points on the surface of the sphere that can be triangulated. When planar triangles are constructed from point-to-point chords, their centers are very useful in deriving other geometry or in comparing one set of triangles to another. The four most important centers of planar triangles used in subdivision are the centroid, *incenter*, *orthocenter*, and *circumcenter*. They are defined at various intersections of angle bisectors or the edge midpoints opposite an angle (*medians*). Figure 4.36 shows how each center is defined for planar triangles. Here we highlight their main characteristics. Table 4.5 provides a quick comparison as well.

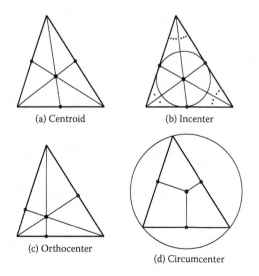

(a) Centroid   (b) Incenter

(c) Orthocenter
(d) Circumcenter

Figure 4.36. Planar triangle's centroid, incenter, orthocenter, and circumcenter.

Centers of Planar Triangles					
Center	**Intersection of**	**Equilateral**	**Acute**	**Right**	**Obtuse**
Centroid	medians	mid-triangle	inside	inside	inside
Incenter	angle bisectors	mid-triangle	inside	inside	inside
Orthocenter	altitudes perpendicular	mid-triangle	inside	on right angle vertex mid-point of	outside
Circumcenter	Bisectors of sides	mid-triangle	inside	hypotenuse	outside

Table 4.5. Centers of planar triangles.

Centroid

- The point of intersection of the medians of the triangle;

- The point always falls inside the triangle;

- The areas of all six sub triangles are equal;

- The *center of gravity* (COG) of the triangle (triangle will balance at its centroid as well as balance along any of its medians); the COG is also called the *barycenter* of the triangle;

- Divides each median, the line from a vertex to the midpoint of the opposite side, in a ratio of 2:1.

- The ratio of the length of sides to one another remains the same, if the triangle is scaled about the centroid;

- A triangle continuously scaled down converges on its centroid;

- The point's coordinates are the average of the vertex coordinates:

$$xc = (x^1 + x^2 + x^3)/3,$$
$$yc = (y^1 + y^2 + y^3)/3,$$
$$zc = (z^1 + z^2 + z^3)/3.$$

Incenter

- The point of intersection of the angle bisector of the triangle;

- The point always falls inside the triangle;

- The point is equidistant from the sides;

- The center of the triangle's *incircle*, the largest circle that can be inscribed and touch all three sides;

- Tangent points of the incircle and triangle sides are the same as angle bisector intersections only when the triangle is equilateral.

Orthocenter

- An altitude is the line that passes through a vertex and is perpendicular to the opposite side;

- Point of intersection of the altitudes of the triangle;

- Similar triangles are formed;

- Orthocenter for *acute triangles* falls inside the triangle and outside for *obtuse triangle;*

- For right triangles, the orthocenter is the vertex, which is the right angle.

Circumcenter

- The circumcenter is the center of the triangle's circumscribing circle, the circle that passes through all three of the triangle's vertices;

- The point is equidistant from the vertices of the triangle;

- The point of intersection of the perpendicular bisectors of the sides of the triangle;

- The circumcenter of acute triangles fall inside the triangle and outside for obtuse triangles;

- For a right triangle, the circumcenter is at the midpoint of the hypotenuse;

- The radius of a circumcenter given the sides of a triangle a, b, and c:

$$r = abc / \sqrt{(a+b+c)(b+c-a)(c+a-b)(a+b-c)}.$$

Figure 4.37 shows a simple application of incenters in spherical work. A circle is drawn inside each of the planar triangles. The circle's radius is 80% of the distance between the incenter and the edge of its triangle. The effect is to give a visual cue about the relative size and shapes of triangles. In these side-by-side comparisons, it is immediately clear that subdivision (a) is more uniform than (b) and that within (b) there is a central

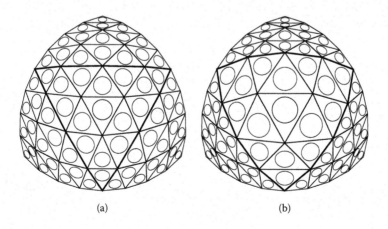

(a)   (b)

Figure 4.37. Incenter circles illustrate triangle shapes and relative sizes.

equilateral triangle (emphasized) whose vertices are at the midpoint of the outermost triangle. This triangle is much bigger than the other three. We will use incenter circles a great deal in Chapters 8 and 10.

4.17.3   Euler Line

We have already seen the orthocenter, circumcenter, and centroid of triangles, where they are the intersections of the altitudes, perpendicular bisectors of the edges, and medians of a triangle. Leonhard Euler (1707–1783), the prolific Swiss mathematician and physicist, observed that in all planar triangles, except an equilateral, the orthocenter, centroid, and circumcenter all lie on an imaginary line (in that order). In his honor, this line is called the *Euler line*. Other centers of a triangle lie on this line also, but these three are the most important for our applications. Note also that a triangle's centroid is always located between the circumcenter and the orthocenter.

Figure 4.38 shows the Euler line segment between the circumcenter and orthocenter for acute (a) and isosceles (b) triangles, similar to the ones created by the points on subdivided spheres. The line segment between the outermost orthocenter and circumcenter is accented. A good feature about this segment is that its length is quite sensitive to the ratio of the triangle's sides. A slight change in the sides causes a significant change in the length of the line segment, and this is what makes it so useful in geodesics because it gives us a way to compare the triangles produced by two subdivision schemas. Note, too, that there is no Euler line for the equilateral triangle (c). This is because all three centers are coincident.

We can use the length and direction of the Euler line segment as a measure of how triangles produced from spherical subdivision deviate from perfect equilaterals. If the schema produces all equilateral triangles (only the vertices of the tetrahedron, octahedron, and icosahedron on their circumsphere do this), the total length of the Euler line segments would be zero. But spherical subdivision produces acute and isosceles triangles as well, so the variation of this small length and its direction on the face of the triangle tells us how much the triangle deviates from equilateral.

Figure 4.39 shows the Euler line segments for every subdivision triangle of a polyhedron's face and shows examples of equilateral, isosceles, and acute triangles. Even though the overall equilateral spherical triangle's apex angles are all 90°, there are no perfectly planar right triangles. With just a quick glance, you can tell what kinds of triangles resulted from the subdivision.

First, only one triangle is equilateral—the one in the middle, because it is the only one without a Euler line. By inspecting the figure, it is also clear there are many isosceles triangles because the Euler line is perpendicular to one of the triangle's edges. If the

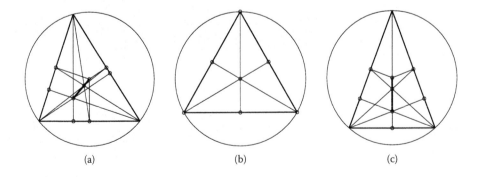

(a)   (b)   (c)

Figure 4.38. Euler line segments for (a) acute, (b) equilateral, and (c) isosceles triangles.

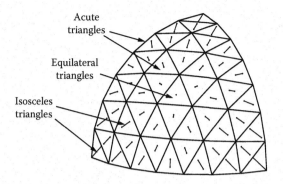

Figure 4.39. Euler line segments for a subdivision triangle.

triangle is acute, the Euler line will not be perpendicular to any edge. Some are labeled in the figure. The subdivision triangles that share one of the polyhedron's face vertices are almost right triangles because their circumcenter point, one of the end points of the Euler line segment, is almost tangent to the triangle's hypotenuse at its midpoint.

We will make use of Euler lines in the next chapter to see which schemas produce the most uniform spherical triangles, an important criterion in some geodesic applications.

4.17.4   Surface Normals

A surface normal is a vector, a mathematical entity that has direction and magnitude in 3D space. The normal to a surface is a direction perpendicular to that surface. If the surface is a plane, the *normal vector* is perpendicular to the plane and everything that lies on it. If the surface is curved or spherical, the normal direction is perpendicular to a plane tangent to the surface at that point. There is only one normal to a plane, but the surface of a sphere has an infinite number of normals. If we visualized the normals to a sphere as arrows, they would all point outward from or inward towards the origin.

Normals are very useful in geodesics calculations, even though they are rarely listed or displayed. Instead, normals are used in various computations. In images generated by computer graphics, they are used to determine whether a viewer can see the faces of an object just by inspecting the normals to the object's faces. If the angle between the viewer's eye direction and the normal (a perpendicular) to the object's face is more than 90°, the viewer can see just the edge of the object or nothing at all. If the angle is less than 90°, the viewer can see the face, although additional checks must be done to see if it is obscured by some other object between it and the viewer's eye.

Another use (unique to spherical subdivision) is to use normals to compare face orientations. If a subdivision results in perfectly equilateral triangles, their normals would have a uniform pattern and the angle between adjacent normals would be the same. But most spherical subdivisions result in a mix of isosceles, acute, or obtuse faces, and the pattern of their normals is also mixed. The angles between normals can be analyzed and displayed with stereographic projection, a topic we will cover later. The analysis tells us how uniform the subdivision faces are.

We do not draw normals when evaluating subdivisions, but we will show them here as arrows to make it clear that a normal is a direction and the direction is perpendicular to each plane of a face that results from a subdivision of the sphere. Figure 4.40(a) shows the normals of a set of subdivision triangles. The base of each arrow is located at the point where the normal's direction line intersects perpendicular to the plane of the triangle and passes through the origin of the sphere. Most of their bases are not

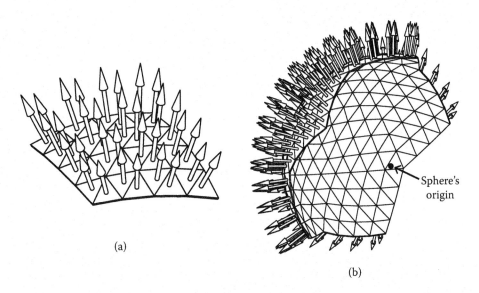

(a)

(b)

Figure 4.40. Surface normals for the planar surfaces of a subdivided sphere.

in the exact center of their respective triangle, an indication that the triangle is not equilateral but slightly isosceles or acute.

Figure 4.40(b) is a larger section of the same subdivided sphere. The cutaway image exposes the sphere's origin, which is represented by a small black dot. Every normal radiates outward from the origin and is perpendicular to the plane of its respective triangle. The angle between any normal and its neighbors is uniform in these figures, suggesting that the subdivision schema produces uniformly oriented faces. Figure 4.40(b) also shows examples of normals and hidden lines. Notice that where normals point away from your view, you cannot see the outside face of the triangle; instead, you see its inside face.

In addition to hidden line computations and evaluating face orientations, normals are very important to the subdivision process itself. Earlier we described how points on a sphere are derived from intersecting great circles. To find the two antipodal points, where two noncoincident great circles intersect, we use normals of the great circle's planes. Figure 4.41 shows how this works. Points A_1 and A_2 define a great circle, and its normal A_n is shown as an arrow perpendicular to the plane of the great circle. Likewise, for points B_1 and B_2 and normal B_n to their plane. Notice that these two normals, A_n and B_n, now define a third plane labeled $A_n B_n$. As you might have guessed, this new plane has a normal, too, and it is labeled N_{AB} in the figure. This third normal also defines the line of intersection of the original two great circle planes. Although the line is infinite in length, it pierces the surface of the sphere at two antipodal points, N_1 and N_2 and these are the points we are looking for. Points N_1 and N_2 are the points of intersection of great circles, A and B, on the surface of the sphere.

To review, we have created a normal to the normals of two great circles and defined the line of intersection of these two great circles. The intersection line pierces the surface of the sphere at two antipodal points defining two points on the sphere where the two great circles intersect. In most subdivision cases, we will use one of the two points as a subdivision point and ignore the other. There are applications, however, where we use both. We will review this technique again in Chapter 8 and show the computation details in Appendix C.

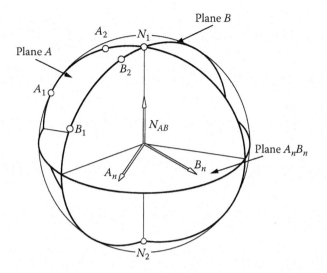

Figure 4.41. Defining points where great circles intersect.

4.18 Summary

We have come full circle in our reasoning. Our tool kit includes primitive geometry such as circles, points, arcs, and chords between points. We can locate them in three dimensions in several different ways, and we can use these points to define convex polygons, some of which can cover the surface of a sphere without gaps or overlaps. Spherical and planar triangles are particularly important, and warrant details about their angles and sides, as well as properties such as excess, area, centers, and orientation.

Evenly distributing points around the circumference of a circle is easy, but doing so over the surface of a sphere is not. In the next chapter, we define this challenge with a series of geometric objectives. We need to be precise about what *"even"* means and have a way of knowing how close we come to achieving an even point distribution on a sphere's surface.

Additional Resources

Aste, Tomaso and Denis L. Weaire. *The Pursuit of Perfect Packing*. Philadelphia, PA: Institute of Physics Publication, 2000.

Critchlow, Keith. *Order in Space: A Design Source Book*. New York: Viking Press, 1969.

Henderson, David W. and Daina Taimiṇa. *Experiencing Geometry: Euclidean and Non-Euclidean with History*. Upper Saddle River, NJ: Pearson Prentice Hall, 2005.

⑤ Distributing Points

In this chapter, we will look at three different but interrelated measures for point distributions on a sphere: covering, packing, and volume. Each provides a metric that allows us to compare one point distribution with another and to know which is more even. We use these measures in Chapter 9 to compare the results of different subdivision schemas.

It is easy to evenly distribute points around the circumference of a circle to any desired precision. But doing the same on the surface of the sphere is very difficult. In this section, we see there are several different perspectives on what an *"even"* point distribution means. We are particularly interested in techniques that can compare the uniformity of one distribution to another.

The essence of spherical subdivision is to define point distributions that meet application needs. The main one we focus on is making every point as equidistant and equiangular as possible to the points immediately around it. Other criteria might include minimum variations in chord lengths or in face areas.

In the strictest sense, the greatest number of points that can be equidistant and equiangular to every other one is four, and these four points are the vertices of a tetrahedron. But we use a less restrictive interpretation in spherical subdivisions and accept distributions that are *"symmetrical"* points on a sphere, where the distance from a point to its neighbors is equal, such as the vertices of the spherical Platonic solids, or nearly equal, like subdivided spherical Platonic solids. Sometimes mathematicians say points are evenly distributed, when the minimum distance between points is maximized.

The point sets that result from our spherical subdivision schemas are not arbitrary at all. The number of points and their distribution depend entirely on the spherical polyhedron we start with, the intensity of subdivision, and the particular subdivision schema used. We are interested in arbitrary point sets because several techniques have been developed to measure how even their distributions are, and we can use these same techniques to evaluate the results of our own not-so-arbitrary subdivisions.

5.1   Covering

The covering problem is sometimes called the *Tammes problem* named for the 1930s Dutch botanist P. M. L. Tammes, who studied pores on pollen grains. We briefly mentioned Tammes' work in Chapter 3. Tammes noticed that pores are distributed on spherical pollen grains in highly symmetric ways and their arrangement was essentially one of packing the largest possible equal circles on a sphere.[1] Tammes posed three questions:

- What is the largest diameter of *n* equal circles that can be placed on the surface of the sphere without overlap?

- How can they be arranged to achieve this maximum?

- Is the arrangement unique?

To meet these conditions, the minimum distance between the centers of the circles must be as large as possible and all distances (circle diameters) must be the same.[2] For a certain small number of circles, such as 2, 3, 4, 5, 6, 12, or 24, there are exact solutions for the Tammes problem. For other distributions of *n* circles, most are numeric (computer) solutions, though they remain unproven *conjectures*.

Circle packing has a large and rich body of literature. In 1964, Fejes Tóth[3] defined the conditions for packing equal circles on a sphere. Theoretic work by H. S. M. Coxeter and applied research by Tibor Tarnai and Zsolt Gáspár also determined the limits of packing equal circles on a sphere. Circle packing remains a key area of research because optimal packing addresses practical challenges, such as locating cell phone repeaters, optimizing satellite orbits, deriving error-correcting codes, or developing antiviral drugs.

There are two broad approaches to distributing *n* points on a sphere—unconstrained and constrained. In unconstrained distributions, the placement of circles is subject only to tangent conditions where a maximum radius is achieved. Any distribution mechanism whatsoever that places them on the surface of the sphere maximizes their angular radius and does not cause any overlaps. Some ingenious techniques have been developed to find good distributions. One technique randomly distributes *n* points in three dimensions and constrains them to lay on the surface of the sphere. From there, the two closest points are moved a little farther apart and the point set is then reexamined. The two closest points are found and moved apart, and the distribution of points is periodically tested for evenness.

When there is no measurable improvement in the evenness of point distribution, the testing of pairs stop. In another technique, *n* points are mathematically modeled

---

[1] (Tammes 1930).

[2] (Radin 2006).

[3] (Tarnai and Wenninger 1985).

as n like-charged particles. Each particle is given the same electrical charge. They are treated as if each is attached to the end of an imaginary spring with the same elastic pull. The other end of the spring is fixed to the origin of the sphere.

Particles with the same charge will repulse each other and when all n points are released, they expand outward from the origin, trying to get away from their neighbors, which are trying to do the same thing. The springs pull them back, keeping the particles from flying off into space. After several expansion-contraction-repositioning cycles, the particles settle down and reach an equilibrium state on the surface of an imaginary sphere. The distribution is stable, because particles are equidistant from the center (equal spring force) and maximally separated from their closest neighbors (electric repulsive forces). The angular radius of a spherical cap centered over each particle is now as large as it will get.[4] We saw an example of this earlier in Chapter 3. These are just some of the many approaches to distributing unconstrained points on a sphere. Except for some relatively low number of points, however, it has been impossible to prove that a particular arrangement is unique, or the best one, or that a particular distribution technique consistently produces better results than another.

In constrained distributions, some type of point-group symmetry is imposed during the distribution process. For example, two or more circles may be clustered side by side or they may have a predefined relationship (i.e., they are always placed two at a time as antipodal pairs or four at a time as tetrahedral quartets).[5] In these last two cases, n is a step function and will always distribute multiples of two or four points on the sphere's surface. Chapter 10 presents techniques for self-organizing grids made of points that are equally spaced.

Another form of constrained distribution is *polyhedral packing*. Circles are centered at the vertices of a polyhedron inscribed in a sphere. Tibor Tarnai has investigated polyhedral packing extensively and we list some of his papers in the additional resources at the end of the chapter.[6] The spherical caps for the Platonic solids, shown in Chapter 4, are the most basic examples of polyhedral packing. In the subdivision schemas we develop later, we constrain points to icosahedral symmetries. Circle packings with icosahedral symmetry have been tabulated for large numbers of circles.[7]

Figure 5.1(a) shows circle packings (caps) for points on a spherical equilateral triangle. All caps have the same radius and the radius is as large as possible without causing overlaps. In this distribution, the distances between any two adjacent edge points of the triangle (dashed) are the same. Thus, every circle over the triangle's edge is tangent. However, the distances between points within the face are close, but not the same. Thus, there is some space between them.

Figure 5.1(b) shows a variation on the packing theme. The points in (b) have the same distribution as (a), and all caps have the same radius. But this time, the cap radius in (b) is determined by the amount of overlap needed to eliminate all gaps and spaces. In other words, in Figure 5.1(b), there are no *interstices*, spaces between each cap, such as in (a). The effect is to tile over the entire surface of the sphere. Tibor Tarnai and Magnus Wenninger have investigated these spherical circle-coverings and shown interesting results where points are defined when the cap's lesser circles intersect

---

[4] Paul Burke maintains a very useful website with many examples of distributing points on a sphere. See local.wasp.uwa.edu.au/~pbourke/geometry/spherepoints.

[5] (Fowler, Tarnai, and Gáspár 2002).

[6] (Tarnai 1980).

[7] (Hardin, Sloane, and Smith 2008).

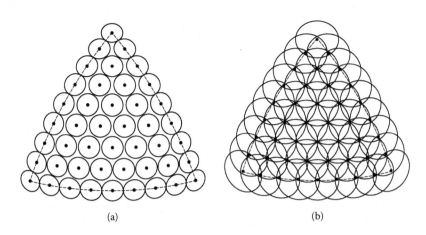

(a)   (b)

Figure 5.1. Tangent and overlapping spherical caps.

(see the bibliography for references). They take on a new hexagonal-pentagonal pattern.[8] When the caps overlap like this, the sphere is *"shingled"* with roofing plates, as it were. Several architects have explored the idea for roofing geodesic domes.[9] They find, however, the caps need more than the minimum radius to ensure enough overlap to avoid rain infiltration.[10] The total area of overlapping caps is not a useful metric of overlap because of double counting, but the common cap radius that just eliminates interstices is a good index. For our subdivision evaluations, we use the nonoverlapping caps when we compare one subdivision schema with another. We include overlapping caps here just for completeness.

What do the caps tell us about point distributions? First, it is easy to see the distribution pattern because the spaces stand out. In Chapter 9, we describe stereographic projection, which is a method for making two-dimensional stereograms of points, arcs, and circles on a sphere that can be measured. Second, we can find the percentage of the sphere's area covered by caps. If two subdivision schemas distribute the same number of points, the one with caps covering more of the sphere's surface has a better distribution of points than the other.

The number of points and their distribution created by our subdivision schemas are not arbitrary, and they are not a continuous function. The total number of points depends on the number of polyhedral faces or edges, the intensity of subdivision, and the particular schema used to subdivide. In all schemas, the maximum radius of a circle or the cap is determined by the arc distance between the two closest vertices in the entire subdivision. Figure 5.2 shows two spherical icosahedra with caps centered at every subdivision grid point. The subdivision in Figure 5.2(a) distributes 642 points on the surface of the sphere. A cap that is centered at every point as large as possible and has no overlapping will have an angular radius of 3.96°. A single cap is about 0.015 steradians and all of them together have a density of 0.77 or cover a little more than three quarters of the sphere's surface. This example has a very good point distribution. The example in Figure 5.2(b) shows both the subdivision grid and spherical caps at every grid vertex. This grid distributes 752 points and its caps cover about 71% of the

---

[8] (Tarnai 1993).

[9] Leonard Spunt developed a patented modular dome structure that essentially uses circular units, somewhat like spherical caps, to replace struts or chord-like structural members when building a dome. The benefit is that fewer different-sized units are needed than traditional strut systems.

[10] See (Spunt 1976, 237) and Fischbeck (2008).

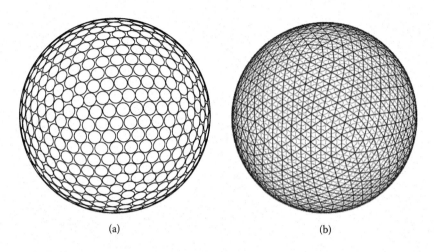

(a)   (b)

Figure 5.2. Spherical caps.

sphere's surface. In Chapter 9 we will use the density of spherical caps to compare one subdivision with another. The goal is to determine which subdivision schema produces the most uniform distribution for the same number of subdivision grid points.

5.2 Packing

Sphere packing is a variation of the Tammes covering problem and arranges identical spheres around points on the surface of a sphere. Sphere packing focuses on maximizing the minimal distance between points. This seems like double talk, but it basically means that for any point distribution, the goal is to make the distance between points as large and nearly equal as possible. One researcher states this objective somewhat philosophically: *"How can you best stay away from other people in a spherical universe?"*[11] In sphere packing, the objective is to find the largest common diameter that a set of equal spheres can have and be placed on another sphere without interference. The centers of the spheres all lie on the surface of the sphere being evaluated.

Let us take the simplest packing case in which just three spheres touch each other. Figure 5.3(a) shows that their centers will be equidistant and the angular separation between one another will be 60°, forming an equilateral triangle between the sphere's centers. We can imagine a fourth sphere cradled on top of these three, with all four centers equidistant and with the same angular separation (see Figure 5.3 (b)). This is the maximum number of points in three dimensions that can be both equidistant and equiangular to one another; they also define the vertices of the tetrahedron.

Within limits, we can place identical spheres around a central sphere where packed spheres will kiss, but not interfere with one another. *Kissing*, a term borrowed from billiards, refers to balls that just touch but do not interfere with each other. The *kissing number* is the largest possible number of equal size spheres that can touch a given sphere in three dimensions. Finding that number of spheres is known as the kissing number problem. The goal is to minimize the gaps between the balls and not have any ball interfere with another. Figure 5.4 shows that the maximum kissing number in three dimensions is 12— that is, 12 spheres surround a central one, with all having the same radius. The centers of the outer 12 spheres form the vertices of an icosahedron.

---

[11] (Bagchi 1997).

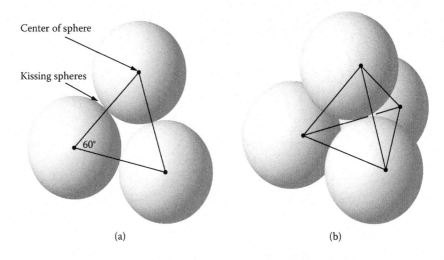

Center of sphere

Kissing spheres

60°

(a)   (b)

Figure 5.3. Angular separation of three tangent spheres.

The kissing number problem in three dimensions can also be described in terms of the angular separation. This angle between the centers of kissing, or near-kissing spheres, will be at least 60°, as we saw in Figure 5.3. Circle packing and the kissing number problem are two views of the same thing.

We can adapt the principles of packing and kissing spheres to spherical subdivision by centering spheres of the same radius at each point on the surface of the sphere that results from subdivision. These packed spheres will have the same radius, but it will be far less than the unit sphere's radius they surround. The largest radius of the packing spheres will be half the chord length between the two closest grid points in the subdivision. If their radius were larger, the spheres would interfere; if less, then the spaces between would increase. Figure 5.5 shows a typical sphere packing for a subdivided spherical icosahedron. All spheres have the same radius, and their centers are all on the surface of the unit sphere. In this schema, only the spheres that are immediately clustered around each of the 12 icosahedra vertices kiss. There are varying gaps between all other spheres.

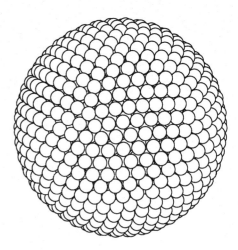

Figure 5.4. Twelve packed spheres.   **Figure 5.5.** Distributed point sphere packing.

If two subdivision schemas distribute the same number of points on the sphere, the one that has the larger radius packing spheres is the one whose points are more uniformly distributed. Thus, the metric of comparison of the two schemas is simply the packing sphere's radius.

5.2.1   200-Year-Old Kissing Puzzle

In three dimensions, the number of spheres that can be packed around a central one is not always obvious. It is easy to arrange 12 spheres to touch a central one, as shown in Figure 5.4. But it's not clear if you can add another. There is enough total space left over, but will one more fit? In the late 1600s, Isaac Newton and astronomer David Gregory argued this point. Both recognized there was enough space left over between the spheres to add another, but would an additional sphere kiss or interfere? Newton thought the kissing number remained at 12 and none could be added, but Gregory argued otherwise. The question was not resolved until 200 years later. In 1874, Newton was proven correct.[12]

5.3   Volume

The volume perspective sees the point distribution problem as one of maximizing the volume of an enclosing surface surrounding the points. This surface is called the *convex hull* and it is a mathematical concept that applies to *n* dimensional space where *n* can be 2, 3, 4, or more dimensions. In two dimensions, it is the minimum convex polygon that surrounds a set of points. The perimeter of this polygon is like a rubber band stretched tightly against the outermost points in a set of points. The polygon is always convex because a line between any two points on the perimeter is always inside the polygon or coincident to its perimeter.

In three dimensions, the convex hull is like shrink-wrapping around the extent of the outermost points. It is the minimum *convex polyhedron* that surrounds the points. The surface resembles a faceted polyhedron that perfectly contains the object or the set of points. A simple example is the Platonic solids. They are convex polyhedra with all points equidistant from the center; therefore, their convex hull is the same polyhedra. This is also the case for our subdivided spheres, where neighboring point sets form planar triangles.

Three-dimensional (3D) convex hulls are important in many applications. For example, in space management, they play a role in packing products and shipping cargo, activities in which the minimum volume needed to accommodate something is important (although this ignores the possibility of nesting objects one within another). In robotics, the convex hulls of objects indicate potential collision zones, where one object can interfere with another, like a robot passing through a maze of objects. In computer graphics, convex hulls are used to distinguish one object from another. The convex hulls of horses do not look like the convex hulls of tea pots. Convex hulls can also be used to find the resting position of objects, or how an object moves or is to be picked up. In our application, the convex hull's volume provides a way to measure just how evenly points are distributed on the surface of a sphere.

The convex hull surface defines a volume and it is affected by the number and distribution of noncoincident points on the sphere's surface. Adding more points or spacing them more evenly will increase the volume. Figure 5.6 shows the effects of both. Figure 5.6(a) and (b) show how by increasing the number of points, it increases the convex hull. The icosahedron (a) and dodecahedron (b) evenly distribute 12 and

---

[12] (Szpiro 2003).

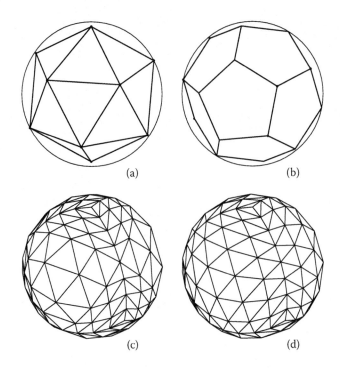

Figure 5.6. The number and distribution of points affect the convex hull.

20 points on the surface of their respective unit circumsphere. The dodecahedron has almost 9.8% more surface area than the icosahedron, simply because it evenly distributes more points on the same unit sphere. When both are inscribed in a unit sphere, the dodecahedron's volume is 66.49%, whereas the icosahedron's volume is less at 60.54% of the sphere's volume.

In Figure 5.6(c) and (d), we see the effects of point distribution on the convex hull. Both subdivided spheres distribute 130 points on the sphere, but the points in schema (d) are more evenly distributed than (c), as demonstrated by the fact that there is less variation in the size of its triangles. The convex hull of (d) is about 1.3% larger as a result. The convex hull of a fully subdivided sphere is simply the smallest convex polyhedron that completely encloses it and tends to look just like a regular faceted sphere, as shown in Figure 5.8. Efficient 3D convex hull algorithms remain a computer science research area, but our spherical subdivision applications greatly simplify calculations. In our applications, all points are on the surface of a sphere. Thus, every point defines part of the convex hull's surface. This is shown in Figure 5.7, where the convex hull for a subdivided spherical polyhedron face is shown. A portion of the sphere's surface is shown, so that it is clear that the points are on the surface. The triangulated mesh that results from connecting the nearest neighboring points is simple to define, and every triangle becomes a surface patch in the convex hull. There are 16 triangular patches and three more surface patches along the sides to the origin of the sphere. The symmetry within our subdivisions reduces calculation as well. Most of our schemas are based on standard geometric subareas of spherical polyhedra such as an equilateral face or a Schwarz triangle. Figure 5.7 is a subdivision of a single face of a spherical octahedron. There is a total of eight octahedral faces subdivided like these in the whole sphere. Highly precise surface and volume calculations need only be performed for these standards and

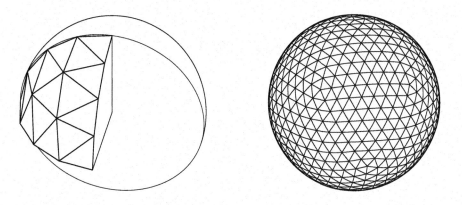

Figure 5.7. Convex hull segment.   **Figure 5.8.** Convex hull of a spherical subdivision.

then multiplied by the number of instances to get the surface area and volume for the whole sphere. Later, in Chapter 9, we will compare the convex hull volumes of different subdivision schemas. The differences tell us which ones produce the best point distributions.

Figure 5.8 shows a complete convex hull surface for a higher order spherical icosahedron subdivision. The surface looks no different than a faceted version. Once again, high-precision calculations are performed on a standard single subdivided face and then multiplied by 20 to get the convex hull volume for the whole sphere.

5.4   Summary

Point distributions on spheres can be defined as covering, packing, or volume problems. Each approach tries to optimize the arrangement of points, maximizing or minimizing a different aspect of their distribution. Cap density tells us what percentage of the sphere's surface is covered by lesser circles (all with the same maximum radius) without causing overlaps. Likewise, the radius of the largest possible sphere centered at every point provides another metric for evaluating point distributions. Spheres, all with the same radius, kiss or just touch each other but do not interfere with each other. The radius is related to the minimum chord distance in the subdivision and indicates how uniformly points are spread out. The convex hull surface provides an area and volume metric for evaluating point distributions. Both increase as more points are placed on the surface of the sphere, and both increase as the same number of points is more evenly distributed.

In the next chapter, we will look at ways to use spherical polyhedra as geometric references to generate and distribute points.

Additional Resources

Aste, Tomaso and Denis L. Weaire. *The Pursuit of Perfect Packing*. Philadelphia, PA: Institute of Physics Publishing, 2000.

Fowler, P. W., T. Tarnai, and Zs. Gáspár. "From Circle Packing to Covering on a Sphere with Antipodal Constraints." *Proceedings of the Royal Society* 458.2025 (2002): 2275–2287.

Szpiro, George. *Kepler's Conjecture: How Some of the Greatest Minds in History Helped Solve One of the Oldest Math Problems in the World.* Hoboken, NJ: John Wiley & Sons, 2003.

Tarnai, T., Zs. Gáspár, and Lidia Szalai. "Pentagon Packing Models for All-Pentamer Virus Structures." *Biophysical Journal* 69 (1995): 612–618.

Tarnai, T. and Magnus J. Wenninger. "Spherical Circle-Coverings and Geodesic Domes." *IASS International Congress* Moscow (1985): 5–21.

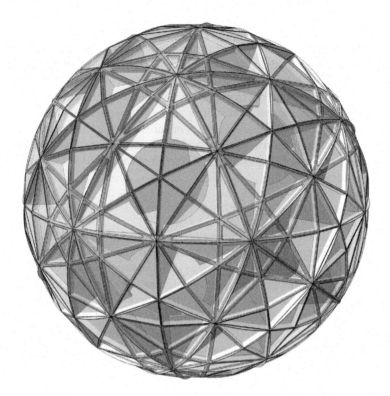

⑥ Polyhedral Frameworks

In this chapter, we cover the characteristics of Platonic solids and Archimedean solids and how to define their geometry precisely, a prerequisite for subdivision. We begin our discussion of polyhedra by defining their most common characteristics. We then detail the advantages each solid offers for spherical subdivision, show their relationships to each other, and highlight the symmetry properties we can exploit when we use them to subdivide spheres. In the next chapter, we use them to jump-start spherical subdivision.

For centuries, polyhedra have fascinated mathematicians, scientists, and artisans alike. They are one of the most intensely studied aspects of geometry. Their 3D faces appear in seemingly infinite arrangements of faceted and star shapes. Their geometry appears in everything from microscopic sea creatures to crystals, from virus cells to soccer balls. Their symmetries apply to mathematics, architecture, art, chemistry, biology, product design, and physics.

Two types of polyhedra, Platonic and Archimedean, are ideal frameworks for subdividing spheres. Platonic solids (we use the terms *solid* and *polyhedron* interchangeably) have faces made of only one type of regular polygon: equilateral triangles, squares, or pentagons. There are five Platonic polyhedra. *Archimedean solids* have faces made of two or more regular polygons and include hexagons, octagons, and dodecagons as well. There are many Archimedean solids, some quite complex. Both types of solids have planar and spherical versions and they are highly symmetrical. Spherical polyhedra give us a huge head start in the subdivision process because their vertices are already evenly distributed on the surface of a sphere that surrounds them and their regular face shapes; with equal edge length, edges give us small and manageable areas to work with as well. We can develop grids within these small symmetrical areas and then replicate the grid repeatedly to cover the rest of the sphere without causing overlaps or leaving any gaps. The same symmetry ensures that the total number of vertices, faces, and edges in a polyhedron will always have a predictable relationship. Spherical subdivision based on spherical Platonic and Archimedean solids are the most common approach, by far.

6.1 What Is a Polyhedron?

Polyhedra are mathematically sophisticated and appear in more than three dimensions. In solid geometry, a polyhedron is a 3D solid whose faces join at their edges. The word derives from the Greek *poly* (many) plus *hedron* from sided. There are many distinct families of polyhedra, the result of differences in their geometry and various mathematical properties. We focus on two types, the Platonic and Archimedean, because they are the most common polyhedral frameworks used in spherical subdivision. The five Platonic solids are regular polyhedra, which means their faces are the same regular type—equilateral triangles, squares, or pentagons. The 13 Archimedean solids are closely related to the Platonic solids, but have two or more regular faces and some have hexagons, octagons, and decagons.

We can define Platonic and Archimedean solids as 3D convex forms with no holes, depressions, protrusions, or penetrations. They are totally defined by their vertices, edges, and faces. All vertices lie on the surface of a surrounding sphere called the circumsphere, whose origin is the same as the center of the polyhedron. The edges of Platonic and Archimedean solids are equal-length straight edges between two neighboring vertices. On spherical versions, edges are equal-length geodesic arcs. The faces of planar polyhedra are regular polygons, and the plane of any one of them, if extended, does not pass through the interior volume of the polyhedron. An individual Platonic or Archimedean solid will have the same number and types of faces sequenced in the same way around every one of its vertices. Faces do not penetrate one another and every face joins another at an edge, leaving no gaps and causing no overlaps. For spherical polyhedra, the sum of the face angles surrounding any vertex is always 360°.

Figure 6.1 shows the planar and spherical version of a dodecahedron, a Platonic solid with 20 vertices, 30 edges, and 12 pentagonal faces. The planar face angles shown in (a) are all 72°, and the dihedral angle, the angle between faces measured perpendicular to their shared edge, is approximately 116.56°. The spherical version shown in (b) has face angles $\beta = 120°$. The arc angles of the spherical edges α are each approximately 41.81°.

Spherical Platonic and Archimedean solids can be derived from their planar versions very easily. All vertices of these solids are the same distance from the center of the polyhedron; thus, they are all on the surface of a circumscribing sphere. The vertices on either end of a chord, a platonic edge, and the center of the solid define the plane of the arc, a geodesic arc, between neighboring points. In Figure 6.1(a), one such plane is shown as a small gray triangle defined by three points: the origin and two vertices.[1] Figure 6.1(b) shows the spherical version of the dodecahedron from the same viewpoint and scale as (a), and every edge in (a) has been transformed into a geodesic arc in (b). The planes of two arcs are emphasized in (b). You can think

---

[1] Three points define a plane.

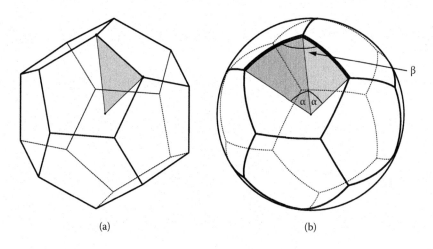

(a)   (b)

Figure 6.1. Planar and spherical polyhedra.

of these geodesic arcs as *central projections* of a planar polyhedron's edges. The projection is on the polyhedron's circumsphere, which acts like a curved projection screen. An imaginary point light source located at the origin produces a shadow of the chord, its arc, on the circumsphere screen.

In Figure 6.1(b), the two emphasized arcs demonstrate another fact about angles on a sphere. The surface angle β between the two arcs is the dihedral angle, between the great circle planes that define them (see Section 4.16.6). The arcs themselves are the sides of the spherical face and are measured by their central angles from the center or origin of the sphere. They are each labeled α. In any given Platonic or Archimedean polyhedron (planar or spherical version), all chords or arcs are of the same length.

With this general introduction, let us take a closer look at the properties of these two families of polyhedra.

6.2 Platonic Solids

The five Platonic solids are *regular polyhedra*. Known since antiquity, regular polyhedra are polyhedra having the same size and type of regular convex face (with equal angles and length sides), and the same number of faces meeting at every vertex. Regular polyhedra are the only polyhedra that enclose a volume of space with a surface composed of only one kind of plane polygon (for example, squares, triangles, or pentagons) and feature one kind of polygon meeting at every vertex. Figure 6.2 shows all five Platonic solids. They are the tetrahedron (1), octahedron (2), cube[2] (3), icosahedron (4), and dodecahedron (5). In Figure 6.2, all solids have the same edge length, and the differences in their enclosed volumes are apparent.

The Platonic solids are the most widely used frameworks for spherical subdivision:

- All vertices are evenly distributed and lie on the surface of a surrounding sphere called a circumsphere;

- Each Platonic solid is composed of equal surface angles, equal dihedral angles, equal chord length, and equal arc lengths;

- Every face, edge, and vertex has rotational and reflection symmetry;

---

[2] A cube is also called *a hexahedron.*

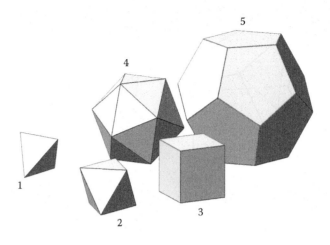

Figure 6.2. Platonic solids.

- A single small Schwarz right triangle totally defines the entire sphere's subdivision;

- A subdivision of a Platonic solid also subdivides its dual, even though its dual's face type is very different. (Duals will be discussed shortly.)

- A spherical equilateral triangle can describe the surface of any Platonic solid, even if its faces are not equilateral;

- Except for the tetrahedron, the planes of faces on opposite sides of the polyhedron are parallel.

Each of these characteristics will be discussed in the remainder of this chapter.

One could say that the tetrahedron is the most equal of polyhedra. It has four vertices, six edges, and four equilateral faces. Not only are its edges equal and all its faces the same, it is the only polyhedron in which every vertex is equidistant from every other one. It's also the only Platonic solid whose vertices have no antipode, which means no two vertices are on opposite sides of the solid or its circumsphere. The tetrahedron can be formed by truncating a cube with two cuts; that is, by slicing off pieces of the solid. Each cut starts along the diagonal of one of the cube's faces and slices through to the opposite side's vertices without passing through the center of the cube. In effect, the opposite vertices of one face and its non-antipodal companion on the cube's opposing face define the tetrahedron's vertices. We discuss truncation in more detail when we describe how some Archimedean solids can be created from Platonic solids.

The octahedron is a familiar form. Many people visualize it as two back-to-back square pyramids. It has 6 vertices, 12 edges, and 8 equilateral faces. Four faces meet at any vertex. Each of the octahedron's six vertices has an antipode and the three axes between antipodal pairs are mutually orthogonal. The octahedron plays a special role in spherical subdivision. Its spherical version is the only spherical polyhedron where its contiguous edges define three mutually orthogonal great circles. Cartographic applications often use octahedral bases when grid systems must have an equatorial and Greenwich great circle. The spherical octahedron also appears in applications where a Cartesian coordinate system is being modeled. From a trigonometric point of view, the faces of the spherical octahedron clearly demonstrate that a spherical triangle can have more than one 90° angle. The octahedron's faces each have six: three 90° surface angles and three 90° edges.

The cube, with its 8 vertices, 12 edges, and 6 faces, is the most familiar polyhedron. It is defined by three parallel pairs of square faces whose midface points form orthogonal axes. Pairs of opposing edges are parallel, and every vertex has an antipode. The cube has a special role in spherical subdivisions because it has subtle symmetry relationships with other polyhedra. These relationships are discussed later in the chapter.

The dodecahedron has 30 edges, 12 pentagonal faces, and 20 vertices, the most vertices of any Platonic solid. The dodecahedron also has the largest volume of any Platonic solid if all the solids have the same length edges. From a symmetry point of view, the dodecahedron's 12 pentagonal faces are six pairs of faces on opposite sides of the polyhedron. Opposing pairs of faces are rotated 36° relative to each other around an axis between their midface points. Every edge of the dodecahedron is parallel to another edge on the opposite side of the polyhedron.

The icosahedron is by far the most frequently used spherical polyhedron in subdivision. The icosahedron has 12 vertices, 30 edges, and 20 equilateral faces, the most faces of any Platonic solid. In its spherical version, each face can be subdivided into six spherical right triangles, which means 120 (6 × 20 faces) of these right triangles tile together to cover the entire sphere. Subdivisions of one right triangle can define the subdivision of the whole sphere. The icosahedron has 15 pairs of opposing edges and its 12 vertices define 6 pairs of antipodal points. Each antipodal pair defines an axis. These symmetries are very important and described in detail later in the chapter.

Max Brückner (1860–1934)

Johannes Max Brückner, a German geometer, was one of the most prolific builders and collectors of polyhedral models. His most famous geometric models were of compounds, stellated non-convex and uniform polyhedra which he published *Vielecke und Vielflache: Theorie und Geschichte*.[3]

Many of Brückner's 146 non-convex polyhedra are stellations that extend the edges or face planes of a regular Platonic polyhedra. They are amazing constructions. Photographs of display models from Table IX and drawings from Table VII of his 1900 treatise appear below. Photography in scientific illustration began as early as 1845 and gained wider spread use by the 1880. Brückner's treatise was a good example of how photography was being used. He was also the first to model the complete icosahedron set with irregular faces. His collection was the most comprehensive demonstration of what was known about polyhedra at the time.

Equally impressive was Brückner's use of projective geometry to develop the polygons for building his models. In some cases, arrangements of non-overlapping edge-joined polygons could be folded in paper, Brückner's primary media, or cut along edges to become the faces and stellations of models.

Brückner's models were intended as research and teaching aids. It's difficult to gauge how his work was received by professional mathematicians at the time but today, his work inspires artists and geometric modelers who continue to build polyhedra models or reproduce their forms with computer programs. See Additional Resources for analysis of Brückner's collection.[4]

---

[3] Title translation "*Polygons and Polyhedra: Theory and History*" (Brückner 1900).
[4] (Mikloweit 2020) and (Hart 2019).

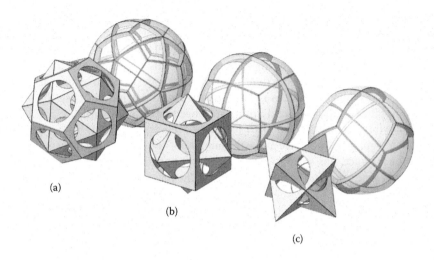

Figure 6.3. Planar and spherical Platonic duals.

6.2.1 Platonic Duals

The *dual* of a polyhedron is another polyhedron resulting from exchanging vertices and faces. A vertex of the dual is located at the center of the original polyhedron's face. The duals of Platonic solids are unique; they are other Platonic solids. Figure 6.3 illustrates the dual relationship for all the planar and spherical Platonics. In the front row, from left to right, are the planar dodecahedron-icosahedron duals (a) and the octahedron-cube duals (b). The tetrahedron (c) is a dual to itself. The back row is the spherical version of the front row. The dual relationship between two polyhedra is especially evident in the planar versions, where you easily see that the faces of one correspond to the vertices of the other. In the spherical versions, the colors of the planar versions are retained to better see how their spherical faces and edges relate. In the spherical versions, you can easily see the edges (great circle arcs) of each dual cross perpendicular to the other and they cross each other at their midpoints. This is also the case in the planar versions, but it's easier to see in the spherical version. Notice also that for any pair of duals, both have the same number of edges, but their numbers of faces and vertices are swapped. That is, if one polyhedron in the dual set, the octahedron-cube dual for example, has six vertices and eight faces (the octahedron), the other (the cube) will have eight vertices and six faces. Another consequence of the symmetry between duals is the spherical subdivision of one of them is also a subdivision of the other.

6.2.2 Shorthand for the Unpronounceable

The Platonic solids have simple names and their characteristics (vertices, edges, and faces) are easy enough to remember. But the Platonic solids are not the only polyhedra. In fact, there are hundreds. Many have catchy titles such as tetrahemihexahedron or the great-inverted-retrosnub-icosidodecahedron. Shorthand notations were inevitable to make it easier to refer to polyhedra and to see their similarities and differences at the same time. There are several notation systems. Most are based on the polyhedron's face types and summarize how faces meet at edges or cluster around vertices. With experience, one can visualize a complex polyhedron simply by looking its notation. In some systems, a rearrangement of notation describes their dual and their *symmetry group*, something we discuss shortly.

Ludwig Schläfli (1814–1895) worked mostly in geometry, arithmetic, and function theory, though some of his best work was in polyhedra, where he invented the notation named after him. *Schläfli Symbols* describe all the Platonic and Archimedean polyhedra. We highlight this system for the Platonic solids because Coxeter, a renowned mathematician, extended this system to describe geodesic subdivisions of spherical polyhedra. The extended system is particularly useful when describing the spherical subdivision schemas. It is a useful notation to become familiar with.

Schläfli symbols take the form $\{p,q\}$ to describe regular polyhedra, with p the number of sides in a regular polygon face and q the number of those faces that meet at any vertex. A cube, for instance, has square faces each with four sides (p) and three of these faces meet at each vertex (q). The Schläfli symbols for the Platonic solids (paired by dual) are:

Icosahedron	$\{3,5\}$	Dodecahedron	$\{5,3\}$
Octahedron	$\{3,4\}$	Cube	$\{4,3\}$
Tetrahedron	$\{3,3\}$		

One useful property of Schläfli symbols is that the reverse of $\{p,q\}$ or $\{q,p\}$ is the symbol for the dual of that polyhedron. In other words, if $\{p,q\}$ *is* a possible solid, so is $\{q,p\}$. You can verify this by looking at Table 6.1, which summarizes the vertices,

Name(s) & Symbols	Vertices	Edges	Faces	{3}	{4}	{5}
tetrahedron $\{3,3\}$	4	6	4	4	–	–
octahedron $\{3,4\}$	6	12	8	8	–	–
cube $\{4,3\}$	8	12	6	–	6	–
dodecahedron $\{5, 3\}$	20	30	12	–	–	12
icosahedron $\{3,5\}$	12	30	20	20	–	–

Table 6.1. Planar and spherical platonics.

edges, and faces for the five Platonic solids we have been discussing. The Schläfli symbols for each polyhedron are given under the polyhedron's name. Notice also that p and q values are simply reversed for duals.

6.2.3   Circumsphere and Insphere

We already mentioned that the vertices of the Platonic solids are tangent to a circumsphere whose origin is coincident with the center of the polyhedron, a point equidistant from each vertex. Less obvious, however, is the fact that another sphere, called the *insphere*, can be inscribed inside each Platonic solid and this sphere is tangent to the midpoint of each Platonic face. These points of tangency define the vertices of its dual.

Figure 6.4 illustrates this relationship by showing the circumsphere and insphere for a dodecahedron. The radius of the insphere, the *inradius*, is the perpendicular distance to the face from the center of the polyhedron. The view in Figure 6.4 shows both spheres and several vertices tangent to the circumsphere. Midface points tangent to the insphere define the vertices of the polyhedron's dual, the icosahedron. Both polyhedra share several systematic relationships that we will investigate next.

The late twentieth-century geometrician H. S. M. Coxeter used the circumsphere-insphere tangent points to define regular polyhedra in a unique way. He states, "*A polyhedron is regular if there exists three concentric spheres one of which contains all the vertices, one contains the midpoints of all the edges, and one that meets all the centers of all the faces.*"[5] The five Platonic solids satisfy these three conditions.

Dual Platonic solids share another subtle relationship. For the icosahedron-dodecahedron and cube-octahedron dual, if either pair's circumsphere has the same radius, they will also have the same radius insphere. This property means the ratio of their volumes is the same as the ratio of their surfaces for either pair of duals.

How can this happen? A polyhedron and its dual are different solids! The answer lies in the fact that when both polyhedra have the same circumsphere, the radius of *circumcircles* around their faces is also the same.

Figure 6.5(a) shows an icosahedron and a dodecahedron inscribed in the same circumsphere. The vertices of both polyhedra are the same distance from their shared center, and the circumcircle of one face from each polyhedron is shown in the figure. The circumcircles around the dodecahedron's pentagonal faces have the same radius as those around the icosahedron's triangular face. Since the circumference of these

Figure 6.4. Polyhedra circumsphere and insphere.

[5] (Coxeter 1973).

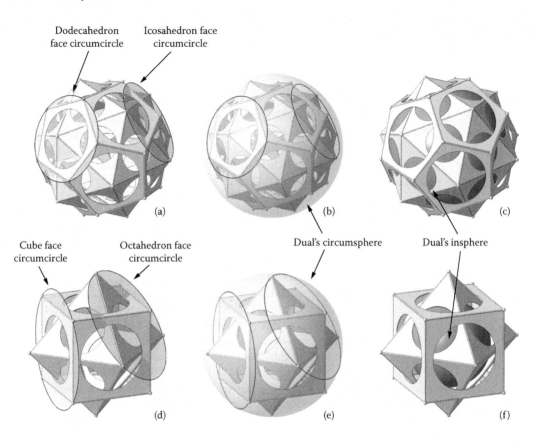

Dodecahedron face circumcircle Icosahedron face circumcircle

Cube face circumcircle Octahedron face circumcircle Dual's circumsphere Dual's insphere

(a)   (b)   (c)   (d)   (e)   (f)

Figure 6.5. Common circumsphere and insphere.

circumcircles pass through their respective face's vertices, which are already on the surface of the sphere, the circumcircles are lesser circles on the surface of the circumsphere. As shown in Figure 6.5(b), they form congruent spherical caps whose base planes are the same perpendicular distance from the origin of the sphere. Thus, both polyhedra have the same insphere radius.[6] The icosahedron's insphere is shown in blue in Figure 6.5(c). It is tangent to the midfaces of both the dodecahedron and icosahedron.

The exact same relationship applies to the octahedron and cube, as shown in Figure 6.5(d)–(f). The faces of both these polyhedra have the same radius circumcircles when they both have the same circumsphere, as shown in (d) and (e). Thus, the ratio of their two volumes is the same as the ratio of their two surface areas. You can verify this by looking at the volume and area ratios for the icosahedron-dodecahedron and cuboctahedron later in the chapter.

6.2.4 Vertex-Face-Edge Relationships

In 1639, René Descartes observed that the total number of a polyhedron's vertices, faces, and edges were systematically related. He noted his discovery in private correspondence and described it in a manuscript that remained unpublished until after Leonhard Euler independently rediscovered the same relationship and published it in 1752.[7] Today, this

---

[6] Note that the polyhedral face circumcircles and resulting spherical caps illustrated here are not the same as the caps surrounding polyhedral vertices illustrated earlier in Chapter 4.

[7] (Adam and Tannery 1996, 265).

Figure 6.6. Vertex, face, and edge examples of Platonics and subdivided spherical Platonics.

formula is called the Descartes-Euler polyhedral formula, but more commonly, *Euler's law*.[8] For convex polyhedra (without any holes), Euler's law states the number of vertices + faces = edges + 2 .

Euler's law applies equally to the Platonic and Archimedean solids and to the highly subdivided spherical versions based on them. Figure 6.6 shows the tetrahedron, cube, and dodecahedron in the front row, and subdivided spherical versions behind them. Each polyhedron is the same scale and is oriented the same way as its spherical version. Although the Platonic and spherical versions (shown in the figure) differ considerably, Euler's law demonstrates that the relationship of their total vertices, faces, and edges remains the same. Table 6.2 confirms Euler's law by listing the vertices, faces, and edges for each of the polyhedra in Figure 6.6.

In the next chapter, we develop techniques for subdividing spheres. We make use of a very handy triangulation number called T to describe the triangulated mesh resulting from a subdivision. With T, it is easy to determine the number of faces, edges, and vertices in a highly subdivided polyhedron. These counts are consistent with Euler's law, no matter how intensely the sphere is subdivided. We will use both T and Euler's law to check our subdivision results.

Euler's Law									
Polyhedra	**Vertices**		**Faces**		**Edges**		**Constant**		**Sum**
Tetrahedron	4	+	4	=	6	+	2	=	8
Subdivided Spherical Tetrahedron	130	+	256	=	384	+	2	=	386
Cube	8	+	6	=	12	+	2	=	14
Subdivided Spherical Cube	194	+	384	=	576	+	2	=	578
Dodecahedron	20	+	12	=	30	+	2	=	32
Subdivided Spherical Dodecahedron	122	+	240	=	360	+	2	=	362

Table 6.2. Examples of Euler's law for Platonics and subdivided spherical Platonics.

---

[8] See (Richeson 2008) for an excellent history of Euler's law and how this law created the field of topology.

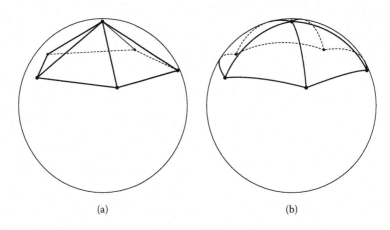

(a)   (b)

Figure 6.7. Five equilateral triangles define space.

Descartes made another important contribution in understanding the relationship between the faces, edges, and vertices of convex polyhedra. He noted that the sum of the face angles around any vertex, an amount he called the *closure deficit*, is always less than 360°. A simple example shows why this is the case. If five equilateral triangles cluster around the same vertex, they form a *pentagonal cap*, as seen in Figure 6.7(a). When each vertex is on the surface of a sphere; the closure deficit is also measuring the sphere's curvature. In this example, the closure deficit is 60° or 360°–5(60°). If the deficit is larger than this, the sphere gets smaller and the curvature is greater. Conversely, if the deficit is smaller, the size of the sphere gets larger and has less curvature. But there are limits. As the deficit approaches 0°, the polygons surrounding the vertex approach a plane. We see this happening if we try to position six equilateral triangles around the same vertex; the deficit would be 0° or 360°—6(60°). They form a perfect plane, which has no curvature. As a result, no space is defined.

In Figure 6.7(b), we see the spherical version of the same vertex points shown in (a). If we extended the pentagonal cap analogy to every vertex that is not already surrounded, we will quickly recreate an icosahedron. This demonstrates that 20, the number of equilateral faces in an icosahedron, is the maximum number of faces possible for a regular polyhedron.

Descartes also observed the sum of all the closure deficits for all vertices in convex polyhedra is always 720°. We use this rather amazing observation later, when we check the results of our spherical subdivisions. Speaking of convex polyhedra (with no holes), Descartes stated: "*If four plane right angles are multiplied by the number of solid angles and from the product are subtracted eight plane right angles, there remains the sum of all the plane angles, which exist on the surface of that polyhedron.*"[9]

What this means is the total face degrees in convex polyhedra equals four right angles (360°) times the number of vertices in the polyhedra less eight right angles (720°).

6.2.5 The Golden Section

The so-called *Golden Section* or *Golden Ratio is* an amazing geometrical construction that has been studied since antiquity. It has been related to human perceptions of balance and beauty and appears in nature, art, architecture, science, and mathematics. We can use the golden section to define very precise polyhedra.

---

[9] From American Mathematical Society's commentary on Descartes' unpublished *Treatise* manuscript, 1999.

The golden section is the result of asking a simple question: what is the result when a line is divided, so the ratio of the lengths of the divisions is the same as the ratio of the division to the length of the line?

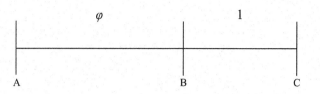

That is, divide a line segment with $AC{:}AB{:}AB{:}BC$. In the language of ancient mathematicians, line AC is divided in an *"extreme and mean ratio."* The result is $AB = 1.618033989... \times BC$. This ratio is universally known as φ.

The value of φ is found from a simple algebraic statement of the original problem. If $BC = 1$, then,

$$\frac{\varphi + 1}{\varphi} = \frac{\varphi}{1}$$

or,

$$\varphi^2 - \varphi - 1 = 0.$$

Among the amazing properties of φ is this result:

$$\varphi - 1 = \frac{1}{\varphi},$$

so $1/1.618033989... = 0.618033989....$

It is also interesting to note that the ratio of two successive terms of the Fibonacci series, where each term is the sum of the preceding two terms $(1, 1, 2, 3, 5, 8, 13, 21,...)$ converges to φ.[10]

The symbol φ occurs often in the geometry of regular polygons. Perhaps the most dramatic illustration is the pentagon, from which the mystical *pentagram* derives, as shown in Figure 6.8. The ratio of the short and long lines (two are accented in bold) is $1{:}\varphi$, the golden ratio.

6.2.6   Precise Platonics

Precise definitions of the Platonic solids are essential when subdividing their spherical versions. We can use unit and φ relationships to define exact vertex coordinates of all five Platonic solids. The face angles of spherical Platonic solids are integral divisions of 360°, and their sides are arc lengths based on trigonometric ratios of the square roots of 2, 3, and phi.

---

[10] See Mario Livio, *The Golden Ratio: The Story of Phi, the World's Most Astonishing Number* (New York: Broadway Books, 2002) for a wonderful review of phi.

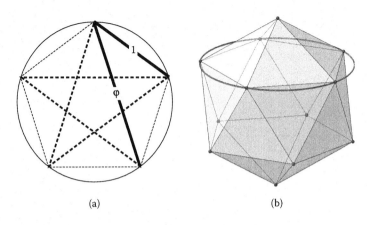

Figure 6.8. The pentagram's golden ratios.

The coordinates of the tetrahedron, octahedron, and cube are simple:

Tetrahedron: $(\pm1, +1, +1)\ (-1, -1, +1)\ (-1, +1, -1)\ (+1, -1, -1)$
Octahedron: $(\pm1, 0, 0)\ (0, \pm1, 0)\ (0, 0, \pm1)$
Cube: $(\pm1, \pm1, \pm1)$

The coordinates of the dodecahedron and icosahedron are related to the golden ratio. The 12 icosahedron vertices are defined by the orthogonal intersection of three golden rectangles; that is, rectangles whose sides have the proportions of 1 to φ. Figure 6.9(a) and (b) shows the arrangement of the golden rectangles to the icosahedron's vertices and to its faces. The coordinates of the icosahedron's 12 vertices corresponding to the figure are

$$\left(0, \pm1, \pm\varphi\right), \left(\pm\varphi, 0, \pm1\right), \left(\pm1, \pm\varphi, 0\right).$$

Figures 6.9(c) and 5.6(d) show the dodecahedron. The coordinates of its 20 vertices are also based on the golden ratio and defined by

$$\left(0, \pm\varphi, \pm\frac{1}{\varphi}\right), \left(\pm\frac{1}{\varphi}, 0, \pm\varphi\right), \left(\pm\varphi, \pm\frac{1}{\varphi}, 0\right), (\pm1, \pm, \pm1).$$

The coordinates given previously for the Platonic solids are precise, but they do not create polyhedra with a unit radius circumsphere. To rescale vertex coordinates to be on the surface of the unit sphere, simply divide each of the coordinate's x, y, and z components by the absolute value of the coordinate:

$$x_{\text{unit sphere}} = \frac{x}{\sqrt{x^2 + y^2 + z^2}}$$

$$y_{\text{unit sphere}} = \frac{y}{\sqrt{x^2 + y^2 + z^2}}$$

$$z_{\text{unit sphere}} = \frac{z}{\sqrt{x^2 + y^2 + z^2}}$$

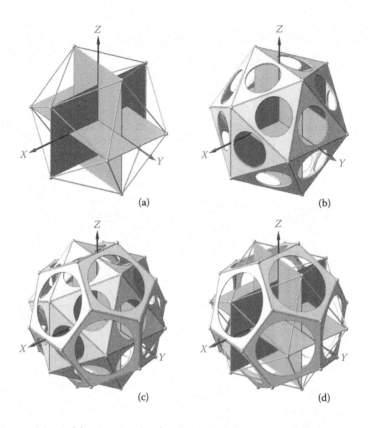

(a)

(b)

(c)

(d)

Figure 6.9. Golden rectangles for the icosahedron and dodecahedron.

6.2.7 Platonic Summary

At this point, we have identified all the key properties needed to use any Platonic solid as a polyhedral framework for spherical subdivision. Table 6.3 provides more specific properties for each one, such as surface and central angles, vertex coordinates, and so forth. The most important facts we need when starting a subdivision are the precise vertex coordinates, face angles, and side angles of the Platonic solid we will use as our framework for spherical subdivision and Table 6.3 provides this information.

The central angles, values, or formulas can be used as is, but once again, the vertex coordinates, shown in the table, are based on unit and φ ratios and must be scaled to place vertices on the surface of a unit circumsphere. With the basics of each Platonic solid established, we now investigate their symmetry. Symmetry shows us how to find common small areas where we can focus our subdivision efforts. Symmetry also shows us how the subdivision of a small area can be made to cover the entire sphere.

6.3 Symmetry

The symmetry of polyhedra is one reason we find them so visually attractive. But symmetry is more than beauty; there are mathematical symmetries at the core of the dual relationships we have seen, and can be exploited to define areas whose subdivision can be replicated over the remaining surface of the sphere.

There are many types of symmetry. Two types are particularly important in spherical work: rotation and reflection. A polyhedron has *rotational symmetry* if its appearance stays the same after being rotated around an axis by some integral fraction of a

Planar Platonics in Unit Circumsphere					
Measure	**Tetrahedron**	**Octahedron**	**Cube**	**Dodecahedron**	**Icosahedron**
Symbol					
Schläfli Symbol	{3,3}	{3,4}	{4,3}	{5,3}	{3,5}
Face					
Edge Length	1.632993162	1.414213562	1.154700538	0.71364418	1.051462222
Circumcircle Radius	0.942809042	0.816496581	0.816496581	0.607061998	0.607061998
Incircle Radius	0.942809042	0.408248291	0.577350269	0.491123473	0.303530999
Vertex Angle	60.0	60.0	90.0	108.0	60.0
Dihedral Angle					
Adjacent Faces	70.52877937	109.4712206	90.0	116.5650512	138.1896851
Adjacent Faces (radians)	$\cos^{-1}\left(\frac{1}{3}\right)$	$\cos^{-1}\left(\frac{-1}{3}\right)$	$\cos^{-1}(0)$	$\cos^{-1}\left(\frac{-1}{\sqrt{5}}\right)$	$\cos^{-1}\left(\frac{-\sqrt{5}}{3}\right)$
Area					
Face r^2	1.154700538	0.866025404	1.333333333	0.87621852	0.478727069
Full r^2	4.618802154	6.92820323	8.0	10.51462224	9.574541383
Volume					
r^3	0.513200239	1.333333333	1.539600728	2.785163863	2.53615071
Vertex Coordinates	(±1,+1, +1) (−1, −1, +1) (−1, +1, −1) (+1, −1, −1)	(±1, 0, 0) (0, ±1, 0) (0, 0, ±1)	(±1, ±1, ±1)	(0, ± φ , ±1/φ) (±1/ φ , 0, ± φ) (± φ , ±1/ φ , 0) (±1, ±1, ±1)	(0, ±1, ± φ) (± φ , 0, ±1) (±1, ± φ , 0)
Reference Spheres					
Insphere Radius	0.333333333	0.577350269	0.577350269	0.794654473	0.794654473
Circumsphere Radius	1.0	1.0	1.0	1.0	1.0

Spherical Platonics in Unit Circumsphere					
Measure	**Tetrahedron**	**Octahedron**	**Cube**	**Dodecahedron**	**Icosahedron**
Face					
Angle	120.0	90.0	120.0	120.0	72.0
Central Angles					
Side	109.47122	90	70.528779	41.8103149	63.434949
Side (radians)	$2\sin^{-1}\left(\frac{1}{\sqrt{3}}\right)$	$2\sin^{-1}\left(\frac{\sqrt{2}}{2}\right)$	$2\sin^{-1}\left(\frac{1}{\sqrt{3}}\right)$	$2\sin^{-1}\left(\frac{1}{\sqrt{3}\phi}\right)$	$2\sin^{-1}\left(\frac{1}{\sqrt{\phi^2+1}}\right)$
Median	125.26439	90	90	69.09484255	58.282526
Arc Factor					
Side	1.9106332	1.5797963	1.2309594	0.729727656	1.1071487
Median	2.186276	1.5797963	1.5707963	1.20593249	1.017222
Chord Factor					
Side	1.6329932	1.4142136	1.1547005	0.71364418	1.0514622
Median	1.7761477	1.4142136	1.4143136	1.134176274	0.983929
Area					
Face r^2	3.1415927	1.5707963	2.0943951	1.047197551	0.6283185

Note: All angles in degrees unless noted.

Table 6.3. Planar and spherical platonic summary.

full rotation of 360°. Rotational symmetry is one of the most used types of symmetry in spherical work. For example, if you hold a cube in your hand and orientate it so one side faces you and you keep that side pointing toward you while rotating the cube in 90° increments either clockwise or counterclockwise around an axis between your eye and the center of that side, the cube will still look the same each time. This is rotational symmetry. The axis of rotation is your line of sight to the center of the cube's side.

A polyhedron has *reflection symmetry* if one or more planes (mirrors) can be passed through the polyhedron's origin, and the image of faces, edges, or vertices on both sides of the plane are identical. For example, if you passed a plane through a pair of opposite edges of an icosahedron, the images on both sides of this plane would be identical. Since an icosahedron has 30 edges, or 15 opposite pairs, we can see that for edges alone there are already 15 reflection symmetries. Almost all polyhedra have this symmetry, but there are a few that don't. These exceptions are called chiral forms because the images on either side of any plane passed through their centers are different. We will discuss chiral forms later in the chapter.

The *identity symmetry*, sometimes called the "*do nothing*" symmetry, is equivalent to rotating a polyhedron 360° around any point in space; the polyhedron simply returns to its original position, without any change taking place. At first, identity symmetry seems useless because no apparent change has taken place. But the identity serves an important mathematical function analogous to multiplying a number by one (unity). We do not use the identity symmetry directly in subdividing spheres, but we include it whenever we enumerate a polyhedron's symmetries.

Figure 6.10 shows examples of rotation and reflection symmetry for a cube. In (a) through (c), a single rotation axis is shown through the centers of two opposing faces, edges, and vertices. All axes also pass through the center of the polyhedron, represented by a small sphere. The cube can be rotated any multiple of 90°, 120°, or 180°, respectively, around these three axes, and the cube will still look the same. This demonstrates rotational symmetry.

Figure 6.10(d) shows two reflection planes, one through mid-edge points, and the other through edges on opposite sides of the cube. These planes also pass through the

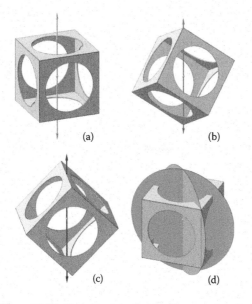

Figure 6.10. Rotation and reflection symmetry.

cube's center. The cube's image on either side of either plane is identical and illustrates reflection symmetry.

6.3.1   Symmetry Groups

A symmetry group is a collection of symmetry-preserving mathematical transformations that when applied to an object, such as a polyhedron, leave it invariant, that is, the object looks the same in shape, position and orientation as it did before the transformation. Subdivisions of spherical polyhedra can retain some, if not all, of the polyhedron's planar symmetry.

Each of the five Platonic solids falls into one of three symmetry groups. The icosahedral group includes the icosahedron-dodecahedron dual. Sometimes the letter *I* is used when making mathematical references to the group. The octahedral group, designated *0,* includes the octahedron and cube. Finally, the tetrahedral group, *T,* includes just the tetrahedron, since it is a self-dual.

We will begin with icosahedral symmetry, which plays a central role in subdivision schemas, because its spherical faces subdivide the sphere into the greatest number of equal areas. This provides a big jump-start in the subdivision process. For this reason, icosahedral symmetry is the most frequently used spherical subdivision framework.

6.3.2   Icosahedral Symmetry

The icosahedron has the highest degree of symmetry of the five Platonic solids. Its 12 vertices define six pairs of antipodes and each antipodal pair defines a rotation axis. Figure 6.11(a) shows all six as blue-gray colored arrows. Each one passes through the center of the polyhedron and around any vertex; five equilateral triangles form a pentagonal cap. If the icosahedron is rotated around any one of these vertex axes any multiple of 72°, the appearance of these caps, or any other part for that matter, will not change.

The icosahedron has 20 faces. The plane of each is parallel to a face on the opposite side of the polyhedron. We can pass an axis through the midface points of opposing pairs of faces and see that each face in the pair is rotated 60° with respect to the other. White arrows in Figure 6.11(b) show all ten of them. The icosahedron can be rotated any multiple of 120° around these axes, and its appearance will not change.

The icosahedron has 15 pairs of opposing and parallel edges. A rotation axis can pass through the mid-edge points of each pair. Thus, for just its edges alone, there are 15 axes of rotation symmetry. The red arrows in Figure 6.11(c) show how the axes relate to pairs of edges. Around any one of these axes, the icosahedron can be rotated any multiple of 180°, and it will still look the same.

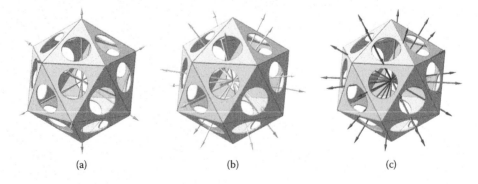

(a)   (b)   (c)

Figure 6.11. Icosahedral rotation symmetry.

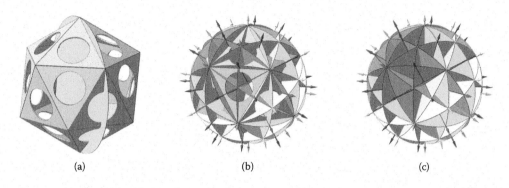

(a)   (b)   (c)

Figure 6.12. Icosahedral reflection symmetry.

To fully describe icosahedral symmetry, we also need to include its reflection symmetries. The icosahedron has 15 planes of symmetry. Each one defines a great circle that passes through a pair of opposing edges and the center of the polyhedron. Figure 6.12 shows these planes in a series of views from the same perspective. A single plane is shown in Figure 6.12(a). It is tangent to four vertices and bisects four of the icosahedron's equilateral faces around their medians. Figure 6.12(b) and (c) shows all 15 symmetry planes and the great circles they create. Figure 6.12(c) shows the relationship of the reflection planes to the vertex, face, and edge rotation axes we saw earlier.

To summarize the icosahedron's symmetries, there are:

24 — four rotations of 72° around centers of 6 pairs of antipodal vertices
20 — two rotations of 60° around 10 pairs of opposite faces
15 — one rotation of 180° around centers of 15 pairs of opposing edges
 1 — identity symmetry

_____

60 symmetries

Reflection symmetry doubles the number of symmetries, for a total of 120.

The dodecahedron and icosahedron are duals and share the same symmetries, though the numbers are rearranged a bit-two rotations of 120° around the centers of ten pairs of antipodal vertices, four rotations of 72° around six pairs of opposite faces, and one rotation of 180° around 15 pairs of opposite edges. The identity symmetry adds one more and reflection doubles the total, equaling the icosahedron's 120 symmetries. Figure 6.13 shows the icosahedron and dodecahedron's dual reflection symmetry with (a) and without (b) symmetry planes.

When we consider all the icosahedron's rotation and reflection symmetries together, 15 great circles are defined in the polyhedron's spherical form. They are shown in Figure 6.14 and they subdivide the icosahedron into 120 Schwarz triangles. The red arcs are icosahedral edges; the blue-gray ones are dodecahedral edges. We will make good use of icosahedral Schwarz triangles later when we develop different schemas for subdividing the sphere and then use CAD techniques to assemble copies of the triangle to make a full sphere.

There are many more great circle combinations for the spherical icosahedron, but none have reflection symmetry, beyond the 15 we have already identified. Figure 6.15 shows a configuration that adds 16 more great circles for a total of 31. Buckminster Fuller discovered this set in the mid-1940s, and the framework greatly influenced his thinking about spheres and inspired his invention of the geodesic dome (see Chapter 2). The additional great circles are the result of adding two new sets. One set is from arcs

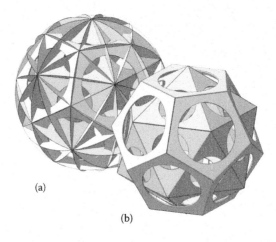

(a)

(b)

Figure 6.13. Icosahedron and dodecahedron dual reflection symmetry.

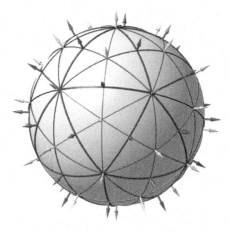

Figure 6.14. Spherical icosahedron and dodecahedron with Schwarz triangles.

Figure 6.15. Icosahedral 31 great circle symmetry.

spanning the mid-edge points of each icosahedral face. These arcs subdivide the icosa-hedron's face into four smaller triangles—one central equilateral surrounded by three isosceles triangles. Since the triangles on either side of the great circle's plane are not the same, these new great circles do not add any new reflection symmetries (see Section 4.16.8). A second set of great circles is added when arcs span from the midedge of one icosahedral face to the nearest non-shared midedge of a neighboring face. These great circles do not add reflection symmetries.[11] Note that while the spherical subdivision, shown in Figure 6.15, contains many more triangles than before, it can still be described by just one left- and one right-handed Schwarz triangle based on the original equilateral icosahedral face.

6.3.3   Octahedral Symmetry

The octahedron and its dual, the cube, have similar rotation and reflection symmetries as the icosahedron, but the octahedron has fewer of them, 48 in all. Figure 6.16(a) shows all the vertex, face, and edge rotation axes together. Three rotational axes pass through three pairs of opposing vertices. The octahedron can be rotated any multiple of 90° around any of these three axes without changing its appearance. In similar fashion, there are four axes through the centers of opposite faces, and any rotation that is a multiple of 120° will not change its appearance. Finally, there are six axes through the midpoints of six pairs of opposing edges. A rotation of 180° around any of these three axes will not change the octahedron's appearance.

Figure 6.16(b) and (c) shows the octahedron's rotation axis along with the nine reflection planes and the nine great circles they create. Figure 6.16(c) shows the relationship of the reflection planes to the vertex, face, and edge rotation axes. Another way to visualize these nine great circles in (c) is to see that three of them follow the edges of the octahedron and six follow the edges of its dual, the cube. We can summarize all the octahedron's symmetries this way:

9 — three rotations of 90° around centers of three pairs of opposite vertices
8 — two rotations of 120° around four pairs of antipodal faces
6 — one rotation of 180° around centers of six pairs of opposing edges
1 — identity symmetry

---

24 Total

Reflection symmetry doubles these symmetries, for a total of 48.

(a)   (b)   (c)

Figure 6.16. Octahedral rotation and reflection symmetry.

---

[11] For a comprehensive analysis of icosahedral great circles, see Robert Gray's website at www.rwgrayprojects.com.

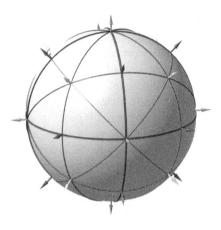

Figure 6.17. Spherical octahedron with Schwarz triangles.

The octahedron and the cube are duals. As in the icosahedral symmetry we saw earlier, the cube retains the same symmetries, but the symmetries are applied in different ways—two rotations of 120° around the centers of four pairs of antipodal vertices, three rotations of 90° around three pairs of opposite faces, and one rotation of 180° around six pair of opposite edges. The identity adds one more symmetry and reflection doubles the total, equaling the octahedron's 48 symmetries.

Figure 6.17 shows a spherical octahedron with the rotation axes and reflection planes we have been discussing. Notice that the reflection planes define great circles that divide the octahedron into 48 Schwarz triangles, 24 left-handed and 24 right-handed. The octahedron-cube dual relationship is evident in this figure. The red great circles outline the octahedron's edges, while the blue-gray ones follow the cube's edges. Since both the octahedron and cube are defined by the same Schwarz triangles, any subdivision of one polyhedron is, in effect, a subdivision of the other.

6.3.4 Tetrahedral Symmetry

The tetrahedron is a self-dual with the fewest vertices, faces, and edges of any polyhedron. It's also the only Platonic solid in which the faces are opposed by a vertex rather than by another face. It is therefore not surprising that the tetrahedron, with fewer vertices, faces, and edges, has the fewest number of symmetries (24 total) of the three groups.

Figure 6.18(a) shows the relationships and all the vertex, face, and edge rotation axes together. Referring to (a), the tetrahedron can be rotated two times, each 120°

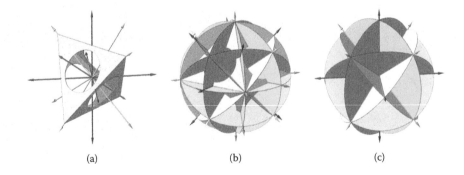

(a)   (b)   (c)

Figure 6.18. Tetrahedral rotation and reflection symmetry.

Figure 6.19. Spherical tetrahedron with Schwarz triangles.

around each vertex-face axis, and one 180° rotation around the centers of three pair of opposite edges without changing its appearance. Figure 6.18(b) shows the tetrahedron's six planes of reflection symmetry, and (c) shows the relationship of reflection planes to rotation axis. Notice that each of the six symmetry planes, shown as great circles, pass tangent to an edge on one side of the tetrahedron and normal to the midpoint of an edge on the opposite side. The tetrahedron's rotation and reflection symmetries combine with the identity symmetry for a total of 12 symmetries in the following way:

8 — two rotations of 120° for each of four axis through a vertex and the middle of an opposing face
3 — one rotation of 180° around the center of three pairs of opposite edges
1 — identity symmetry

12 Total

Reflection doubles the count to 24.

Figure 6.19 shows a spherical tetrahedron and retains the rotation axes and reflection planes (great circles) we have been discussing. Combinations of great circles divide the tetrahedron into 24 Schwarz triangles, 12 left-handed and 12-right-handed. Notice the spherical cube pattern, a result of the tetrahedron-tetrahedron dual. It might seem surprising to see this pattern. However, a tetrahedron can be created from the diagonals of the cube's faces, and this is what is apparent in the figure.

6.3.5   Schwarz Triangles and Symmetry

In Chapter 4 we described the Schwarz triangle as the largest spherical right triangle that can tile a spherical polyhedron's face and cover the sphere without gaps or overlaps. Each pair of Platonic duals or symmetry group has its own Schwarz triangle. But the story does not end there. Symmetry groups also reveal other relationships.

The icosahedron's Schwarz triangles define its dual, as expected; but somewhat unexpected is the fact that the same tiling also defines the spherical octahedron and tetrahedron. Figure 6.20 shows why this happens. Four identical spheres are shown with the same set of icosahedral Schwarz triangles. All views are from the same point and scale. Figure 6.20(a) is the full set of Schwarz tiling for the icosahedron. One face is highlighted in (b) and a dodecahedron face is highlighted in (c). But the same tiling also defines the octahedron's face, as shown in (d). How does this happen? Octahedral faces have three 90° angles and three 90° sides. The 90° angle of the icosahedral Schwarz triangle defines each octahedral face angle, and each octahedral face edge (also 90°) is defined by the three different sides of icosahedral Schwarz triangles. They add up to 90° (31.71747° + 20.90516°

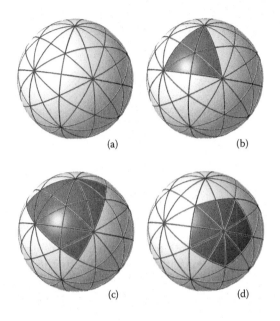

Figure 6.20. Icosahedral Schwarz triangle defines the dodecahedron and octahedron.

+ 37.37737°). We also notice that the combination of Schwarz triangle edges that define octahedral edges form continuous great circles, each orthogonal to one another, a hallmark of the spherical octahedron. The point here is that although the octahedron has its own Schwarz triangle definition and it is larger in area than the icosahedron's (when both have the same circumsphere radius), we can use symmetry relations to define the octahedron with an icosahedral Schwarz triangle. In practice, however, we generally use only the Schwarz triangle for the polyhedral framework we want to use. This maximizes the symmetry we can exploit. It's evident from the octahedral face tiling in (d) that using the icosahedron's Schwarz triangle to define an octahedron's face instead of the octahedron's own Schwarz triangle has cost us some rotation and reflection symmetry.

We can examine the octahedron's Schwarz triangles and see similar relations to its dual, the cube, and to the tetrahedron. Figure 6.21 shows four spheres tiled with the same octahedral Schwarz triangles, as shown in (a). The view of each sphere is the same. Single octahedron and cube faces are shown in (b) and (c), and their dual symmetry is obvious. But since a tetrahedron can be defined by the face diagonals of a cube, it's no surprise that the octahedron-cube triangles also define the tetrahedron's face, as shown in (d). The spherical tetrahedron face angles are each 120° and its edges are 109.47122°. Two back-to-back octahedral triangles, each 60°, fit perfectly in each of the tetrahedron's angles, and the sum of two octahedral triangle hypotenuses (each 54.73561°) equals a tetrahedral edge. Unlike the octahedral face we previously defined with icosahedral Schwarz triangles, the tetrahedron's face defined here with octahedral Schwarz triangles has full rotational and reflection symmetry.

Given the relationships of Schwarz triangles between the three symmetry groups, one could argue that only two different pairs of Schwarz triangles, one for the icosahedron and another for the octahedron, are needed to define every spherical Platonic solid.[11a] But once again, the practice is to use the group's own Schwarz triangles because they are the largest right triangles that will tile their faces and still maintain full rotational and reflection symmetry. Table 6.4 summarizes the characteristics of the Platonic Schwarz triangles.

[11a] Schwarz triangles, though not called such, are basic to the ancient craft of Temari Balls. See Section 3.23.1.

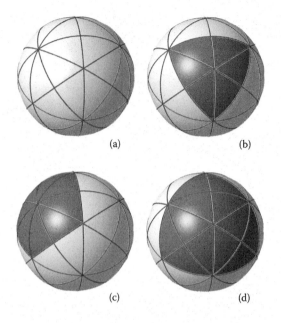

(a)   (b)

(c)   (d)

Figure 6.21. Octahedral Schwarz triangles define the cube and tetrahedron.

Platonic Schwarz Triangles					
Triangular Part	**Tetrahedron**	**Octahedron**	**Cube**	**Dodecahedron**	**Icosahedron**
Surface Angles					
A	60.0	60.0	45.0	36.0	60.0
B	60.0	45.0	60.0	60.0	36.0
C	90.0	90.0	90.0	90.0	90.0
Central Angles					
a	54.73561032	45.0	35.26438968	20.90515745	31.71747441
b	54.73561032	35.26438968	45.0	31.71747441	20.90515745
c	70.5277937	54.73561032	54.73561032	37.37736814	37.37736814
Arc Factors					
a	0.95531662	0.78539816	0.61547971	0.36486383	0.55357436
b	0.95531662	0.61547971	0.78539816	0.55357436	0.36486383
c	1.23095942	0.95531662	0.95531662	0.65235814	0.65235814
Chord Factors					
a	0.91940169	0.76536686	0.60581089	0.36284333	0.54653306
b	0.91940169	0.60581089	0.76536686	0.54653306	0.36284333
c	1.15470054	0.91940169	0.91940169	0.64085182	0.64085182
Area					
Stradians	0.52359878	0.26179939	0.26179939	0.10471976	0.10471976
Spherical Degrees	30.0	15.0	15.0	6.0	6.0

Note: All angles in degrees.

Table 6.4. Platonic Schwarz triangles.

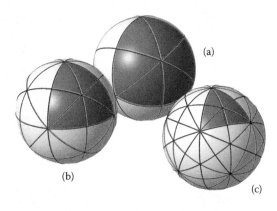

Figure 6.22. Deltahedra equilateral faces.

6.3.6 Deltahedra

Symmetry within the Platonic solids provides another tool for subdivision. Geometrists Martyn Cundy and A. P. Rollett proposed the term *deltahedra* to describe any polyhedron whose faces were all equilateral triangles.[12] The name reflects the Greek capital letter *delta* (Δ), which has the shape of an equilateral triangle. There are many different deltahedra. Three are Platonic solids, but no Archimedean solids are deltahedra. Figure 6.22 shows the three Platonic deltahedra: the tetrahedron (a), octahedron (b), and icosahedron (c). As we have already seen, they also define symmetry groups. What's important is the fact that any subdivision technique we develop for a general spherical equilateral will subdivide any deltahedra and its dual, even though the dual's face type is some other type. For example, subdividing an octahedron's equilateral face also subdivides a cube's square face. The same applies to the spherical icosahedron's equilateral face and the dodecahedron's pentagonal face. Although we can develop subdivision techniques for any regular spherical polygon that tile a sphere, we benefit most from techniques that subdivide equilaterals or a right Schwarz triangle. In Chapter 8, we will show how to do this.

6.4 Archimedean Solids

The Archimedean solids, so named to honor Archimedes of Syracuse (287-212 BC), are the second-most important polyhedral frameworks for spherical subdivision. Known long before the mathematician's lifetime, these 13 solids are like Platonic solids in that they are convex polyhedra. But unlike Platonic solids, they have two or more types of regular polygon faces. For this reason, they are classified as *semiregular* polyhedra. Faces join at their edges, they do not penetrate one another, and they always sequence the same way around every vertex. Figure 6.23 shows all 13 of them with the same edge length.[13] Referring to the numbered solids in the figure, they are the truncated tetrahedron (1), truncated cube and truncated octahedron (2 and 3), lesser and greater rhombicuboctahedron (4 and 5), cuboctahedron (6), icosidodecahedron (7), snub cube and snub dodecahedron (8 and 9), truncated icosahedron and truncated dodecahedron (10 and 11), and finally, the lesser and greater rhombicosidodecahedron (12 and 13).[14]

---

[12] (Cundy and Rollett 1961, 78).

[13] Two Archimedeans, the snub cube and snub dodecahedron, come in a left- and right-handed version. One could argue there are really 15 Archimedeans, not 13.

[14] Most of the Archimedean solids were named by Johannes Kepler (1571–1630), the German mathematician, astronomer, and astrologer who formulated the laws of planetary motion.

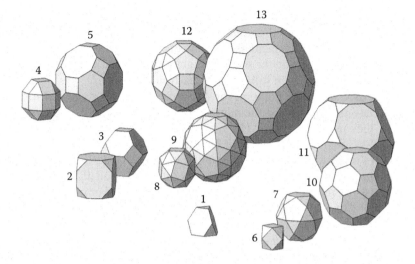

Figure 6.23. Archimedean solids.

There are important differences between the Platonic and Archimedean solids that affect their use in spherical work. The Platonic solids distribute points (vertices) evenly on the surface of their circumsphere. The distance between any vertex and its neighbors, and the angle between any two neighboring points and every dihedral angle for a solid, is the same. Platonic solids also have only one insphere that's tangent to the midface point of their faces. A pair of Schwarz triangles, left- and right-handed, can describe the entire surface of spherical versions.

Archimedean solids do not have every one of these properties. While their vertices are on the surface of their circumsphere and the distances between any vertex and its neighboring vertices are the same, the angle between adjacent edges at a vertex and the dihedral angle between faces are not always the same. Archimedean solids have two or more inspheres, each tangent to the midface points of one set of like faces but not tangent to the midface points of the other face types. The *midsphere* of an Archimedean solid does, however, intersect every edge of that solid at its midpoint. Multiple spherical triangles are needed to describe the surface of spherical versions. In some spherical applications, it is important to have a subdivision grid that has at least one sequence of grids that defines a complete great circle. Some subdivision techniques result in grids that form great circles, but two Archimedean solids, the icosidodecahedron and cuboctahedron, have sequences of edges that naturally form great circles in their spherical version. We will discuss them in a moment. The octahedron, a Platonic solid, is the only other polyhedron we have discussed having this property.

Archimedean solids have duals just like Platonic solids. They are called *Catalan solids*, and are quite beautiful. However, they do not offer any special advantages over Platonic or Archimedean solids as frameworks for spherical subdivision. We do not cover them in this text. Instead, we include several excellent references in the additional resources at the end of this chapter.

6.4.1 CundyRollett Symbols

Individual Archimedean solids can have two, three, or four different face types. The types of regular polygons and their order around each vertex of a solid is always the same for a given solid. Thus, an Archimedean solid can be fully specified by simply listing the order of its polygon faces around a vertex. Within an Archimedean solid, the vertex face orders are all the same and the solid is said to be *vertex-transitive*,[15] which means the face arrangement around any one of its vertices can be transformed to any other vertex position on the solid through some rotation or reflection symmetry. All Platonic and Archimedean solids are vertex-transitive.

Schläfli devised symbolic notation for Archimedean solids, as he did for Platonic solids. His Platonic symbols list the number of sides in the regular polygon and how many times they meet at a vertex. His symbol for a cube is {4,3}, which means four-sided polygons (squares) meet three at a time at a vertex. But Archimedean solids have more complex face orders than Platonic solids, and as a result, their Schläfli symbols are not as intuitive. Geometrists H. M. Cundy and A. P. Rollett, modified Schläfli's symbols to list the regular face types in their vertex order.[16] For example, the *Cundy-Rollett* symbol {3.5.3.5} means every vertex of this solid is surrounded by alternating triangles and pentagons. It could also be written as {3.5}². This Archimedean solid is called the icosidodecahedron and we will see its details shortly. To distinguish Cundy-Rollett symbols from Schläfli symbols, we use periods between parameters in the former and commas in the latter.

6.4.2 Truncation

Truncation is the way some Platonic solids become semiregular Archimedean solids, and their names reflect this fact. When a Platonic solid is cut, new and regular (equal edge and angle) polygonal faces result. *Uniform truncations* pass the cut plane less than halfway along the edges that meet at any vertex. *Complete truncations* cut every edge midway. Figure 6.24 shows the differences between the two. Some Archimedean solids require cutting a Platonic solid several times. There are a couple of Archimedean solids, called *snubs*, which cannot be made by simple vertex or edge cuts alone.

We introduce the Archimedean solids, both planar and spherical, by symmetry groups.

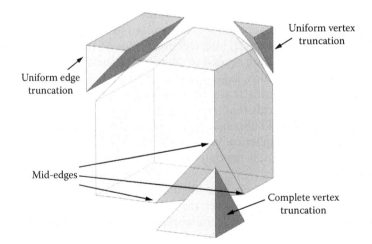

Figure 6.24. Vertex and edge truncations.

---

[15] A polyhedron is vertex-transitive if any vertex can be carried to any other by a symmetry operation (Cromwell 1997, 369).
[16] (Cundy and Rollett 1961, 84).

6.4.3   Archimedean Solids with Icosahedral Symmetry

The first six Archimedean solids are part of the icosahedral symmetry group and can be inscribed in either a dodecahedron or icosahedron. On average, this series distributes (not evenly) nearly twice as many vertices on the sphere as any other symmetry group. Table 6.5 shows the planar and spherical versions together from the same viewpoint scaled to the same unit circumsphere. All six solids in this group can all be inscribed in a dodecahedron or icosahedron.

The icosidodecahedron combines the 12 pentagonal faces of a dodecahedron with the icosahedron's 20 equilateral faces; it can be created by complete vertex truncations of either a dodecahedron or an icosahedron. The result is alternating face types

Name(s) & Symbols	Vertices	Edges	Faces:	{3}	{4}	{5}	{6}	{8}	{10}
icosidodecahedron {5.3.5.3} or {5.3}²	30	60	32	20	–	12	–	–	–
truncated icosahedron {5.6.6} or {5.6²}	60	90	32	–	–	12	20	–	–
truncated dodecahedron {3.10.10} or {3.10²}	60	90	32	20	–	–	–	–	12
small or lesser rhombicosidodecahedron {3.4.5.4}	60	120	62	20	30	12	–	–	–
great rhombicosidodecahedron {4.6.10}	120	180	62		30	–	20	–	12
snub dodecahedron or snub icosidodecahedron {3.3.3.3.5} Two opposite-handed versions	60	150	92	80	–	12	–	–	–

Table 6.5. Archimedean solids with icosahedral symmetry.

around each vertex. Polyhedra with two pairs of alternating regular face types at each vertex are called *quasi-regular*.[17] Perhaps the most important property of this solid, and the reason it is often used in spherical subdivision, is that its edges form six hemispheric planes, each of which is a *decagon* and will define a great circle in its spherical version. In Chapter 8 we show a subdivision technique called Mid-arcs. When applied to a spherical icosahedron, the technique creates the spherical icosidodecahedron framework.

The truncated icosahedron is widely recognized as the traditional design for soccer balls. In physics and chemistry, this solid resembles the newly discovered carbon 60 (C_{60}) molecule and has been named the *"Fullerene"* or *"Buckyball"* after Buckminster Fuller.[18] Both are shown on the right. The truncated icosahedron can be created by uniform vertex truncation of an icosahedron or dodecahedron. Thus, its faces are tangent to either the dodecahedron or the icosahedron. We will see variations of the truncated icosahedron in several spherical subdivision schemas. Much of what has been said about the truncated icosahedron applies equally to the truncated dodecahedron, which we will review in a moment. It can be created by uniform vertex truncation of the dodecahedron or icosahedron; thus, it can be inscribed in either solid.

The small rhombicosidodecahedron can be created by truncating the edges and vertices of other Archimedean solids. Each of its faces belongs to one of three circumscribing solids: the dodecahedron, the icosahedron, and the rhombic triacontahedron. The small rhombicosidodecahedron can also be made by rotationally expanding the faces of a dodecahedron along an axis perpendicular to their face centers. The great rhombicosidodecahedron is the most complex and the largest of the Archimedean solids when all polyhedra have the same edge lengths. It has more regular face types (squares, hexagons, and decagons) than any other Archimedean solid and, as a result, the most vertices and edges. The great rhombicosidodecahedron distributes twice as many vertices on the surface of the sphere of any Archimedean solid: 120 in all. When compared to other Archimedean solids with the same edge lengths, it is the largest by far. Like the others in this symmetry group, its faces also belong to a circumscribing dodecahedron and the icosahedron.

The snub dodecahedron, sometimes called the snub icosidodecahedron, places 12 pentagonal faces on the surface of a circumscribing dodecahedron. The snub dodecahedron and another Archimedean solid, the snub cube, are *chiral polyhedra*; that is, they do not have reflection symmetry. No plane passing through the center of the polyhedron produces equal images on both sides. There are two versions of this polyhedron: one is left-handed, while the other is right-handed. Neither version can be made by simple truncations of any Platonic solid. One of the spherical subdivisions schemas that we will develop in a later chapter creates an entire class of skewed subdivisions like the snub cube. We will look at skewed subdivisions in Chapter 8.

---

[17] Quasi-regular polyhedra are sometimes called combinational solids because they combine the faces of two other polyhedra.

[18] A new kind of carbon, C_{60}, was discovered in 1985 by scientists Robert Curl, Harold Kroto, and Richard Smalley. They were awarded the Nobel Prize in Chemistry in 1996 for discovering this new class of compounds. The molecular model resembles a truncated icosahedron and is named the Fullerene or Buckyball, in honor of Buckminster Fuller.

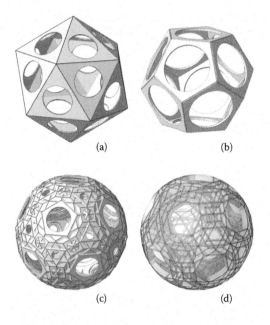

(a)   (b)

(c)   (d)

Figure 6.25. Superimposed Archimedean polyhedra with icosahedral symmetry.

We can see the symmetry of this group by superimposing both their planar and spherical versions at the same scale. Figure 6.25(a) and (b) shows the Platonic duals that define this symmetry group. They are positioned as duals—the dodecahedron's vertices in (b) correspond to the same position as the icosahedron's midface points in (a). Figure 6.25(c) and (d) shows the Archimedean planar and spherical polyhedra in this symmetry group superimposed and from the same viewpoint as (a) and (b). The snub dodecahedron is omitted because it cannot be made by simple truncations. In (c) it's easy to see symmetrical patterns that result from the various vertex and edge truncations of the icosahedron and dodecahedron. Figure 6.25(d) is simply the spherical version of (c) and emphasizes the arc edges. The symmetry of all four figures is evident.

6.4.4   Archimedean Solids with Octahedral Symmetry

The next six Archimedean solids are related to the octahedron and its dual, the cube. They are the truncated octahedron and cube, cuboctahedron, the lesser and greater rhombicuboc-tahedron, and the snub cube. Table 6.6 shows them along with their basic characteristics. Every planar and spherical version is shown from the same viewpoint scaled to the same unit circumsphere. Except for the snub cube, five of the six solids can be formed by truncating the octahedron's vertices and/or edges to a certain depth. Each cut creates new face types and vertices. All six solids in this group can be inscribed in an octahedron or cube.

Referring to Table 6.6, the truncated octahedron and truncated cube are simple uniform vertex truncations of the vertices of an octahedron or cube. All cut planes are uniform vertex truncations of the octahedron and cube at a distance between a vertex and neighboring edge midpoints. Sets of like faces (squares or hexagons) in the truncated solid are parallel to each other and axes normal to a pair of faces are mutually orthogonal, like Cartesian axes. The spherical forms of both solids are often used in

Name(s) & Symbols	Vertices	Edges	Faces	{3}	{4}	{5}	{6}	{8}	{10}
truncated dodecahedron {4.6.6} or {4.6²}	24	36	14	–	6	–	8	–	–
cuboctahedron {3.4.3.4} or {3.4}²	12	24	14	8	6	–	–	–	–
truncated cube {3.4.3.4} or {3.4}²	24	36	14	8	–	–	–	6	–
small or lesser rhombicuboctahedron or rhombicuboctahedron {3,4,4,4} or {3,4³}	24	48	26	8	18	–	–	–	–
great rhombicuboctahedron or truncated cuboctahedron rhombitruncated cuboctahedron omnitruncated cuboctahedron {4.6.8}	48	72	26	–	12	–	8	6	–
snub cube or snub cuboctahedron {3.3.3.3.4} Two opposite-handed versions	24	60	38	32	6	–	–	–	–

Table 6.6. Archimedean solids with octahedral symmetry.

applications, where *x*-, *y*-, and *z*-axes are needed, or in cartographic applications where great circles are required for the equator and Greenwich Meridian.

The cuboctahedron is unique among polyhedra. It can be formed several ways. First, it can be made by complete vertex truncations of either a cube or octahedron. It can also be made by combining the cube's six faces with the octahedron's eight faces or by combining eight tetrahedra, each sharing one point in the center of the polyhedron. No matter how the cuboctahedron is made, every one of its edges is equal and every vertex is equidistant to its neighbors, as well as to the center of the polyhedron. It's the only polyhedron like this. Another unusual property of the cuboctahedron is it has only two different face types, and each type surrounds the other at every vertex. Despite having two types of faces, every dihedral angle for the cuboctahedron is the same. The icosidodecahedron is the only other quasi-regular Archimedean solid. We will return to these two solids later in the chapter. They offer great circle advantages in spherical work.

The small rhombicuboctahedron is another unique polyhedron. Each of its faces belongs to one of three circumscribing solids: the cube, the octahedron, and the rhombic dodecahedron. What is unusual here is the same number of faces and types can be rearranged to make a different polyhedron that isn't a Platonic solid or an Archimedean solid. A polyhedron with this property is said to be *isomeric*. Few polyhedra have this property. We will take a closer look at this characteristic later in the chapter. The great rhombicuboc tahedron can be made by truncating the cuboctahedron's vertices. Its faces belong to one of two circumscribing solids, the cube and the octahedron.

The snub cube, sometimes called the snub cuboctahedron, cannot be made by simple truncations of either an octahedron or a cube. Although its equilateral faces are skewed relative to the edges of the octahedron, its square faces lie on the surface of the circumscribing cube and eight of the triangular faces lie on the circumscribing octahedron.

Although each Archimedean solid in the Octahedral symmetry group is quite different, their group symmetry shows through when we superimpose them. The octahedron-cube dual that defines this symmetry group is shown in Figure 6.26(a) and (b). They are positioned as duals: the octahedron's vertices in (a) correspond to the cube's midface points in (b). The edges of both solids would be perpendicular to one another if both were overlaid. Figure 6.26(c) and (d) shows the Archimedean planar and spherical polyhedra in this symmetry group superimposed and from the same viewpoint as (a) and (b). The snub cube is left out because its skewed faces obscure the others. It cannot be made by simple truncations like the others. The highly symmetrical patterns, clearly seen in Figure 6.26(c), are the result of symmetrical vertex and edge truncations of the octahedron and cube. Figure 6.26(d) is simply the spherical version of (c) and emphasizes the relationship of the arc edges to one another.

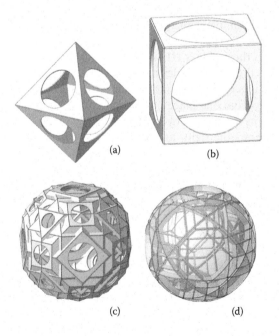

(a)

(b)

(c)

(d)

Figure 6.26. Superimposed polyhedra with octahedral symmetry.

Name(s) & Symbols	Vertices	Edges	Faces	{3}	{4}	{5}	{6}	{8}	{10}
truncated tetrahedron {3.6.6} or {3.6²}	12	18	8	4	–	–	4	–	–

Table 6.7. Archimedean solid with tetrahedral symmetry.

6.4.5 Archimedean Solids with Tetrahedral Symmetry

This is the smallest symmetry group and it contains only one solid, the truncated tetrahedron. It is the only polyhedron with just hexagonal and triangular faces. It can be created by making a uniform vertex truncation of a tetrahedron or by truncating all the vertices of a cube to different depths. The first technique is the most common. Either way, the result is a polyhedron bounded by four hexagons and four triangles, as shown in Table 6.7.

As far as we know, the truncated tetrahedron has not been used for spherical subdivisions in industrial products or architecture. It is included here to complete the Archimedean symmetry examples.

6.4.6 Chiral Polyhedra

Earlier in the chapter, we said two Archimedean solids, the snub cube and snub dodecahedron, were chiral polyhedra, but until now, we did not ascribe any significance to this. A chiral polyhedron is one that has two distinct forms or *enantiomorphs, a laevo* (left-handed) and *dextro* (right-handed) version. Your hands are nearby examples of enantiomorphs. No combination of rotations or translations of one hand will produce the shape of the other. In a sense, you must turn one inside out to make the other. You can demonstrate this by turning a glove inside out. A chiral polyhedron cannot be made by simple truncations of a Platonic solid.

We care about chiral polyhedra because we will develop an entire family of subdivisions related to them. Class III subdivisions are chiral, and their lack of reflection symmetry means the geometry developed for a local area can only cover the remainder of the sphere, if it is rotated around one or more axes. We lose the convenience of reflecting geometry from one place on the sphere's surface to another place, something we can easily do in some cases by simply changing vertex coordinate signs.

Figure 6.27 shows the right- and left-handed versions of the snub cube (a) and (b) and the snub dodecahedron (c) and (d). Neither set of polyhedra can be created from a Platonic solid by vertex or edge truncations. Both sets of figures have been scaled, so the most extreme vertices are tangent to a unit circumsphere. Try as you may, you cannot pass a plane through any of them (and their origin) and produce identical images on both sides.

Although none of the vertices of chiral polyhedra are antipodal pairs, they retain symmetric relationships with their Platonic octahedron and icosahedron cousins. In Figure 6.28(a), we see them superimposed—the octahedron and snub cube on the left, the icosahedron and snub dodecahedron on the right. Even though the snub's chiral form appears "*twisted*" relative to its Platonic counterpart, the systematic relationship between their faces, edges, and vertices is easy to see. The square faces of the snub cube and pentagonal faces of the snub dodecahedron are rotationally symmetric to the vertices of the octahedron and icosahedron.

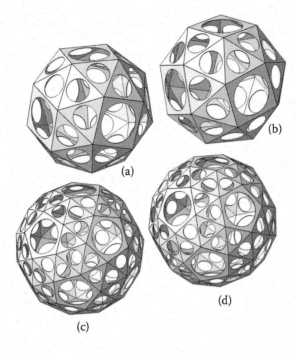

Figure 6.27. Snub cube and snub dodecahedron.

Figure 6.28. Chiral polyhedra and subdivisions.

Figure 6.28(b) shows a Class III spherical subdivision using an octahedral (left) and icosahedral foundation (right). The bold spherical Platonic edges in the figures are *not* part of the subdivision grid; they are displayed in the same orientation as the polyhedra in (a) to show their relationships to their Platonic frameworks. No matter how finely we subdivide the sphere using this schema, the vertices, edges, and faces produced by the grid will retain their symmetry relationship to their foundation polyhedron. Moreover, they will always remain chiral and will lack reflection symmetry.

6.4.7   Quasi-Regular Polyhedra and Natural Great Circles

The cuboctahedron and icosidodecahedron are classified as semiregular and *quasi-regular polyhedra*. It turns out that these two are also the only Archimedean solids whose edges form natural equatorial planes that translate to great circles in their spherical versions. They are also quasi-regular polyhedra. The cuboctahedron {3.4.3.4} alternates equilateral and square faces around each vertex, while the icosidodecahedron {3.5.3.5} alternates equilateral triangles and pentagonal faces. Quasi-regular solids have equal dihedral angles, equal-length sides, and all vertices tangent to their circumsphere. These are also characteristics of the Platonic solids, but most important, they have continuous sets of edges that form planes through the center of the polyhedron and form great circles in their spherical versions. It's not surprising to see quasi-regular polyhedra used in spherical applications where at least one great circle is required. For example, a geodesic dome needs to sit evenly on the ground, and a mold for a golf ball needs to have a hemispheric parting line in order to mold it. The octahedron is the only Platonic solid, without further subdivision, that also has natural hemispheric planes. Figure 6.29 shows the two quasi-regular polyhedra and the octahedron (not classified as quasi-regular) in applications that exploit their great circles. In the figure, the polyhedra and application have approximately the same orientation and scale.

The cuboctahedron has four equatorial planes. Figure 6.29(a) shows it alongside a computer model of Buckminster Fuller's famous octet truss connector. It was detailed in Section 3.13.

The octahedron shows up in many spherical applications, not only because it offers great circles, but it also offers three mutually orthogonal hemispherical planes that naturally fit applications where a Cartesian *xyz*-reference system is needed. Figure 6.29(c) and (d) shows its use in mapping where the equator and Greenwich references are required. The cartographic grid shown is called the *quaternary subdivision*, a meshing system that hierarchically subdivides equilateral triangles into four smaller ones by adding new edges that span between the mid-edge points of the original triangle.[19] In Chapter 8, we develop this schema in detail; it's called the Mid-arcs schema.

The icosidodecahedron offers six great circles; they are easily seen in Figure 6.29(e) and (f).[20] This application shows the arrangement of dimples on a golf ball. The designer needed at least one great circle plane as a parting line for the molding process. Using an icosidodecahedral foundation, he needed to develop two dimple arrangements to cover the entire ball, one for the triangle, another for a symmetrical part of the pentagonal face. In Hwang, the inventor of this design, chose to place six large dimples totally within each of the equilateral face. The icosidodecahedron's pentagonal faces are further subdivided into five slightly isosceles triangles. Three dimples

---

[19] (Dutton 1991).
[20] (Hwang November 1996, sheet 5/9, fig. 8).

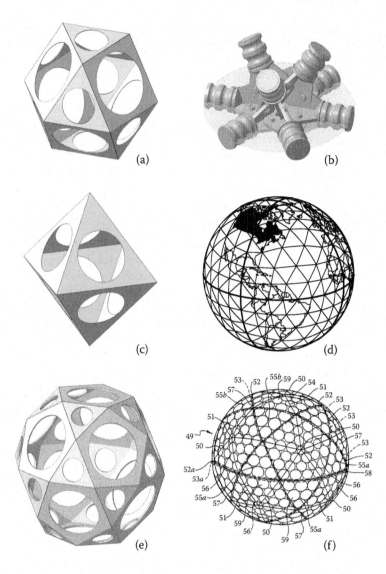

Figure 6.29. Applied quasi-regular polyhedra.

are totally within this triangle, while the remaining ones straddle edges or center over the pentagon's midface point. Most golf balls are molded and, until recently, have required a hemisphere parting line, an equatorial line that is not crossed by dimples. Golf balls are one of the most creative and diverse applications of spherical subdivision. We discuss their spherical subdivision geometry shortly.

6.4.8 Waterman Polyhedra

We have discussed Platonic and Archimedean solids in detail because we will use them in the next chapter to jump-start the spherical subdivision process. For completeness, however, we want to mention the Waterman polyhedra because some are nearly spherical and already highly subdivided. Unlike spherical Platonic and Archimedean solids, Waterman polyhedra do not evenly distribute points on a sphere or even distribute points on the same radius sphere. Waterman polyhedra are important additions to a spherist's tool kit because they exhibit many of the properties we have been discussing, such as symmetry, great circles, antipodal points, and so forth, and they demonstrate a rich way to develop spherical geometry that can solve certain application problems.

In 1988, Steve Waterman began a series of investigations into gridded maps. We saw an example of his butterfly map projection alongside Cahill's map when we discussed Fuller's Dymaxion maps in Chapter 2. Waterman's projection studies led to the development of entirely new families of polyhedra and some resemble highly subdivided spheres. The way they are constructed with sphere packing is totally unique.

Sphere Packing and Kepler's Conjecture

Waterman Polyhedra are derived from sphere packings. Packing applications are intensely studied today at the theoretic and practical level. Besides green grocers who might want to stack the most produce in a crate or neatly arrange fruit for display, sphere packing plays an important part in atomic level chemistry and new materials science. Higher-dimensional dense packings play a key role in error-correcting codes used by cell phones, space probes and the Internet to send signals through noisy channels.

In 1611, the German astronomer Johannes Kepler wrote an essay entitled **On the Six-cornered Snowflake** in which he made the famous statement "*no packing of congruent balls in Euclidean three-space has density greater than that of the face-centered cubic packing.*"[21]

The problem Kepler and subsequent mathematicians focused on was the density of congruent spheres in infinite space. Density is the total volume of the spheres divided by the total volume of the container into which they are packed.

Unproven for 400 years, the *Kepler Conjecture* is considered the oldest problem in discrete geometry. Then in August of 1998, mathematician Thomas Hales announced a proof so complex that it required many more years of computer programming to validate it before the mathematics community would accept it.[22] Hales proved Kepler right, Face-centered cubic packing does yield the highest possible density.

What is Face-centered cubic packing and how does it compare? Figure 6.30 illustrates the three cubic packings: Simple cubic (a), Body-centered cubic (b) and Face-centered cubic (c).[23]

In *Simple cubic packing* (*SC*), spheres are aligned so that their centers are at the corners of squares and then stacked. Each successive layer is the same (a). As the stacking repeats, rows and columns maintain their orthogonal orientation. Simple cubic spheres occupy 52% of the space; this packing is not very efficient.

In *Body-centered cubic* (*BCC*) packing, each sphere layer has eight nearest neighbors (b). In this respect, it is like *SC*. However, the next layer atop is different. Spheres on this layer occupy the square-shaped "*holes*" formed in the first layer and the third layer now occupies the square holes formed by the second layer and so forth. In *BCC*, every other layer of spheres is the same. Spheres packed *BCC* occupy 68% of the volume, an improvement over *SC* packing.

---

[21] (Kepler and Hardie 1966).
[22] (Hales 1998).
[23] Sphere packing models courtesy of Christopher Kitrick.

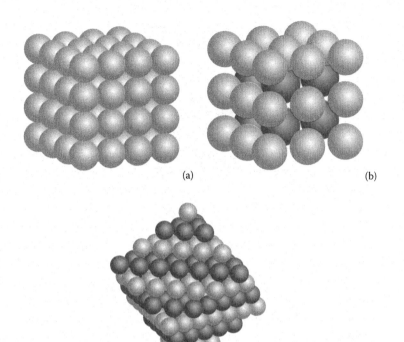

(a)

(b)

(c)

Figure 6.30. Three cubic sphere packings.

Face-centered cubic (*FCC*), also called Cubic-closest packing (*CCP*), produces the highest density, see (c); it's also the most complex cubic packing. Each sphere is surrounded by six others in the same layer. Above any set of seven spheres are six depressions arranged in a hexagon pattern and they support the second layer of spheres nested in every other of these hollow spaces. A third plane of spheres is placed above the "*holes*" of the first plane not covered by the second plane. A fourth layer of spheres is directly above and repeats the first layer. In *FCC*, every third layer repeats. The result is 74% of the space is occupied. Kepler was right, it's the maximum density possible.

Buckminster Fuller also studied spherical packing as a way to visualize the efficiency of subdivision grids and as clusters he called vector equilibriums. Figure 6.31(a) shows him in studio surrounded by *Synergetic* models. In the lower left of the photo is an example of *BCC* sphere packing. In the lower right is his model called "*omnidirectional closest packing of spheres around a nucleus.*" It is defined by the vertices of the cuboctahedron and it's *FCC* packing. The layering sequence is shown in (b).[24,25]

---

[24] Fuller in studio and Synergetics Geometry images by permission of the Estate of R. Buckminster Fuller.
[25] Mathematician H.S.M Coxeter wrote a paper in 1974 with a simple proof of the packing theorem, which was (very nearly) stated by R. Buckminster Fuller (Marks 1960, 35): When congruent balls are arranged in cubic close-packing so as to fill a tetrahedron, square pyramid, octahedron, cuboctahedron, truncated tetrahedron, or truncated octahedron, with n+1 balls along each edge, the total number of peripheral balls is bn^2+2, where b = 2, 3, 4, 10, 14, or 30, respectively. See section Additional Resources for references to Coxeter's analysis.

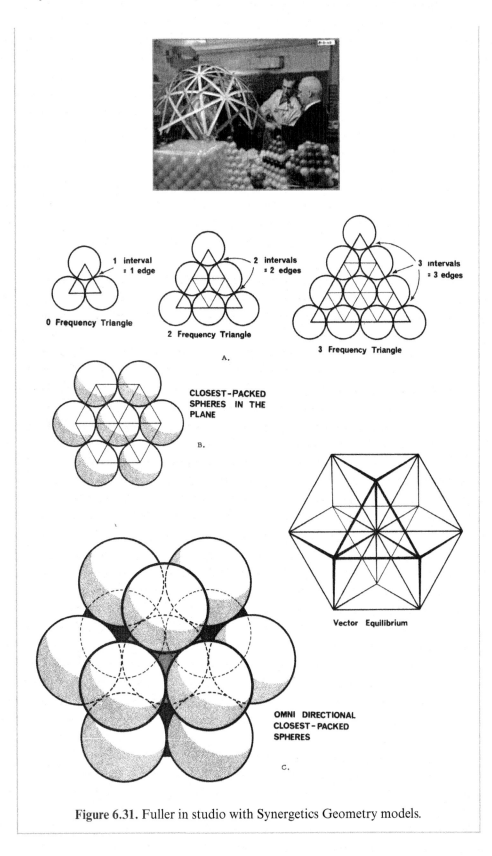

Figure 6.31. Fuller in studio with Synergetics Geometry models.

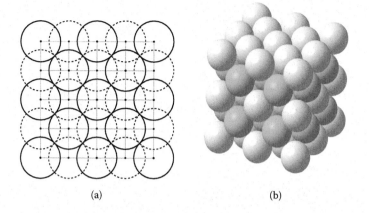

(a)   (b)

Figure 6.32. Cubic closest packing.

Waterman polyhedra begin with a 3D lattice of points and pack spheres in *Cubic close packing (CCP)*, also called face-centered cubic (*FCC*) packing. An infinite number of tangent spheres are packed in layers one layer on top of another. Each layer of spheres occupies the inter-grid spaces of their adjacent layers. If all spheres have a diameter of $\sqrt{2}$, their centers will fall only on integer coordinate values, if one of the sphere's center occupies the origin of the Cartesian space at coordinates (0, 0, 0). Figure 6.32(a) shows how *CCP* spheres are arranged on alternating layers. The grid spacing is one unit in the *x, y,* and *z* (perpendicular to the page) direction. The solid spheres are on even *z*-planes, including the *z* = 0 plane.

In *CCP*, 12 spheres surround every sphere in the pack, and their centers are equidistant from the center of the sphere they surround. If the centers of these 12 spheres are connected, they define the vertices of a cuboctahedron, Fuller's vector equilibrium (see Section 2.4).[26]

A Waterman polyhedron is defined by the convex hull of all the lattice points (the centers of stacked spheres) that are within a defined radius from the lattice's origin. Points outside this radius are ignored. A convex hull is the minimum convex polyhedra (faces, points, and edges) that surround the points. In a sense, the convex hull is like a shrink-wrap surrounding the most extended set of points.

The Waterman Polyhedra, called W1, are defined by a radius of $\sqrt{2} \times 1$. This radius selects only the 12 points surrounding the origin, and their convex hull is the same as the cuboctahedron or Fuller's VE. As the defined radius increases, more lattice points are selected and their convex hull, the Waterman polyhedron, becomes more and more complex and is made up of many more faces, points, and edges. When the defined radius has a root greater than 300, Waterman polyhedra resemble spherical subdivisions. Figure 6.33 shows two examples with selected lattice points (with spheres around them) and the resulting convex hull Waterman polyhedron.[27] These Waterman polyhedra are referred to by their square root numbers. Notice that a higher root number results in more points, edges, and faces on the sphere.

---

[26] In synergetic geometry, Buckminster Fuller makes extensive use of *CCP* spheres and refers to some of their configurations as the *isotropic vector matrix* (IVM). See (Edmondson 1987, 127) for a good explanation of how IVM extends Fuller's concept of vector equilibrium (VE) and how it relates to his octet truss design. Steven Waterman believes that Fuller was not aware that *FCC* (he called it *CCP*) had square root integer distance between any two-sphere centers.

[27] Images courtesy of Rene K. Mueller, SimplyDifferently.org.

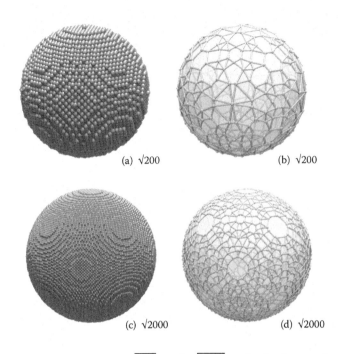

(a) √200   (b) √200

(c) √2000   (d) √2000

Figure 6.33. Waterman polyhedra $\sqrt{200}$ and $\sqrt{2000}$ point lattices and convex hulls.

Waterman polyhedra exhibit high degrees of edge, face, and vertex symmetry. For some cases, it is possible to pass planes through the Waterman sphere stack and define a lattice of points that create hemispherical or copula polyhedra that resemble geodesic domes.[28] The *FCC* packing and positioning (one sphere at the origin of the Cartesian space) produces one family of Waterman polyhedra; however, other families result when the origin of the sphere cluster is repositioning to other symmetrical positions within the lattice. There are seven possible origins for *FCC* packings,[29] and corresponding Waterman Polyhedra can be created with sphere stacks having a $\sqrt{2}$ diameter, or any other consistent diameter. If their diameter is not $\sqrt{2}$, their centers will not necessarily fall on integer *xyz*-coordinates.

Waterman polyhedra names are simple. They are labeled *Wn,* with *n* being the defined radius integer parameter.

The concept of Waterman polyhedra has been generalized to include point lattices created by packed (stacked) polyhedra in addition to spheres. Packs of polyhedra, restricted to certain integer *x, y, z* grids, will produce more than 60 new families of polyhedra. In these cases, the Waterman polyhedra naming convention is necessarily more complex because their names must account for the type of polyhedron in the stack, the origin location for the stack lattice, and the defined radius parameter. For more information on Waterman polyhedra, see Mark Newbold's website.[30]

To summarize, basic Waterman polyhedra are developed from Waterman spheres with diameters of $\sqrt{2}$ and packed in a *CCP* or *FCC* manner. Their centers are located

---

[28] See Paul Bourke, "*Waterman polyhedra data generator,*" local.wasp.uwa.edu.au/~pbourke/geometry/waterman/gen/, 2004.

[29] Seven possible point lattice origins: (1) sphere center, (2) spheres touch point, (3) spheres triangle center (there are two variations), (4) spheres forming tetrahedron, (5) spheres forming square-based pyramid, and (6) spheres.

[30] dogfeathers.com/java/ccppoly.html.

at integer x-, y-, and z-coordinates, and one sphere in the pack is centered at the Cartesian origin, at coordinates (0, 0, 0). A defined radius from the origin, a multiple of $\sqrt{2}$, selects a set of lattice points, the centers of spheres. A Waterman polyhedron is the convex hull of the selected set of lattice points. The number of points, edges, and faces in a Waterman polyhedron (the convex hull of the selected points) obey Euler's law, which states the number of vertices + faces = edges + 2. The number of vertices in a Waterman polyhedron is a fixed set. In other words, if one Waterman polyhedron has 2,000 points, another will not have an arbitrary 2,001 points. Chords between points define polygons, most of which are not regular (equilateral triangles, pentagons, hexagons, etc.). A general definition of Waterman polyhedra include *FCC* lattices created by both polyhedral and spherical packs. Different Cartesian origins for these lattices in combination with defined radii producing different families of Waterman polyhedra. Not all points of a Waterman polyhedron are on a circumscribing sphere, such as the Platonic and Archimedean solids already described; Waterman polyhedra exhibit high degrees of point, edge, and face symmetry. As the defined radius increases, the number of different face types, edge lengths, angles, and areas also increase. These may be limiting factors in some spherical applications.

6.5 Circlespheres and Atomic Models

In this chapter we have seen polyhedra enclose space, define great circles, and distribute points on their circumsphere. For the most part, polyhedra have been portrayed as static, highly symmetric bodies with predictable topological relationships among their faces, edges, and vertices. Artist Kenneth Snelson, who discovered the principle of *tensegrity* (see Section 2.7.1), found that a subset of Platonic and Archimedean solids possess adjacent face relationships and axial frameworks that can make new polyhedral forms he calls *circlespheres*. Snelson observed that certain polyhedra can have two-color face mappings; their faces can be colored one of two colors, and no face will be adjacent to another of the same color. Using this base, he then replaces two-color faces with dual-polarized (north/south) magnetic disks. This second form demonstrates how forces interact to create a stable body, while allowing systematic motion within the body. In so doing, the circlesphere acts as a model of the atom.

Figure 6.34 shows some of the polyhedra that can be two-color mapped. In these examples, four faces always meet at a vertex. Alongside each two-color polyhedron is

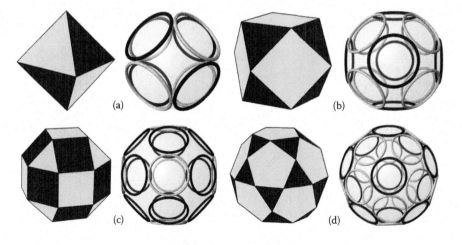

Figure 6.34. Circlespheres.

their circlesphere version. Magnetic rings with dual-polarization replace certain faces of the polyhedron. Each lesser-circle ring is made of polarized material, one side of the ring is polarized north or south, while its other side is polarized in the opposite direction.[31] Figure 6.34 shows that the octahedron (a), cuboctahedron (13), lesser rhombicuboctahedron (c), and icosidodecahedron (d) meet the two-color and dual polarization criteria. In each case, no two faces or rings touch another face with the same color or polarization. In each circlesphere example, all lesser circles are on the same circumsphere, and all are the same diameter. What is unique here is the fact that every ring only touches other rings that have opposite magnetic polarization. Notice, too, that the points of tangency between rings suggest the mid-edge points of the polyhedron on which they are based.

6.6 Atomic Models

Snelson recognized that circlespheres could be exploited in yet another way. He could replace each magnetic ring with a magnetic disk. If the disks were mounted on axes that ran through the center of the polyhedron's faces and the middle of the polyhedron, the disks would form a spherical gear set—that is, turning one disk would cause all the others in the system to turn synchronously and at the same number of revolutions per minute.

Figure 6.35 shows examples of circlesphere disk configurations for the icosahedron (8 magnets), cuboctahedron (14 magnets), small rhombicuboctahedron (26 magnets), icosi-dodecahedron (32 magnets), and lesser rhombicuboctahedron (62 magnets).[32] Notice that for any of the configurations shown, it is possible to rotate one disk either clockwise or counterclockwise, which would cause all the other disks to rotate synchronously. If a north-polarized face is rotated clockwise, all other north-polarized

Figure 6.35. Models of atomic forms.

---

[31] Circlesphere images courtesy of Kenneth Snelson.
[32] (Snelson 1966).

faces on the circlesphere will rotate clockwise, while south-polarized faces will rotate in the opposite direction.

Snelson made another surprising discovery with circlespheres. He noticed that circle-spheres behave much like the charged fields of atoms, and thus provide a visualization aid for how the atom works. Circlespheres with magnetic disks are kinetic models of forces (magnetism) and motion (rotary), and the number of magnets that can fully link together in circlespheres is uncannily close to the numerical sequences by which electrons fill "*shells*" or energy levels of atoms according to the periodic table of elements. In a sense, circle-spheres had become models for atomic forms by representing the electronic structure of atoms and molecules with interacting magnetic rings and discs.

Snelson's circlespheres and magnetic polyhedra extend a subset of Platonic and Archimedean solids to new dimensions and offer new ways to visualize and model atomic structures?[33] It is somewhat ironic that Fuller's synergetic geometry, a design cosmology based on natural systems, geometric relationships, strength of materials (in compression and in tension), and charged particles, completely missed the synergetic system Snelson discovered in magnetic circlespheres and force fields that model the atom.

Additional Resources

Polyhedra General

Coxeter, H. S. M. *Introduction to Geometry*. New York: John Wiley & Sons, 1961.

_____.*Regular Polytopes*. Mineola, NY: Dover Publications, Inc., 1973.

Cromwell, Peter R. *Polyhedra*. Cambridge [Eng.], Cambridge: Cambridge University Press, 1997.

Cundy, H. M. and A. P. Rollett. *Mathematical Models*. Oxford: Oxford University Press, 1961.

Hart, George W. 2008. "The Encyclopedia of Polyhedra." www.georgehart.com/virtual-polyhedra/vp.html.

___. "Max Brückner's Wunderkammer of Paper Polyhedra." Bridges Linz 2019, Linz, Austria, 2019.

Holden, Alan. *Shapes, Space and Symmetry*. New York: Columbia University Press, 1971.

Mikloweit, Ulrich. "Exploring Noble Polyhedra with the Program Stella4D." Bridges 2020, Helsinki, Finland, 2020.

Pugh, Anthony. *Polyhedra: A Visual Approach*. Berkeley: University of California Press, 1976.

Richeson, David S. *Euler's Gem: The Polyhedron Formula and the Birth of Topology*. Princeton, NJ: Princeton University Press, 2008.

---

[33] See (Hearney 2009, 110–118) for a detailed explanation of modeling atomic shells.

Wenninger, Magnus J. *Dual Models*. Cambridge [Eng.], New York: Cambridge University Press, 2003.

____. *Polyhedron Models*. Cambridge [Eng.], New York: Cambridge University Press, 1971.

Sphere Packing

Aste, Tomaso and Denis L. Weaire. *The Pursuit of Perfect Packing*. Philadelphia, PA: Institute of Physics Publishing, 2000.

Conway, John and N.J.A. Sloane. *Sphere Packings, Lattices and Groups*, New York: Springer-Varlag, A Series of Comprehensive Studies in Mathematics, 1993.

Coxeter, H.S.M. *Polyhedral Numbers*, Boston: Boston Studies in the Philosophy of Science, Vol 15, 1974, 15–25. (also available from Springer).

Tarnai, Tibor. "Polymorphism in Multisymmetric Close Packings of Equal Spheres on a Spherical Surface." *Structural Chemistry* 13.3–4 (2002): 289–295.

Spherical Polyhedra

Wenninger, Magnus J. *Spherical Models*. Cambridge [Eng.], New York: Cambridge University Press, 1979.

Wenninger, Magnus J. and Peter W. Messer. "Patterns on the Spherical Surface." *International Journal of Space Structures* 11.1–2 (1996): 183–192.

Polyhedra in Art, Design, and Nature

Loeb, Arthur L. *Space Structures: Their Harmony and Counterpoint*. Reading, MA: Addison Wesley Pub. Co., Advanced Book Program, 1976.

Pearce, Peter. *Structure in Nature Is a Strategy for Design*. Cambridge, MA: MIT Press, 1980.

Pearce, Peter and Susan Pearce. *Polyhedra Primer*. Palo Alto, CA: Dale Seymour Publications, 1978.

Posamentier, Alfred S. and Ingmar Lehmann. *The (Fabulous) Fibonacci Numbers*. New York: Prometheus Books, 2007.

Williams, Robert. *The Geometrical Foundation of Natural Structure: A Source Book of Design*. New York: Dover Publications, 1979.

Other Polyhedra

Hart, George W. and Henri Picciotto. *Zome Geometry: Hand-on Learning with Zome Models*. Emeryville, CA: Key Curriculum Press, 2001.

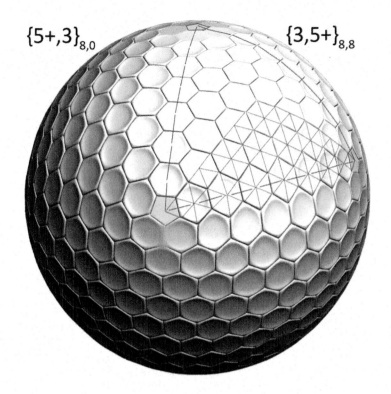

$\{5+,3\}_{8,0}$ $\{3,5+\}_{8,8}$

⑦ Golf Ball Dimples

We introduced golf ball dimples in Chapter 3 and pointed out how dimples have improved the distance and control golfers can hit a ball. Here we look at the basic ways designers distribute dimples on spherical polyhedra. We will look at patented designs and see how they exploit the symmetry and great circles of various polyhedral frameworks. We will see very creative arrangements of dimples that optimize the packing of dimples with different diameters. The approach, used in many designs, is to define an arrangement within a small working area, such as a Schwarz triangle, and then replicate the arrangement to cover the rest of the ball without causing overlaps or gaps. United States Golf Association (USGA) regulations require that golf balls be spherically symmetric. While dimple shapes and sizes can vary a great deal, the more times a fixed pattern is replicated to cover the ball's surface the better, provided of course the dimples are large enough to be effective, and are evenly spaced to eliminate bald spots. Most designers develop their dimple patterns within standard design areas based on a polyhedral foundation.[1]

Most designers start with one of the five spherical Platonic solids or one of their 13 Archimedean relatives. The chapter head image is a hexagonal/triangular dimple pattern based on a spherical icosahedron.[2] Usually, the designer will subdivide the

---

[1] (Solheim 1987, sheet 3/5, fig. 7).
[2] See Section 8.1 for explanation of the $\{5+,3\}_{8,0}$ hexagonal or $\{3,5+\}_{8,8}$ triangular notation. Overlay grid by Christopher Kitrick.

polyhedron's faces into smaller working areas, including Schwarz, equilateral, and isosceles triangles. The number of different working areas determines how many dimple patterns a designer needs to develop; each area requires its own pattern. For example, each of the icosahedron's 20 equilateral faces might be divided into six Schwarz triangles (three left-handed ones and three right-handed ones); a dimple pattern for one of these would be replicated 120 times, covering the entire ball through combinations of rotation or reflection. This is a good return for such a small designed area.

In the last chapter we described three symmetry groups that characterize the Platonic solids: the icosahedral (icosahedron and dodecahedron), the octahedral (octahedron and cube), and the tetrahedral groups. Polyhedral-based dimple patterns usually fall into one of these symmetry groups and we can use golf ball dimples to show how these symmetries are applied.

Why are these groups so important? There are two main reasons. First, the higher the number of rotation symmetries there are, the more spin axes a ball has and the more consistently the ball will fly. Of course, this assumes that the dimple pattern is uniform. Second is layout flexibility. Platonic solids share all their symmetries with their Archimedean relatives. For example, a designer could develop a dimple pattern within the equilateral faces of the icosahedron or its Schwarz triangles. If he chooses an icosidodecahedron, he has equilateral and isosceles triangles to work with, all of which could be further subdivided. The icosahedron and icosidodecahedron are different polyhedra, but they belong to the same symmetry group; a symmetrical dimple arrangement in one will share symmetries with the other. The icosahedral group offers more choices than the other two groups.

7.1   Icosahedral Balls

The icosahedron and its dual, the dodecahedron, share the same 120 rotation and reflection symmetries around their vertices, edges, and faces (see Section 6.3.2). The icosahedral group has the most symmetries of any group. For this reason, it is the framework of choice among ball designers.

Figure 7.1 shows a series of icosahedral group designs. We have already seen pure icosahedron examples, so here we start with its dual, the dodecahedron, which is also in the icosahedral group. Figure 7.1(a) shows a creative dimple arrangement for a spherical dodecahedron, which normally has no complete great circles. Brent Emerson's design arranges three different dimple sizes in such a way as to effectively subdivide each of the pentagonal faces into five diamonds, five isosceles triangles, and a center pentagon. In so doing, he creates ten dimple-free great circles.[3] Only one of the dodecahedron's 12 faces is shown in the figure. Note that the innermost pentagonal area, defined by five dimples labeled 60, surrounds the vertex of the dodecahedron's dual, the icosahedron. Figure 7.1(b) shows a design by In Hwang based on a spherical icosidodecahedron.[4] This Archimedean solid combines the 20 triangular faces of the icosahedron with the 12 pentagonal faces of the dodecahedron. They are clearly visible in the drawing, which emphasizes the six great circles formed between faces. The large triangles are equilateral, while the ones that subdivide the pentagonal face are isosceles.

Within these two types of triangles, the designer makes a straightforward dimple arrangement. Once again, the icosahedral symmetry is apparent. The center dimple

---

[3] (Emerson et al. 2003, sheet 4/8, fig. 4).
[4] (Hwang November 1996, sheet 5/9, fig. 8).

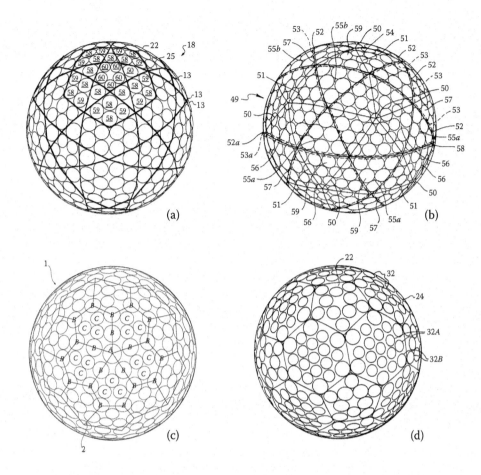

Figure 7.1. Golf balls with icosahedral symmetry.

in the middle of each pentagonal face is a vertex of the underlying icosahedron. In similar fashion (a), each of these vertices is surrounded by five equal-sized dimples. The icosidodecahedron is a popular design base because it offers all the symmetry axes of the icosahedron and has six natural great circles that can be used for parting lines, something a pure icosahedral base does not offer.[5]

Figure 7.1(c), is an unusual Catalan base. The 13 Catalan solids are duals of the 13 Archimedean solids,[6] and this one is called a pentagonal hexecontahedron.[7] It is the dual of the snub dodecahedron and has two distinct forms, which are *mirror images* (enantiomorphs) of each other. The distinct forms preclude reflection symmetries within a form, but some rotational symmetry remains. The dimple marked *A* is a vertex of the icosahedron and, once again, it is surrounded by five equal-diameter dimples marked *B*. We show this design to make the point symmetries, such as icosahedral, octahedral, or tetrahedral, are constant between a Platonic solid and its related Archimedean and Catalan solids.

---

[5] This advantage is becoming less important as molding technology improves. Titleist's *Staggered Wave* golf ball molding parting line actually weaves in and out of the dimples that cross the hemispheric boundaries.
[6] The Catalan solids are named after Belgian mathematician, Eugène Catalan, who was the first to write about them in 1865.
[7] (Winfield and Aoyama 2003, sheet 1/6, fig. 1).

Figure 7.1(d) uses another Archimedean base, the truncated icosahedron (resembles a soccer ball), in which 12 pentagonal and 20 hexagonal faces combine.[8] Dimple diameters do not vary much in this layout, which is one reason for the large land areas, the undimpled areas, between some clusters. Although this polyhedral base is quite different than previous ones, it still carries over some of the icosahedron's symmetry. The five equal-diameter dimples in the center of each pentagonal face surround an icosahedral vertex.

Commercial balls with icosahedral symmetry include Acushnet Co. Pinnacle *Platinum Distance*, Wilson *Ultra*, Ben Hogan *Hawk*, Dunlop Sports *DDH II*, Dicks Sporting Goods Maxfli *HT-9*, and Callaway/Spalding Top-Flight *Complete Distance*.

7.2   Octahedral Balls

Until the 1980s, octahedral layouts were very common. The octahedron's trirectangular work areas are easy to work with, and the base offers three orthogonal great circles for parting lines that facilitate molding and make it easier to make tools and dies. The octahedron and its dual, the cube, have only 48 rotation and reflection symmetries around their vertices, edges, and faces—less than half the number of the icosahedral group. Figure 7.2(a)[9] shows a classic dimple design, called the Atti pattern,[10] like the Class I subdivision schema we will see in the next chapter. Within each trirectangular face, dimples are arranged in three concentric "*rings*." Around the edges of each triangle, 21 dimples enclose an inner ring of 15 dimples. These 15 dimples, in turn, surround the six mid-face dimples.

Each face will have 42 dimples; the entire ball has 336, covering more than 80% of its surface. The Archimedean solid cuboctahedron combines the cube's six quadrilateral faces with the octahedron's eight equilateral faces to form the polyhedral foundation in the design shown in Figure 7.2(b).[11] The designer further subdivides each equilateral into six Schwarz triangles and each quadrilateral face into four isosceles triangles. Dimples within the quadrilaterals have a grid-like arrangement.

The octahedron's dual, the spherical cube, is shown in Figure 7.2(c).[12] Each quadrilateral face is subdivided into four congruent isosceles work areas. Only one dimple pattern is required; it will be replicated 24 times over the ball. Dimple numbering indicates each triangle has seven 101-size and eight 102-size dimples, or 60 total for each of the cube's faces. The whole ball is covered by a total of 360 dimples. The octahedral vertices coincide wherever two great circles meet at 90°. One is labeled 103 in the drawing.

Figure 7.2(d)[13] shows the six squares and eight hexagonal faces that make up this solid. The truncations do not add any great circles to the octahedron's original three. The dimple numbers indicate their sizes. It is impressive that just two sizes can get this level of coverage and leave so little land area between them. Once again, the octahedral symmetry shows through; every vertex of the octahedron is surrounded by four equal-sized dimples.

These four examples illustrate the flexibility of the octahedral group. When great circles intersect at small angles, it is difficult to fill the corners without resorting to dimples too small to contribute to the ball's aerodynamics. If they are omitted, how-

---

[8] (Mackey 1991, sheet 2/2, fig. 3).

[9] (Morgan 2005, sheet 3/13, fig. 3).

[10] Named after Ralph Atti who received many patents for golf ball construction and manufacturing.

[11] (Morell 1990, sheet 1/4, fig. 2).

[12] (Gobush 20 Sept. 1988, sheet 6/10, fig. 11).

[13] (Gobush 23 Aug. 1988, sheet 4/6, fig. 6A).

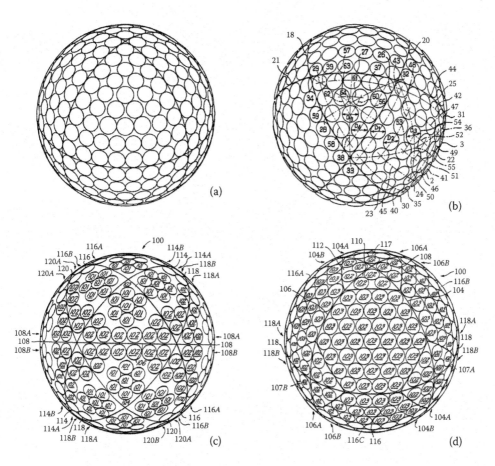

Figure 7.2. Golf balls with octahedral symmetry.

ever, bare patches result. Notice how well dimples pack in the great circle corners in Figure 7.2(a), as compared to the other layouts.

Commercial golf balls with octahedral symmetry include the Slazenger *Raw Distance Practice*, Nike *Tour Accuracy DD*, Spalding *Super-Flight*, and MacGregor *Tourney*.

7.3 Tetrahedral Balls

Tetrahedral symmetry is not used often for golf balls, perhaps because it offers the fewest number of symmetries (24 in all), and its large equilateral faces have considerable spherical excess (see Chapter 4). We show some designs for completeness. The symmetry principles we have been discussing—repeatable design areas, great circle parting lines, and dimple arrangement—still apply. Figure 7.3 shows two golf ball designs based on the tetrahedron. Their equilateral faces are so large it is difficult to see their edges in these illustrations. We have, therefore, outlined one face in each design to make both the faces and their design areas easier to read. Both designs subdivide the tetrahedron's faces into six right Schwarz triangles. Dimples are arranged within these areas, and none cross any of the six great circles. One great circle serves as a parting line. In Figure 7.3(a),[14] two rows of small dimples labeled *S* parallel two sides of the Schwarz triangle; the longer run of these dimples is along the polyhedral face edge.

---

[14] (Stiefel and Nesbitt 1999, sheet 1/2, fig. 2).

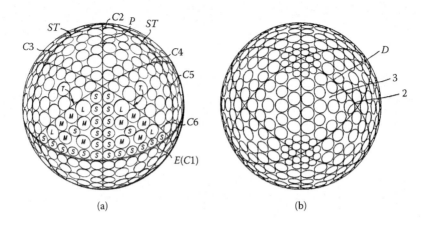

(a)   (b)

Figure 7.3. Golf balls with tetrahedral symmetry.

The design in Figure 7.3(b)[15] also places same-sized dimples along the polyhedral edge and the face medians within each Schwarz triangle. Clusters of three small dimples in two of the Schwarz triangle's corners yields the distinctive dimple pattern in the middle of each of the tetrahedron's faces and around its vertices.

7.4 Bilateral Symmetry

Although we have emphasized dimple designs with polyhedral bases, we present one final dimple layout based on loxodromes. We described this unique spherical spiral earlier in Section 4.9. It is not immediately obvious that it could be used to distribute points on a sphere. A loxodrome design was first proposed for a golf ball in 1924 by Jarvis Hunt.[16] Here we show a modern, sophisticated use of this spiral, which produces *bilateral symmetry* instead of the multi-axis symmetry examples we have been discussing. Figure 7.4(a) shows a design by Donald Bunger and associates, with four left-handed and four right-handed spirals between the North and South Poles of the

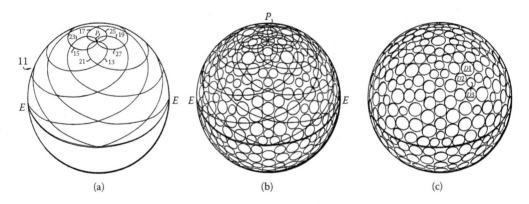

(a)   (b)   (c)

Figure 7.4. Golf balls with loxodrome design.

---

[15] (Yamada 1990, sheet 4/6, fig. 5).
[16] (Hunt 1924).

golf ball.[17] These spirals define a single point at each pole, 24 in each hemisphere and four along the equator. Since none of the spirals are great circles, the four-sided areas formed by their intersections technically are not spherical polygons. The points where the spirals intersect each other are rotationally symmetric around the poles. Loxo-drome spirals naturally continue to the lower half of the sphere, but the designer de-cided to break the curves at the equator and rotate the lower hemisphere's spirals 45° relative to the upper one; thus, the equatorial plane is not a mirror plane. Great care is needed to find spirals that form useful intersections and design areas. If the spi-rals intersect at too shallow an angle, very small dimples result, and they have less of an aerodynamic effect. Note how the designer centers some dimples on the loxo-drome or at their intersections, while others are arranged totally within four-sided areas. Figure 7.4(b) shows the relationship of the spirals to the dimple arrangement, while (c) shows the completed design, minus all reference lines. This design distributes 410 dimples of three sizes over the ball.

7.5 Subdivided Areas

At this point, we have a good sense of the three main symmetry groups a golf ball designer can use as a polyhedral foundation. Assuming the design is not a random layout, his choice of polyhedral foundation determines the regular spherical faces he will work with, either equilateral triangles, quadrilaterals, pentagons, or hexagons. In many cases, the designer will further subdivide the polyhedral face to gain additional great circles for parting lines. These subdivisions also may be used to define standard areas, so dimple arrangements developed within will cover the rest of the ball through rotation and reflection symmetries.

Archimedean solids and subdivided Platonic solids will have at least two, if not more, work areas. An icosidodecahedral base, for example, has pentagonal and equi-lateral triangular faces, and both may be further broken down into Schwarz triangles or nested equilaterals. If the ball is a hybrid, it may combine some standard work ar-eas with entirely different ones. We have already seen hybrid cases where some of the regular faces of a spherical icosahedron are preserved, yet other areas are developed as bands of dimples around the equator.

Figure 7.5(a)[18] demonstrates an important subdivision, where nested equilateral triangles define the center points of dimples. The approach is quite different from the Mid-arc subdivision of the equilateral triangle shown in (b),[19] which can also be viewed as a series of right Schwarz triangles. Figure 7.5(c) is an innovative subdivision of an equilateral triangle. The schema results in three congruent spherical quadrilaterals.[20] The subdivision, however, does not result in adding great circles to the polyhedral base. The pentagon within a pentagon, shown in Figure 7.5(d),[21] is similar in approach to the triangle within a triangle in (a). In Figure 7.5(e) the pentagonal area is subdi-vided, so two standard areas, a diamond and isosceles triangle, surround a central pentagon. [22] This calls for three dimple arrangements.

A number of designers use icosidodecahedral bases. Figure 7.5(f)[23] demonstrates one way to subdivide the 12 pentagons into isosceles work areas adjacent to the

---

[17] (Bunger et al. 1991, sheets 1/9, 4/9, and 5/9; figs. 2, 8, and 9, respectively).

[18] (Aoyama 1999, sheet 9/9, fig. 10).

[19] (Woo 1993, sheet 1/5, fig. 1).

[20] (Aoyama 1995, sheet 3/7, fig. 3).

[21] (Emerson et al. 2003, sheet 4/8, fig. 5).

[22] (Shaw and Haines 1979, sheet 3/3, fig. 3).

[23] (Hwang November 1996, sheet 4/9, fig. 7).

(a)

(b)

(c)

(d)

(e)

(f)

(g)

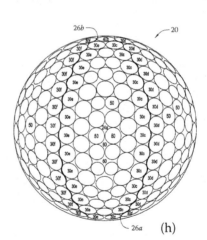

(h)

Figure 7.5. Subdivision areas.

20 equilateral triangles. Spherical quadrilaterals are not used often as work areas, but the arrangement in Figure 7.5(g)[24] shows that any number of triangular subareas could be defined, provided the patterns within are symmetrical. In Figure 7.5(h), the standard design area is an entire lune, six of which cover the entire sphere.[25] Given the symmetry of dimple arrangement within, it would be possible to define a *birectangular* triangle as a subunit.

[24] (Aoyama 1990, sheet 5/8, fig. 8).
[25] (Veilleux, Simonds, and Shannon 2007, sheet 2/11, fig. 2).

7.6 Dimple Graphics

Dimples made it easier for us to read the underlying spherical subdivision of golf balls. We are not designing golf balls in this book, but we can use graphic styles that look like dimples to make it easier to read the subdivision schemas we create. Each schema we present in the next chapter results in a triangular grid. We can place dimple-style graphics at grid vertices or in the middle of the grid's triangular faces. In the first case, we use circles or small balls centered at each grid point. In the second case, we place a hole centered at the triangle's centroid. Figure 7.6 shows five graphic styles for the same spherical grid. A typical subdivision grid in (a) is shown with circles at each vertex in (b) and shaded in (c). These last two styles resemble golf balls. Figure 7.6(d) shows the same subdivision, but this time, holes are located at the centers of each triangular face and then shaded in (e). Each style emphasizes a feature. Some make it easier to see how uniform the spacing is between grid points; others indicate differences in the triangles shape and size. We will make extensive use of these graphic styles in the next chapter.

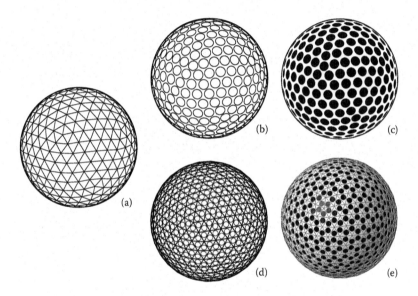

Figure 7.6. Grid accents, dimples, and holes.

7.7 Summary

We have defined what a polyhedron is, detailed the characteristics of the Platonic and Archimedean solids, and shown precise ways to define their planar and spherical geometry. Schläfli and Cundy-Rollett's modified Schläfli symbols provide simple and useful notations for specifying a polyhedron and seeing the similarities and differences between them. The most important polyhedral frameworks for spherical subdivision are the Platonic deltahedra—the tetrahedron, octahedron, and icosahedron. They have equilateral faces that break down into Schwarz triangles, and each defines a symmetry group. Any subdivision of a deltahedron's Schwarz or equilateral face triangle also subdivides its dual, even though the duals have a very different face type. While the Platonic solids, especially the icosahedron, are the most common frameworks for subdivision, we will see instances where quasiregular Archimedean solids, the cuboctahedron and icosidodecahedron, solve special cases where a natural great circle of edges is needed. We have also seen that two of the Archimedean solids, the snub cube and snub dodecahedron, are chiral and do not have reflection symmetry. Chiral subdivisions are a recent and distinct class of spherical subdivisions. The rotational and reflection symmetry inherent in the icosahedral, octahedral, and tetrahedral groups gives us methods for replicating geometry from one part of the sphere to another. We can subdivide a small area, such as a Schwarz triangle, and then replicate it to cover the entire sphere without gaps or overlaps.

Golf ball dimples are comprehensive demonstrations of the principles of spherical subdivision. Dimple layouts are practical solutions to difficult geometry problems. Most golf ball dimple arrangements are developed from a polyhedral base. Designers exploit symmetry, great circles, and local working areas that can be replicated over the sphere to pack different-sized dimples. The variety and creativity of these spherical designs are amazing, not only in their creativity, but as solutions to difficult spatial problems.

In the next chapter, we explore various subdivision techniques based on the polyhedral frameworks and symmetry groups we have just covered.

Additional Resources

Naruo, Takeshi and Taketo Mizota. *"The Influence of Golf Ball Dimples on Aerodynamic Characteristics."* The 2014 Conference of the International Sports Engineering Association, Elsevier: Procedia Engineering 72 (2014): 780–785.

Stewart, Ian. "Crystallography of a Golf Ball." *Scientific American* (Feb. 1997): 96–98.

Tarnai, Tibor. "Geodesic Domes and Golf Balls." *Space Structures* 4.2 (1993): 1176–1183.

Tarnai, Tibor and András Lengyel. *"Dimple Patterns of Golf Balls and Node Arrangement of Geodesic Domes."* International Association for Shell and Spatial Structures (IASS) 53.3 (2012): 185–195.

⑧ Subdivision Schemas

Spherical polyhedra provide a convenient starting point for subdividing spheres because they jump-start the process by evenly distributing an initial set of points on the surface of the sphere. This chapter describes how to define more points between this initial set. When each point is connected by an arc or chord to the points immediately around it, a triangular grid results. The top four figures above show this sequence. Some spherical applications require design features, such as openings or panels or different face shapes. The geometry can become quite complex. The lower two figures above suggest added features, such as holes and panels that are based on the subdivision grid. These features are typically added with Computer Aided Design after the base subdivision geometry is established.

In the mid-1960s, Joseph Clinton developed a class system for categorizing spherical grids.[1] Each class refers to triangular grids that have the same basic orientation relative to the edges of the spherical polyhedron they subdivide. Class I grids run parallel to the polyhedron's edges. The subdivided spheres shown on the previous page are all Class I. Class II grids run perpendicular to the polyhedron's edges, while Class III

---

[1] In 1966, Joseph Clinton and Mark Mabee, under NASA contract, undertook a study of spherical geometries and developed the spherical node identification and grid topological class system we use today. Their Classes I and II were first published in *Dome Book 2*; see (Clinton 1971). Magnus Wenninger added Class III for skewed grids in 1979 in his book *Spherical Models*; see (Wenninger 1979).

grids are skewed and run at an angle to the edges. We will cover each class in this chapter, highlighting techniques unique to each and showing examples of how the subdivision can be implemented.

In 2002, Clinton introduced the totally new group of spherical tessellations having edges of equal length. These tessellations are distinct from the Class system presented in this chapter. Sometimes called self-organizing grids, they can solve challenging configurations that Class techniques cannot. They are covered in detail in Chapter 10.

The presentation of each Class method is identical. Each starts with orienting one of the polyhedron's faces in a way that makes calculations easier. This face is called the *principal polyhedral triangle (PPT)*, and it serves as our subdivision workbench. Points are located on the PPT's edges, and great circle arcs span from one edge point to another. Subdivision grid points are defined where these arcs intersect over the face of the PPT or at some point near their intersection. We will illustrate each of these steps, noting the mathematical principles. We will also present the key formulas and algorithms in Appendix C.

We have tremendous flexibility in spherical subdivision. We can start with any number of spherical polyhedra and can develop different types of grids covering them. By selectively spanning points with either great circle arcs or chords, we can make triangles, diamonds, hexagons, pentagons, or other polygons. With so many choices, it is natural to ask which combination is best. Of course, the answer depends on the application. In the next chapter we present a series of metrics that make it easy to see their differences and similarities. In the end, you will be equipped to answer this question for your own applications.

8.1   Geodesic Notation

In Chapter 6 we saw examples of Schläfli symbols. They summarize polyhedral geometry with two numbers $\{p,q\}$, which are read as p-gons (pentagons, hexagons, etc.) that meet q times at a vertex. The Schläfli symbol for a dodecahedron, for example, is $\{5,3\}$, indicating that three pentagonal faces (five edges) meet at each vertex. This notation is extended to include symbols for grids that cover spherical polyhedra.[2] *Geodesic notation* is very useful because, at a glance, it states the type of tessellation, how many grid members intersect at a vertex, the intensity of the grid, and the grid's orientation with respect to the base polyhedron's edge.

The general form of geodesic notation is $\{p, q+\}_{b,c}$ where p indicates the tessellation shape and q is the *valence* (number of grids) that meet at a subdivided polyhedron vertex. When three subdivision grid members meet at a vertex, its valence is called *trivalent*, and when four, five, and six grid members meet at a vertex, they are called quadravalent, *pentavalent*, and hexavalent, respectively. All of our subdivision examples develop triangular grid, thus, $p = 3$. In an icosahedron, five grid members (pentavalent) meet at each polyhedral vertex. Thus, $q = 5$ (for tetrahedra or octahedra, $q = 3$ and 4, respectively). The "+" sign indicates that more than q grid members meet at other vertices; these cases occur at points other than the base polyhedron's vertices. For an icosahedron with a triangular tessellation grid, $\{p,q\}$ would be $\{3,5+\}$ or 3-gons (triangles) with five grid members meeting at each of the polyhedron's vertices. The + sign, indicated by 5+, means that at points other than the polyhedron's vertices, more than five grid members meet.

Perhaps the most useful aspect of geodesic notation is the b,c subscript because it indicates the frequency of the grid, how course or fine it is, and the orientation of the grid with respect to the sides of the polyhedron's faces.[3] The frequency (designated by

---

[2] (Goldberg 1937, 106).

[3] H. S. M. Coxeter used Michael Goldberg's formula of $T = b^2 + bc + c^2$ with geodesic notation $\{p,q\}b,c$ to describe Buckminster Fuller's geodesic grids; see (Coxeter 1971).

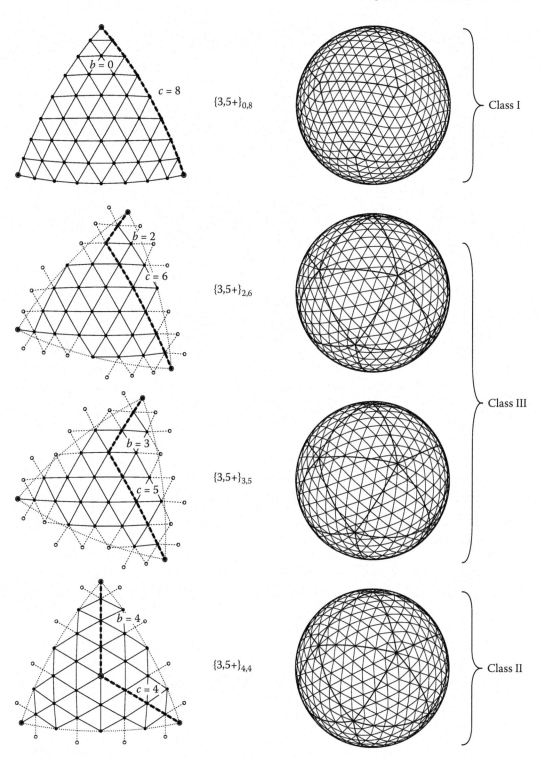

Figure 8.1. Examples of geodesic notations for Class I, II, and III tessellations.

the superscript Greek letter *nu* or ⱽ) of a subdivision grid is the sum $b + c$, and both b and c can be any nonnegative integer.[4] Either b or c can be 0, but not both in the same subdivision.[5] Most of the examples in this chapter are frequency 8^v, thus, b and c will have values, such as 0,8 or 4,4 or 2,6, respectively.

Figure 8.1 shows three classes, or families, of grids for a single PPT—Class I, II, and III. A class is simply a group of subdivision methods or schemas that result in the same grid orientation with respect to the edges of the polyhedron's face. Within a class, there will be small variations in grid points within the PPT from one schema to another, but the general orientation of the grid remains the same. Each left-right pair of illustrations in Figure 8.1 shows one grid type covering a single PPT and covering a full sphere. We have said that Class I grids run parallel to the PPT's edges; a typical one is shown in the topmost pair in Figure 8.1. Class II grids are the other extreme and run perpendicular to the PPT's edges, as shown in the bottom pair. The Class III grids, shown in Figure 8.1, are placed between Class I and II in the figure to emphasize the fact that their grids are oriented somewhere in between Class I and Class II and are skewed with respect to the PPT's face.

Figure 8.1 also shows the geodesic notation and an example of a fully subdivided sphere. Each of these classes has the same {3, 5+} specification because all the subdivisions shown are for a spherical icosahedron. Only their b, c values change because they indicate how the grid is oriented. To see how b, c coordinates work, start at any PPT apex and follow the shortest path of grid lines to reach the next PPT apex to the left of the direction you are traversing in the grid. In the figure, b directions are down and to the left, while c directions are down and to the right. If you start at the uppermost PPT vertex, count how many grids you traverse in the two directions; this is the b, c value that appears in the geodesic notation.

8.2   Triangulation Number

Goldberg used geodesic notation to investigate dual grids (grids based on other grids), and the symmetries of grids on a sphere. In the process, he found a direct relationship between b, c and the topology of a grid, the number of parts on a fully subdivided sphere. The relationship $T = b^2 + bc + c^2$, sometimes called the triangulation number, or simply T, is very useful.[6] It provides a coefficient that summarizes the total faces, edges, and vertices in a completely tessellated sphere.[7] Figure 8.2 shows how to find the triangulation number and the formulas that use it to enumerate the number of faces, edges, and vertices in a fully subdivided sphere. The example uses T to find the number of parts in a 4^v Class I spherical octahedron.

$$T = b^2 + bc + c^2$$
$$\text{Faces} = T \times \text{number of polyheadra face,}$$
$$\text{Edges} = T \times \text{number of polyheadra edges,}$$
$$\text{Vertices} = T((2 \times \text{valence})/(6 - \text{valence})) + 2$$

---

[4] The Greek letter nu (v), a symbol for subdivision frequency, first appeared in early worksheets developed by Synergetics, Inc. and Geodesics, Inc. in the 1950s. It is still used in this context today.

[5] It is not often seen, but when either b or c is 1 and the other is 0, there is no tessellation and the notation describes an unsubdivided polyhedron. A basic tetrahedron polyhedron, for example, would read {3,3+}0,1.

[6] (Goldberg 1937, 104–108).

[7] The triangulation number was important in defining the possible icosahedral surface lattices possible in the pioneering virus research of Donald Caspar and Aaron Klug. See Section 3.2.

Example

Spherical octahedron:
Faces—8
Edges—12
Vertices—6

Geodesic notation: $\{3,4+\}_{0,4}$

Subdivided octahedron:
$T = 4^2 = 16$
Faces $= T(8) = 128$
Edges $= T(12) = 192$
Vertices $= T[(2(4))/(6-4)]+2 = 66$
Euler's law: vertices + faces = edges + 2 = 194

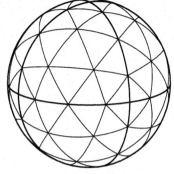

Figure 8.2. Triangulation number.

The triangulation number is also an indication of the possible patterns that can be made from the triangular grid. If the number is evenly divisible by three, you can make diamond and hexagonal-pentagonal patterns by selecting different combinations of points from the triangular grid. All Class II subdivisions can make these other patterns, but not every Class I or III combination can do so.

8.3 Frequency and Harmonics

In geodesic notation, the frequency of a tessellation, $b+c$, is the number of segments into which the grid divides the polyhedron edge. Some subdivision schemas can produce any even or odd frequency. Others are limited to even ones or to powers of 2. A grid may also contain subgrids, or harmonics. A harmonic is a sub-frequency that is mathematically and geometrically related to another sub-frequency. If a tessellation has harmonics, some of its grid points define other tessellation grids. These might be useful in some geodesic applications, for example, where grid overlays are used for trussing or harmonics that define smaller units within the PPT that facilitate construction or prefabrication.

The easiest way to visualize both frequency and harmonics is to see them side by side. Figure 8.3 shows an example of doubling frequencies for icosahedron Class I, II, and III tessellations. The frequencies in (a) and (b) are 2^v, 4^v, 8^v, and 16^v for Class I PPTs and Class II diamonds (described later). They are shown in clockwise order around a common icosahedra vertex. In a similar manner, (c) shows frequencies 3^v, 6^v, 12^v, and 24^v for Class III tessellations. Notice in each figure that the lower frequency grid is also part of the higher one. The lower frequency grid is a harmonic of the higher frequency grid.

Harmonics follow rules. All frequencies are positive integer numbers and the sum $b+c$. Class I and III can have even or odd frequencies. Class II and its harmonics are always even frequencies because the tessellation grid is symmetrical about the polyhedron's face vertical axis. A frequency can have a harmonic, if it is not a prime number. The numbers, 2, 3, 5, 7, 11, 13, 17, 19, 23, and 29 are prime, and therefore, any Class I subdivision with these frequencies will not have harmonics. Class III subdivisions are skewed, and both b and c specifications must be nonprime for there to be a harmonic.

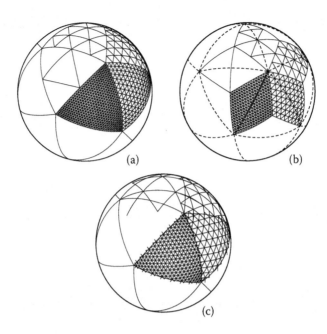

Figure 8.3. Class I, II, and III tessellation harmonics.

Tessellation Harmonics						
v	**Prime Factors**	**Harmonic Frequencies**				
2	2 (Prime)					
3	3 (Prime)					
4	2,2	2^v				
5	5 (Prime)					
6	2,3	2^v	3^v			
7	7 (Prime)					
8	2,2,2	2^v	4^v			
9	3,3	3^v				
10	2,5	2^v	5^v			
11	11 (Prime)					
12	2,2,3	2^v	3^v	4^v	6^v	
13	13 (Prime)					
14	2,7	2^v	7^v			
15	3,5	3^v	5^v			
16	2,2,2,2	2^v	4^v	8^v		
17	17 (Prime)					
18	2,3,3	2^v	3^v	6^v	9^v	
19	19 (Prime)					
20	2,2,5	2^v	4^v	5^v	10^v	
21	3,7	3^v	7^v			
22	2,11	2^v	11^v			
23	23 (Prime)					
24	2,2,2,3	2^v	3^v	4^v	6^v	8^v	12^v
25	5,5	5^v				
26	2,13	2^v	13^v			
27	3,3,3	3^v	9^v			
28	2,2,7	2^v	4^v	14^v		
29	29 (Prime)					
30	2,3,5	2^v	3^v	5^v	6^v	10^v	15^v

Table 8.1. Tessellation harmonics.

Figure 8.3(c) shows Class III examples of frequencies 3^v, 6^v, 12^v, and 24^v. Their b,c specifications are doubled in each successive tessellation. Their geodesic notations are $\{3,5+\}_{1,2}$; $\{3,5+\}_{2,4}$; $\{3,5+\}_{4,8}$; and $\{3,5\}_{8,16}$, respectively. Note that both b,c values double each time.

To determine if a frequency has one or more harmonics, first find its prime factors. The prime factors of a positive integer are the prime numbers that divide into it exactly, leaving no remainder. If the frequency has prime factors, the product of any combination of them is a subdivision grid harmonic, if it also satisfies the rules above. Table 8.1 shows the harmonics for frequencies 2^v–30^v. Column 1 is the base frequency, and column 2 lists its prime factors, if it is not a prime. Columns 3 onward are the frequency's harmonics. Referring to Table 8.1, a 24^v tessellation, for example, has prime factors 2, 2, 2, 3. Thus, a 24^v tessellation will have 2^v and 3^v tessellation harmonics as well as tessellations for the product of various combinations of these prime factors, six in all: 2^v, 3^v, 4^v, 6^v, 8^v, and 12^v. Keep in mind that for Class III subdivisions, both b and c must be nonprime for there to be harmonics.

To review, the frequency of a subdivision tessellation is $b+c$, or the number of times the grid divides up the polyhedron's edge. If the frequency is a prime number, the tessellation has no prime factors, and therefore, no harmonics. If the frequency is not prime, the product of any combination of its prime factors will be a harmonic grid in the overall tessellation.

8.4   Grid Symmetry

In Chapter 6, we defined symmetry as any combination of rotations and reflections that leave the polyhedron unchanged. Tessellation grids exhibit these symmetries. Most of the subdivisions developed in this chapter are subdivisions of the equilateral faces of the tetrahedron, octahedron, or icosahedron. Although these polyhedra are rotationally symmetric around their vertices, edges, and faces, the subdivision grids developed over them are not necessarily symmetric. Figure 8.4 shows some examples of three classes of grids we develop later in this chapter. The center of each face is marked.

The Class I and II grids, shown in Figure 8.4(a) and (b), are symmetric by reflection and by rotation. That is, the face can be rotated around its center (indicated by a $\oplus$ symbol) any multiple of 120°, and its appearance does not change. Likewise, the grid can be reflected around any of the principal triangle's altitudes or edges and remain unchanged. Class I and II tessellations are generally symmetric rotationally and by reflection, but Class III tessellations are not.

Class III grids run at an angle to the PPT's edges. As a result, their grids are *enantiomers*, which means the grid's reflection is an opposite-handed version.[8] You cannot pass a plane through a Class III subdivided sphere and produce equal images on either side. This is analogous to the images of your left and right hands. They are enantiomorphs—opposite-handed reflections of each other. Notice in Figure 8.4 that even after reversing the b,c coefficients for the Class I and II tessellations (a) and (b), the grid would still look the same. But when b,c is reversed in Class III, the grid becomes opposite-handed, as (c) and (d) demonstrate. Class III grids do have rotational symmetry; however, to review, subdivisions of spherical polyhedra are grouped into three classes, each distinguished by the orientation of their subdivision grid with respect to

---

[8] The Latin words *laevo* and *dextro* are sometimes used to designate the left- and right-handed image or to indicate a rotation direction.

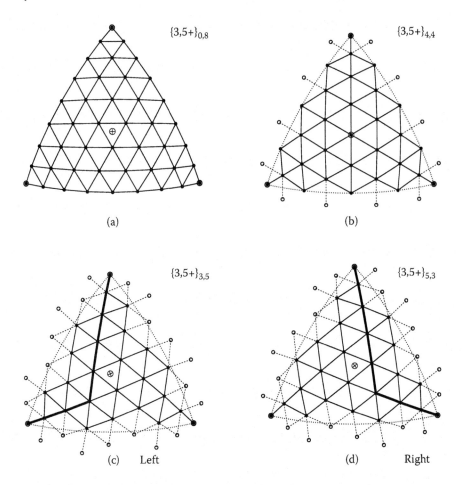

Figure 8.4. Class I, II, and III tessellation rotation and reflection symmetry.

the polyhedron's edges.[9] In Class I and II subdivisions, tessellation grids run parallel and perpendicular to the polyhedron's edges. In Class III, grid chords run at angles to the edges, never parallel or perpendicular to them. Class I and II subdivision schemas generally produce grids that are symmetric by rotation and reflection. However, Class III grids are only rotationally symmetric. Their reflections are not symmetric; they are opposite-handed.

For a subdivided spherical polyhedron, a geodesic notation has evolved that summarizes the type of tessellation grid, the number of grid members that meet at a typical polyhedron's vertex, its *valence*, and the coordinates of the tessellation's grid. Coordinates b,c define the orientation of the grid with respect to the polyhedron's edges, as well as the grid's frequency. The frequency of a subdivision grid is the sum $b + c$, or the number of times the grid subdivides a polyhedron's edge. Most Class I and III subdivision techniques can produce any even or odd frequency, but some produce frequencies that are only powers of 2. Class II frequencies are always even frequencies. A subdivision grid may contain harmonic grids (grids within grids), if the frequency is not a prime number. Harmonics, if present, offer more flexible implementations

---

[9] (Wenninger 1979, 98–100).

of geodesics in some applications. You can derive a triangulation number, T, from a subdivision's geodesic notation and use it to calculate the total number of faces, edges, and vertices in a fully subdivided sphere.

8.5   Class I: Alternates and Ford

This section shows the step-by-step process for subdividing a sphere with Class I tessellations. We demonstrate four different subdivision schemas. Figure 8.5 shows examples of Class I tessellations developed on several different polyhedral foundations. Although Class I subdivisions are typically used with deltahedra—polyhedra with only equilateral faces, such as the octahedron, tetrahedron, and icosahedron—the schema can be used with isosceles areas as well. Figure 8.5 shows examples of subdivided deltahedra. From left to right they are icosahedron 8^v, icosahedron 4^v, tetrahedron 8^v, and octahedron 11^v.[10] Although some of their grids look very similar, there are subtle but important differences that may influence which Class I schema you use in your own applications. We highlight these differences in the next chapter and make side-by-side comparisons of all the schemas.

The earliest Class I schemas were developed by Buckminster Fuller, Don Richter, Jeffrey Lindsay, T. C. Howard, and Duncan Stuart in the late 1940s and early 1950s in Buckminster Fuller's various startup companies. Early Class I subdivisions were called *alternates* and *Ford* breakdowns. The name *alternate* most likely evolved to mean this subdivision was the alternate method to the triacon method, another popular subdivision technique at the time, or from the way the method subdivides alternating sides of a spherical triangle. The *Ford* moniker is likely a tribute to the grid developed for the 1953 Ford Rotunda dome project in Dearborn, Michigan.[11] Many geodesic domes have been built using Class I grids. Today, this classic grid finds new uses in product design, cartographic reference systems, computer graphics, scientific instruments, sports balls, and other industrial applications.

It is easy to recognize Class I subdivisions. Their three-way grid "*parallels*" the three edges of the polyhedron's triangular faces, as shown in Figure 8.5. As a result,

Figure 8.5. Examples of Class I subdivision schemas.

---

[10] Geodesic notations are: $\{3,5+\}_{0,8}$, $\{3,5+\}_{0,4}$, $\{3,3+\}_{0,8}$, and $\{3,4+\}_{0,11}$, respectively.
[11] See Section 2.7.5.

the polyhedron's edges, though subdivided, remain part of the final grid. This makes Class I subdivisions easy to recognize, and makes it easy to see which polyhedron is being used as its base. None of the other classes have this characteristic. Class I can be any even or odd frequency, and one of its geodesic b, c specifications is always zero. The tessellation frequency is $b + c$, and the grid may contain harmonic grids, depending on their prime factors.

Each Class I subdivision schema we develop in this chapter begins by properly orienting the PPT. The PPT becomes our spherical workbench. Reference points are located along its edges. Pairs of reference points on opposite edges of the PPT define great circles that cross and intersect over the PPT's surface. Each great circle intersection defines two points. One point is within the PPT's face, while the other is on the opposite side of the sphere and is ignored. We repeat this process again and again, each time using a different combination of edge reference points and corresponding great circles. The process ends when this single PPT is completely tessellated. We copy the resulting PPT grid and replicate it to cover the rest of the sphere. The Class I schemas developed here differ only in the way the edge reference points are defined and how great circles between them intersect to define subdivision grid points. All Class I subdivision schemas start by orienting the PPT and defining reference points along its edges. We start with general considerations for the PPT.

8.5.1 Defining the Principal Triangle

A single face from a spherical deltahedron—a tetrahedron, octahedron, or icosahedron—serves as our subdivision workbench. Figure 8.6 shows three deltahedra above, each oriented vertex-zenith, along with one of their faces subdivided with a 4^v grid and its geodesic notation below.

It is evident that the icosahedron's PPT (far-right illustration in Figure 8.6) has the smallest face area and the most uniform grid of the three. Its chords, triangle face sizes, and shapes show less variation than the other two. The spherical icosahedron

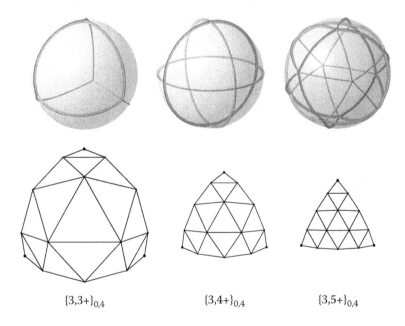

$\{3,3+\}_{0,4}$ $\qquad$ $\{3,4+\}_{0,4}$ $\qquad$ $\{3,5+\}_{0,4}$

Figure 8.6. Class I principal triangles for tetrahedron, octahedron, and icosahedron.

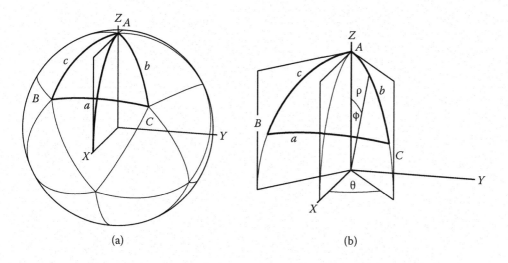

(a)   (b)

Figure 8.7. Principal triangle orientation and vertex coordinates.

is the best reference polyhedron of the three because its PPT has the smallest surface area of the Platonic solids and, of the three deltahedra, distributes the most evenly spaced vertices on its circumsphere. Its PPT has only one-twentieth of the area of a full sphere, whereas the tetrahedron and octahedron's PPTs are larger and cover one-fourth and one-eighth of the area of the sphere, respectively. The spherical icosahedron distributes 12 evenly spaced vertices on its circumsphere, while the tetrahedron and octahedron only distribute four and six vertices, respectively, on theirs. For these reasons, the icosahedron is the most frequently used reference polyhedron for spherical subdivision, by far. We will use the icosahedron in every subdivision schema we demonstrate and include an occasional tetrahedron or octahedron to make the point that the same techniques apply equally well to them.

With a spherical icosahedron as our base, the next task is to define precisely the geometry of one of its PPTs. We orient the PPT vertex-zenith and one of its faces symmetric to the xz-plane, as shown in Figure 8.7(a). This PPT orientation simplifies our calculations. We already know the coordinates of the PPT's zenith point, $(0, 0, 1)$, and we know that any point we calculate along edge b can be mirrored to edge c simply by reversing its y-coordinate sign. The vertex-zenith orientation does not affect any of the PPT's face or edge angles.

To find the coordinates of vertices B and C, it is easiest to first define them in the spherical coordinate system and then convert them to Cartesian values because we already have all the spherical information we need. Figure 8.7(a) shows the overall orientation, while (b) shows the relationship of the two coordinate systems to the PPT. Referring to Figure 8.7(b), the great circle planes that define edges b and c are tangent to the z-axis. The dihedral angle between these planes is simply the PPT's surface angle, or 72° (for the tetrahedron and octahedron, it is 120° and 90°, respectively). In the spherical coordinate system, a point is defined by three values: *rho* (ρ), *pi* (ϕ), and *theta* (θ), where ρ is the radius of the unit sphere, ϕ is the PPT's edge (angle), and θ is the PPT's face angle on the xy-plane, measured around the z-axis from the $+x$-axis (either clockwise or counterclockwise).

In Chapter 6, we used the golden ratio to precisely define the face and edge angles of the spherical Platonic solids. Using these angles, along with the radius of the unit sphere (1.0), we have all the information we need to define the spherical coordinates

Deltahedra PPT Coordinates and Chord/Arc Factors							
	Vertex C Spherical Coordinates			Vertex C Cartesian Coordinates			Chord Factor	Arc Factor
Deltahedron	ρ	φ°	θ°	X	Y	Z	Chord Factor	Arc Factor
Tetrahedra	1.0	109.471220634	60.0	0.471405	0.816497	−0.333333	1.632993	1.910633
Octahedra	1.0	90.0000000000	45.0	0.707107	0.707107	0.000000	1.414214	1.570796
Icosahedron	1.0	63.434948823	36.0	0.723607	0.525731	0.447214	1.051462	1.107149

Table 8.2. Deltahedra PPT coordinates and chord/arc factors.

of PPT vertex C. Table 8.2 lists the spherical coordinates for apex C. The chord and arc factors are based on the sphere's radius and the PPT's edge (angle). We can find the Cartesian coordinates of vertex C by simply converting the spherical values to Cartesian values, using the following formulas. Most scientific calculators can do this automatically.[12]

$$x = r\cos\theta\sin\varphi,$$
$$y = r\sin\theta\sin\varphi,$$
$$z = r\cos\varphi.$$

To summarize the Cartesian coordinates of the icosahedron's PPT's apices are:

$$A = (0.000000, 0.000000, 1.000000),$$
$$B = (0.723607, -0.525731, 0.447214), \quad \text{and}$$
$$C = (0.723607, 0.525731, 0.44724).$$

We have now established all the basic facts about the principal triangle: surface and edge angles, PPT vertex coordinates, and chord/arc factors. We have all the information we need to begin subdividing the PPT.

8.5.2 Edge Reference Points

Class I subdivision begins at the PPT's edges. A series of reference points is located along the edges, dividing the edges into segments equal in number to the subdivision frequency. Reference points have a regular pattern. Their spacing may be equal or unequal, but all three PPT edges will have the same pattern regardless of the spacing used.

There are two approaches to defining reference points: *Equal-chords* and *Equal-arcs*. Each approach can produce several different grids. For Equal-chords, the PPT's apex-to-apex chords (all three are the same length) are subdivided with equally spaced reference points. These reference points are on the PPT's edge chord; thus, they are inside the sphere. They must be projected to the surface of the sphere before they can be used as subdivision references. For Equal-arcs, the PPT's arc edge (not its chords)

---

[12] Two spherical coordinate conventions are in common use: (ρ,φ,θ) and (ρ,θ,φ). Be sure to use the convention consistent with your calculator.

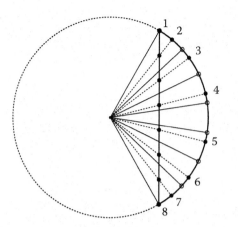

Figure 8.8. 7^v Equal-chords, Equal-arcs subdivision.

is evenly subdivided with points, which are used as references, as is. No projection is needed because these points are already on the surface of the sphere.

Figure 8.8 illustrates the difference between these two approaches with a 7^v subdivision of a tetrahedron's chord. We use an odd frequency so that reference points from the two approaches will not be superimposed on one another. This makes the figure easier to read. We also use the chord of a tetrahedron, the longest of any Platonic with a unit circumsphere, to make comparisons between Equal-chords and Equal-arcs more vivid. Referring to the figure, points 1 and 8 are the chord's end points on the surface of the circumscribing sphere represented in the cross-section by the circumscribing circle.

For grids based on Equal-chord reference points, the chord between points 1 and 8 is divided into seven equal segments (7^v) with points located between them. The dashed lines show chord points projected from the center of the sphere to its surface. They are numbered 2 through 7 (1 and 8 are vertices of the tetrahedron's PPT edge and are already on the Sphere's surface). By inspecting the figure, you can easily see the arc spacing between projected chord points is not equal. The difference between the largest and smallest arc length is more than 2:1 in this example. To put some numbers to this, the tetrahedron's overall edge arc is approximately 109.5°, and the seven arcs resulting from Equal-chords subdivision are 9.4°, 14.1°, 19.8°, 22.8°, 19.8°, 14.1°, and 9.4°, in that order. It is also evident that the arc lengths are symmetrical; that is, arc 1-2 equals 7-8, arc 2-3 equals 6-7, and so on.

The Equal-arc points in Figure 8.8 are marked with small hollow circles on the surface of the sphere. They do not have to be projected. Each is spaced at 15.6°. Note that lines between them and the center of the sphere divide the chord into unequal segments. Appendix C lists simple formulas for dividing chords and arcs between any two points on the sphere into equal segments and projecting points to the surface of the sphere.

8.5.3   Intersecting Great Circles

Intersecting great circles is the most intuitive and easiest way to define points on the surface of the sphere. We make good use of this technique in our Class I schemas. Recall that a great circle is defined by three points: the origin and any two non-tangent, non-antipodal points on the sphere. Figure 8.9 shows two great circles defined by two

Figure 8.9. Defining points by intersecting great circles.

pair of points (red) and the origin (not shown). Two new antipodal points are created where these great circles intersect. One of them (brown) is visible in the figure; the other is on the opposite side of the sphere and cannot be seen.

The procedure for defining Class I grid points on the PPT's face is very simple. We use the origin and pairs of edge reference points on opposite edges to define the plane of each of the two great circles. The line of intersection between any two great circles pierces the surface of the sphere at two antipodal points. One of these points will be within the PPT's surface; we will ignore the other one, which is on the opposite side of the sphere. This new point will become a tessellation vertex in the final grid, or it will be used as an intermediate point to find a tessellation vertex, depending on which variation of grid we are making. We use this technique extensively in Class I subdivisions because it is extremely accurate, can be applied in many geometrical situations, and is easily programmed into calculators or computers.

Let's review the Class I technique. Class I tessellations run parallel to the edges of the spherical polyhedron, and the spherical polyhedron's edge remains a part of the final tessellation grid. The initial choice of polyhedron determines the size of the PPT to be subdivided. Only one PPT is subdivided and the tessellation is then replicated to cover the rest of the sphere. We position the polyhedron vertex-zenith with the selected PPT's edge symmetrical to the xz-plane. Its vertex angle, chord, and edge arc lengths are found along with the coordinates of each vertex. Next, we locate reference points along each PPT edge, which divides the edge into segments equal to the frequency. Two approaches to dividing the PPT edge are used: Equal-chords or Equal-arcs. The Equal-chord approach produces unequal, but symmetric, points along the PPT's edges, while Equal-arc produce equally spaced points and Equal-length arc segments. All three edges of the PPT will have the same pattern of reference points. Based on these edge reference points, great circle planes are defined and intersected. The intersection of two great circles is a line that intersects the sphere's surface at two antipodal points. One of these points falls within the edges of the PPT's face and we ignore the other one. We can use at least four different combinations of edge reference points and great circle intersections to define a spherical grid. Each combination results in a slightly different grid.

8.5.4   Four Class I Schemas

Two approaches for defining PPT edge reference points—Equal-chords and Equal-arcs—have been described, along with a general technique for defining points on a sphere at the intersection of great circles. It is now time to combine these techniques and subdivide the PPT's face. Figure 8.10 shows how subdivision grid points result from different combinations of intersecting great circles. Each grid point will be a little different, depending on the choice of Equal-chords or Equal-arcs references and the intersection of great circles. An octahedron PPT is properly oriented with 4^v reference points located along its edges. The PPT in Figure 8.10(a) uses Equal-chord reference points along its edges, and defines a new point within the PPT's face by intersecting three great circles. The planes of the three great circles are approximately parallel to a PPT edge.

In Figure 8.10(a), notice that all three great circles based on Equal-chord edge reference points intersect at a single point. This means we could have intersected any two of them, and their intersection would have created the same tessellation point. We show three to emphasize the fact that for Equal-chord reference points, all three will intersect at the same point. For reference purposes, we will call this subdivision schema Equal-chords. Figure 8.10(b) and (c) shows two more schemas; both are based on Equal-arc reference points. In (b), a grid point is defined by intersecting just two great circles. Two will always intersect and define a point within the PPT. As before, the planes of the great circles run approximately parallel to the PPT's edges.

Intersecting two great circles is certainly easier than intersecting three of them, but this simpler approach has one subtle, but significant disadvantage—the tessellation

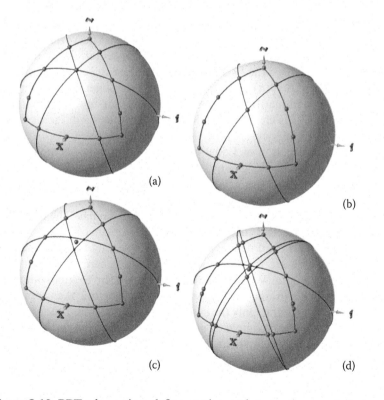

Figure 8.10. PPT edge points define vertices at intersecting great circles.

vertex created is not rotationally symmetric around the center of the PPT. This asymmetry is too subtle to appear in Figure 8.10(b) with only one vertex, but it will become quite noticeable when we subdivide the whole PPT. This schema demonstrates how asymmetry occurs, even when all other references are symmetrical. There are applications where grid symmetry is not a concern and this technique is useful. We call this schema Equal-arcs (two great circles).

Despite the asymmetry problem just uncovered, Equal-arc reference points yield superior results when we intersect three great circles, rather than just two. Figure 8.10(c) illustrates a surprising example of such situations. Unlike the great circles in the Equal-chords approach, three great circles based on Equal-arc reference points do not intersect in a single line. They intersect in three lines instead, creating three pairs of antipodal points, three of which are on the PPT's face. These three points (brown) form a small triangle, which Duncan Stuart aptly called a *window*.[13] The planar version of the small window triangle is just inside the sphere; only its three vertices are on the surface.

Although the window appears to be a small equilateral triangle, most windows are acute triangles. We can use the triangle's centroid, a single point, to summarize the three window vertex points. We use the centroid because if the window was continuously scaled down until it was so small it approximated a single point, and if the ratio of its sides were preserved while doing this, it would get smaller and smaller and converge to its centroid. Thus, the centroid—the intersection point within the triangle of its three edge bisectors—is the single point that best summarizes the window's *apices*. The centroid, however, is not on the surface of the sphere, it is on the plane of the window triangle and must be projected to the surface before we can use it as a tessellation grid point. Figure 8.10(c) shows the final tessellation vertex point as a red vertex centered over the window. Although more computation is involved in this schema, it is well worth the effort. The vertices are uniformly distributed, rotationally symmetric about the PPT's center, and mirror symmetric around the altitude axis of the PPT's vertices. We call this schema Equal-arcs (three great circles). Figure 8.10(d) superimposes all three schemas to demonstrate the influence various combinations of edge reference points and great circles have on defining grid points. It is clear that each schema produces a different grid point for the same PPT and frequency.

There is another variation of Equal-arcs. It simply subdivides the equilateral PPT into four smaller spherical triangles by spanning great circle arcs between the mid-edge points of the PPT. We call this subdivision schema Mid-arcs and we will develop it later in the chapter.

We have outlined four different Class I tessellation schemas. The first defines vertices by intersecting three great circles based on selected pairs of Equal-chord reference points. The second and third define vertices by intersecting two and then three great circles based on Equal-arc reference points. The last schema is a recursive triangle within a triangle technique. We are now ready to apply them to subdividing a complete spherical icosahedron. Although our examples are icosahedral, they apply equally well to tetrahedra and octahedra. Only the characteristics (angles and coordinates) of their PPTs change. We will show completed examples of them at the end of the Class I section, though we do not develop their grids.

---

[13] The term *"window"* was first coined by Duncan Stuart in a technical report in 1952 but it did not pass into general usage until Joseph Clinton (1971, I-24), Shoji Sadao, Magnus Wenninger (1979, 93–95), and others popularized it in the literature.

8.5.5 Equal-Chords

The Equal-chords schema, as the name implies, starts with Equal-chord reference points along the PPT's edge. The Equal-chords approach results in unequal arc lengths between edge reference points, when they are projected to the surface of the sphere. This is just noticeable along the PPT's edges in Figure 8.11(a) and (b). Pairs of reference points (and the origin) define great circle planes that are approximately parallel to the PPT's edges. Selected great circles intersections are shown in Figure 8.11(a) and their resulting tessellation vertices in (b). For example, great circle arcs $a1$-$b1$, $a6$-$c2$, $b3$-$c3$ intersect at a single line, creating two antipodal points. The ones that fall on the surface of the PPT are selected and shown in (b). Two other example vertices show how the process repeats itself. This process continues until every edge reference point combination has defined a PPT grid point. The completed grid, shown in Figure 8.11(b), is quite uniform. It is rotationally symmetric about the PPT's center, its apices, and its mid-edge points. The grid is also mirror symmetric about the PPT's three altitudes. All grid points are on the surface of the sphere and define either planar or spherical triangles.

A great number of surface treatments are possible. In Figure 8.12(a), grid triangles are shown as simple planar surfaces. Panels are suggested in (b) with holes centered at each triangle's incenter. Their radius of the incircle is 80% of the incenter-to-edge distance; thus, each hole is equidistant to its triangle's three edges. Holes make it easier to compare face sizes and, to some extent, their shapes. An equilateral face would have a hole perfectly centered inside it, while the hole is more off-center if the triangle is skewed. Holes also tell us something about triangle sizes. The largest triangle in this schema is the equilateral triangle at the center of the PPT, while the smallest are found next to the three PPT apices; these four are highlighted in (b). Barely noticeable, the triangles along the edge increase in size slightly, as they approach midedge and move towards the center of the PPT. We will see this pattern in other schemas as well. We will take a closer look at face sizes and shapes in the next chapter.

The subdivided PPT, shown in Figure 8.12, is only one face of the spherical polyhedron. In Figure 8.13 the PPT has been replicated to cover the rest of the sphere. The PPT just developed is highlighted, and the total parts on the sphere are listed.

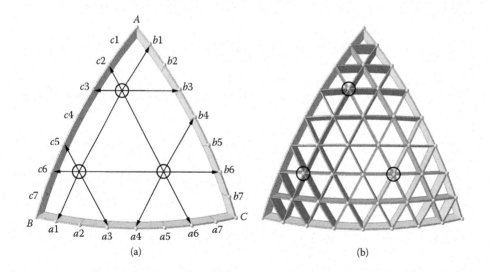

(a)   (b)

Figure 8.11. Equal-chords tessellation schema.

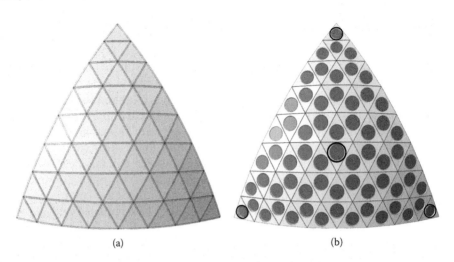

(a)   (b)

Figure 8.12. Equal-chords tessellated PPT.

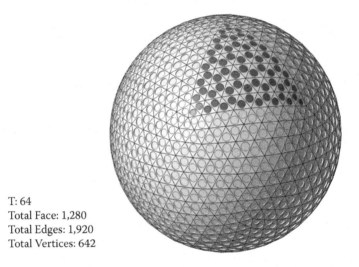

T: 64
Total Face: 1,280
Total Edges: 1,920
Total Vertices: 642

Figure 8.13. Equal-chords tessellated sphere.

The sphere is covered by rotating copies of the subdivided PPT to the other 19 face positions on the spherical icosahedron. We describe how this is done in Appendix B.

8.5.6   Equal-Arcs (Two Great Circles)

We have said that two non-tangent great circles always intersect at two antipodal points. When we use points along the edge of the PPT to define those intersecting great circles, one of the intersection points is always on the surface of the PPT. In Equal-chords, we used three great circle intersections to define one tessellation vertex. We could have created the same tessellation by intersecting just two great circles. However, when Equal-arc reference points along the PPT's edge are used rather than Equal-chord reference points, the great circles intersect in unexpected ways. We will show this by first defining grid points with just two great circle intersections, and then with three. The differences between the two sets of points are surprising.

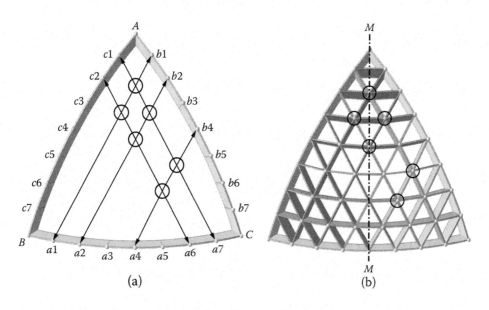

(a)

(b)

Figure 8.14. Equal-arcs (two great circles) schema.

Figure 8.14 shows the icosahedron PPT with Equal-arc reference points along the edges. Their uniform spacing is evident. In this schema, a grid point on the face of the PPT is defined where two great circle arcs intersect. Figure 8.14(a) shows examples of intersecting great circles: $c1$ -$a7$, $c2$-$a6$ with $a1$-$b1$, $a2$-$b2$, $a4$-$b4$.

Using just two great circles is an attractive option. The intersection is easy to calculate, and is a straightforward procedure. We repeat the process for each pair of great circles between reference points along edges c-a and b-a. As simple as this process is, there is an unexpected consequence, however, when you only use two great circles to define a grid point. The edge reference points in Figure 8.14(b) are labeled the same as (a). Notice that the grid points closest to edge a are closer to the edge than the corresponding grid points are to edges b and c. Every horizontal row of grid points seems somewhat displaced towards the bottom of the PPT. In fact, they are.

The biggest consequence of intersecting just two great circles (based on Equal-arc reference points) is that the resulting grid points are not rotationally symmetric about the PPT's center. They are not symmetric about the PPT's edges or apices, either. There is only one axis where they are mirror symmetric, and that is the one PPT altitude line marked M-M in Figure 8.14(b). These symmetry artifacts are a direct result of using just two great circles to define a vertex. For Equal-chord reference points, three great circles will intersect at the same point as two great circles, but not when Equal-arc reference points are used. For some applications, Equal-arcs (two great circles) asymmetry is not an issue, but for others this schema will not serve.[14]

Figure 8.15 shows the surface treatments for a fully tessellated Equal-arcs (two great circles) PPT. They are the same ones used in Equal-chords. However, this time, the panel holes tell a different story. Just like the previous subdivision schema, each hole's radius is 80% of the distance between the incenter of the triangle and the closest edge of its triangle. First, the largest and smallest triangles are no longer in the center

---

[14] Rodolfo Aguilar made excellent use of this schema in "*Study of the Stability of Framed, Triangulated Geodesic Domes under the Action of Concentrated Loads*" (Aguilar 1964).

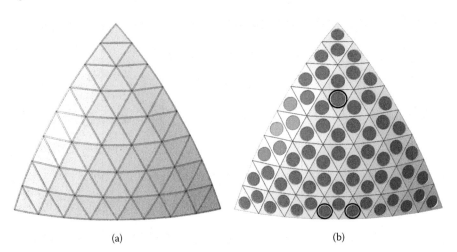

(a)   (b)

Figure 8.15. Equal-arcs (two great circles) tessellated PPT.

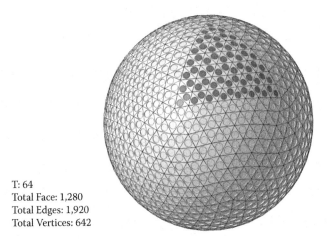

T: 64
Total Face: 1,280
Total Edges: 1,920
Total Vertices: 642

Figure 8.16. Equal-arcs (two great circles) tessellated sphere.

or at the PPT's apices, as they were in Equal-chords. The largest triangle now is above center, and the smallest ones are the two triangles along the PPT's bottom edge. They are circled in Figure 8.15(b).

Figure 8.16 shows a fully tessellated Equal-arcs (two great circles) sphere, along with the total part counts. This schema creates the same number of parts as Equal-chords. Although it appears quite uniform at this scale, an analysis of face size in the next chapter will show it is not as uniform as the previous Equal-chords schema. If rotational symmetry is not a requirement, Equal-arcs (two great circles) is an acceptable schema.

8.5.7 Equal-Arcs (Three Great Circles)

The previous example cast some doubts on the advantages of using Equal-arc reference points to define great circle intersections. However, when three intersecting great circles are used to define grid points instead of two, the results are outstanding, yielding the most uniform tessellation gird of the three schemas thus far. Like the previous Equal-arcs (two great circles) schema, this one is not without its surprises, but they are easy to deal with.

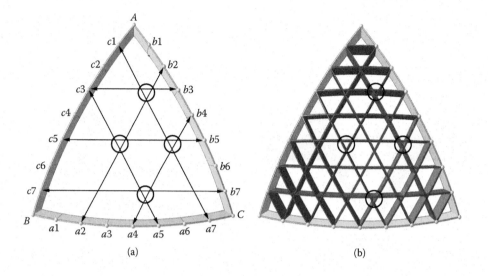

Figure 8.17. Equal-arcs (three great circles) schema.

In this schema, Equal-arc reference points along the PPT's edges are selected so that pairs of them define three great circle arcs approximately parallel to their respective PPT edge. Unlike Equal-chords schemas where great circles reference Equal-chord points, those based on Equal-arc references do not intersect at a single line. Each pair of great circles intersects at a different line and defines its own pair of antipodal points. With three intersecting great circles, six antipodal points are created and three of them (one from each antipodal pair) fall on the surface of the PPT. Figure 8.17(a) illustrates the result.

In Figure 8.17(a), examples of four three-way intersections are circled. Taking just one of these examples, the great circle arcs between $c3$-$b3$, $c1$-$a7$, and $b2$-$a2$ intersect at three different places, creating three different points within the PPT. These points are on the surface of the sphere but they form a small window triangle rather than the single tessellation vertex we need. Figure 8.17(b) shows the 21 window triangles that result when the entire PPT is subdivided. Notice that the windows become larger and larger towards the center of the PPT; the smallest ones are the ones nearest the PPT's three apices.

Each window is actually a small triangle whose planar surface is just inside the sphere; only the apices of the window triangle are actually on the surface of the sphere. Figure 8.18(a) shows a detail of the window and its vertices. The red arcs represent the great circle arcs, and the small balls represent their intersection points. What is needed is a single point that summarizes the three window vertices. The triangle's centroid is the most appropriate choice because if the triangle were continuously scaled down while maintaining the proportions of its sides, its geometry would converge at a single point at its centroid. The centroid of a triangle is defined as the intersection of its medians. The average of its three vertex coordinates produces the same point we call C. To find the window's centroid point:

$$Cx = \frac{(x_1 + x_2 + x_3)}{3},$$

$$Cy = \frac{(y_1 + y_2 + y_3)}{3},$$

$$Cz = \frac{(z_1 + z_2 + z_3)}{3}.$$

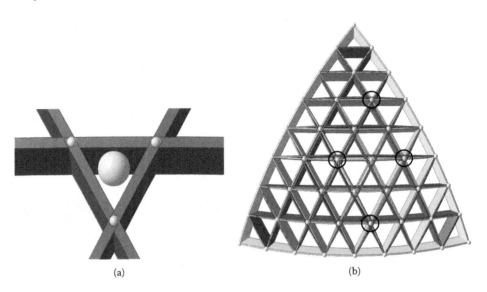

(a)   (b)

Figure 8.18. Equal-arcs windows and tessellation vertices.

We are almost done; only one small detail remains. The window's centroid is on the plane of the triangle, which is inside the sphere. The point must be projected from the origin to the surface of the sphere to create the final tessellation vertex we need. This is a simple matter of normalizing—creating a unit vector length—for the point. Appendix C explains this simple procedure. Figure 8.18(b) shows the tessellation vertices that result when all the centroids of the window triangles are found and projected to the sphere's surface. The original great circle arcs are shown in red, the final grid in blue-green. This figure highlights intersections that were used as examples in previous figures.

Figure 8.19 shows the completed PPT with the surface and panel treatments we have been using. The results are highly uniform, and are the best yet. The final grid resembles Equal-chords to some extent. The largest and smallest triangles are in the same positions as Equal-chords. The mid-PPT face triangle is also equilateral, and just

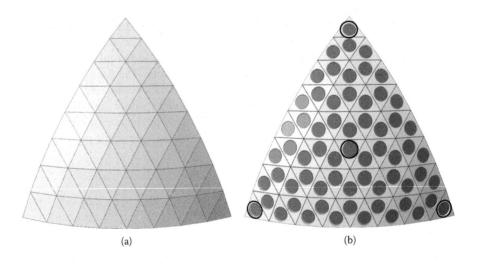

(a)   (b)

Figure 8.19. Equal-arcs (three great circles) tessellated PPT.

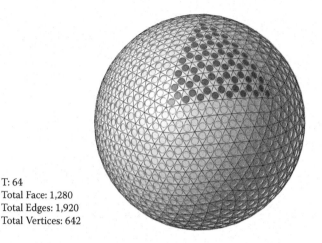

T: 64
Total Face: 1,280
Total Edges: 1,920
Total Vertices: 642

Figure 8.20. Equal-arcs (three great circles) tessellated sphere.

as in Equal-chords, triangle areas increase as you move from the PPT apex to midedge, and from midedge to midface. The main differences between the Equal-chords and Equal-arcs (three great circles) schema are chord lengths and face angles. We compare the two schemas in the next chapter.

Figure 8.20 shows a complete sphere based on the PPT just created. Notice that its part counts are the same as the other schemas. The subdivision method did not change this fact, since the schemas all have the same b,c specification. Certainly, Equal-arcs (three great circles) involves more computation than either of the previous two schemas, but the results are worthwhile. Analysis will show that it has the best distribution of points and the least variation in chord lengths, angles, and face angles of any Class I schema.

8.5.8 Mid-Arcs

The earliest work with Mid-arcs was done by Jeffrey Lindsay in the early 1950s. At that time, the technique was referred to as the alternate method. We call the technique Mid-arcs because it is more descriptive of how the technique works. Deltahedra spherical faces are all equilateral. Mid-arcs takes advantage of this fact and subdivides a PPT by creating one Mid-arc reference point on each of its edges. Each pair of Mid-arc points defines a great circle arc across the triangle and these three arcs subdivide the triangle into four smaller ones. There is no need to intersect great circles. We use the same method of defining four new triangles within an existing one until we have subdivided the entire PPT. Each time we subdivide all the triangles in a PPT this way, we double the PPT's frequency. Mid-arcs is very easy to understand and to calculate. It is popular with cartographers and astrono-mers[15] who have devised ways to reference individual triangles on the sphere by taking advantage of the schema's hierarchical nesting of triangles within triangles.[16] Often, they subdivide a spherical octahedron rather than an icosahedron because the octahedron results in natural great circles that can be aligned with the Earth's equator and Greenwich Meridian.

Figure 8.21 shows how Mid-arcs work. The pattern of triangles within triangles is evident. Each successive layer of triangles is based on the Mid-arc points of an outer triangle. This recursive nesting of triangles within triangles is the essence of Mid-arcs.[17]

---

[15] (Dutton 1996, 1991) and (Szalay et al. 2005).

[16] (Fekete 1990).

[17] Cartographers sometimes use quaternary subdivision meshing schemas that hierarchically subdivide equilateral triangles similar to Mid-arcs (Dutton 1996). See also Section 3.3.

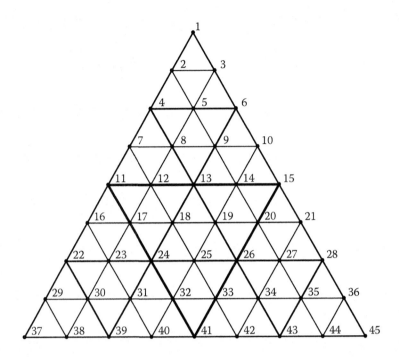

Figure 8.21. Mid-arcs vertex grid reference numbers.

Figure 8.22 applies the schema to the same icosahedral PPT we have been using all along. We number the vertices from top to bottom, left to right so that Figures 8.21 and 8.22 can be compared. As usual, subdivision begins at the PPT edges, where each has a single Mid-arc reference point.[18] In Figure 21(a), they are numbered 11, 15, and 41. These points can be found by projecting the mid-chord point between 1 and 37 to the surface of the sphere or finding the middle of the arc between the PPT's apices. Either way, great circle arcs between these points define four new subtriangles (1-11-15, 11-37-41, 11-41-15, and 15-41-45), each of which has Mid-arc points (4, 6, 13, 22, 24, 26, 28, 39, and 43); see Figure 8.22(b). These new Mid-arc points define 12 more edges and 30 new vertex points when all edges are considered (c). This process repeats itself once again until a total of 42 tessellation vertices and 108 edges are defined. Figure 8.22(d) shows the complete tessellation. Keep in mind that all of these vertices and edges were derived from the apices of the original PPT and the Mid-arc reference point along each edge. Calculations have to be precise; otherwise, errors will accumulate. Our example stops at 8^v, but we could keep on subdividing each triangle indefinitely, doubling the frequency each time.

The Mid-arcs schema is straightforward and easy to calculate. However, there are some limitations. Unlike the other Class I schemas, Mid-arcs can only tessellate frequencies that are powers of 2 (i.e., 2^v, 4^v, 8^v, 16^v, 32^v, etc.). Any even or odd frequency is possible with the other Class I schemas, but only even ones are possible with Mid-arcs. The second drawback concerns the tessellation faces themselves. Despite the fact that we start with an equilateral PPT face and place reference points only in the middle of each of its edges, the vast majority of subdivision triangles created in the final tessellation are not equilateral. Only those faces whose centroids are tangent to the PPT's own

---

[18] The edge midpoints of the icosahedron (or of the dodecahedron) are the vertices of the icosidodecahedron (20 triangular faces and 12 pentagonal faces). See Chapter 6 for examples.

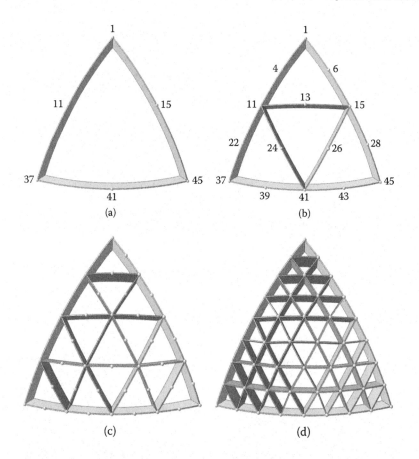

Figure 8.22. Mid-arcs tessellation schema.

centroid are equilateral; all others are slightly isosceles or acute. In this example, only triangles 1-37-45, 11-15-41, 13-24-26, and 18-25-19 are equilateral.

A few numbers illustrate this last point. The icosahedron's PPT edges are each 63.43495°. The first new Mid-arc triangle (11-15-41) is also equilateral with 36° edges (its centroid is tangent to the PPT's centroid), but the other three triangles created (1-11-15, 11-27-41, and 15-41-45) are slightly isosceles. Two of their edges are 31.71747°, and the other edge shared with the mid-PPT triangle is 36°. Although other Class I schemas have this same characteristic, Mid-arcs produce more variation in chord length and triangle area than others do. We will see this later in the next chapter. Figure 8.23 shows a full sphere subdivided by the Mid-arcs technique along with its part counts. The PPT we subdivided is highlighted.

Figure 8.24 illustrates another disadvantage of Mid-arcs tessellation (left) by comparing its grid side by side with an Equal-arcs (three great circles) tessellation for the same PPT. Although the PPT edge reference points are the same in both schemas, Mid-arcs faces increase in size more quickly as they approach the center of the PPT, and the chords change length faster as they approach the middle of the PPT. As a result, sequences of chords that cross the PPT are not as smooth in transition as they are with Equal-arcs. An example run of chords showing this difference is circled. Figure 8.25 makes the same comparison, but uses tetrahedral PPTs instead. Tetrahedron PPTs are four times larger than an icosahedron's face (not to scale in the figure), and this makes it easier to see the differences between the two schemas. The first-level subdivision of

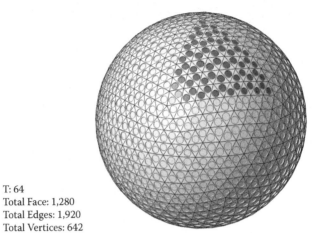

T: 64
Total Face: 1,280
Total Edges: 1,920
Total Vertices: 642

Figure 8.23. Mid-arcs tessellated sphere.

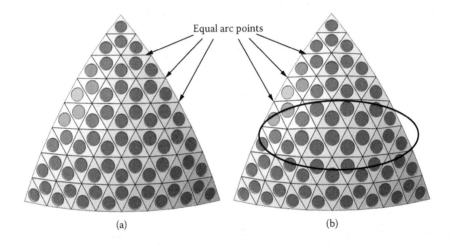

(a)   (b)

Figure 8.24. Icosahedral Equal-arcs versus Mid-arcs PPT tessellations.

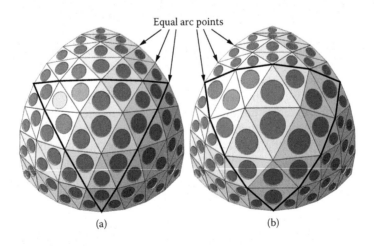

(a)   (b)

Figure 8.25. Tetrahedral Equal-arcs versus Mid-arcs PPT tessellations.

the PPT is highlighted in each illustration, and the differences are pronounced. Mid-arcs central PPT face is noticeably larger and the faces near the PPT apices are smaller than they are in Equal-arcs tessellations. In Figure 8.25(b) the tetrahedron's central equilateral triangle (highlighted) has three times as much area as the ones at each apex. The tetrahedron's equilateral faces have significantly more spherical excess than the icosahedron's and highlight these differences.

Mid-arcs is the simplest subdivision technique presented thus far. The fact that frequencies are always powers of 2 and the repetitive definition of triangles within triangles offers a natural hierarchy that can be exploited to reference individual triangles on the sphere.[19] Mid-arcs also demonstrates a basic spherical principle—there are no similar triangles on a sphere. You cannot subdivide a spherical equilateral triangle around its mid-edge points and define four other equilaterals as you can in plane geometry.

8.5.9 Subdividing Other Deltahedra

We have developed four different Class I schemas in this chapter. Although each example subdivides a spherical icosahedron, the same methodology applies to tetrahedra and octahedra. The only difference is the initial PPT's characteristics (vertex angle, edge arc length, and vertex coordinates) used to start the process. Use the values in Table 8.2 to subdivide the others.

Figure 8.26 shows examples of 8^v Equal-arc schemas for the tetrahedron and octahedron (geodesic notation $\{3,3+\}_{0,8}$ and $\{3,4+\}_{0,8}$, respectively). They are shown in a partially disassembled state and with mixed surface treatments. Both of these polyhedra have larger PPT areas than the icosahedron for the same radius sphere. As a result, their tessellation triangles show much more variation in tessellation triangle area and

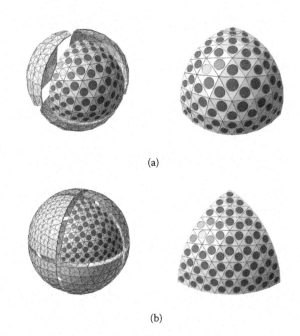

(a)

(b)

Figure 8.26. Equal-arc (three great circles) tessellations of tetrahedra and octahedra.

---

[19] (Fekete 1990).

shape, as the incenter circles clearly show. The tetrahedron shows the greatest variation of the two. These examples clearly demonstrate how important the choice of foundation polyhedra is when the goal is to create the most uniform subdivision possible. It is no wonder the icosahedron is the most frequently used polyhedral base.

8.5.10 Summary

Class I schemas were among the first subdivision methods developed. They apply to any deltahedra. Except for Mid-arcs, which is limited to frequencies that are powers of 2, any even or odd frequency is possible and, depending on prime factors, may contain subgrids at harmonic frequencies. Class I subdivisions maintain the polyhedron's edge as part of the final tessellation grid, though the edge is segmented. The geodesic notation for Class I is distinctive: either the b or c coefficient is always 0.

Class I tessellations begin with Equal-chords or Equal-arcs reference points along the PPT's edges. Pairs of points (and the origin) define intersecting great circles that are approximately parallel to the PPT's edges. Grid points result where their line of intersection pierces the sphere's surface:

- *Equal-chords.* Based on Equal-chord reference points, great circles parallel to the PPT's edges are intersected and their intersection points define points on the tessellation grid within the PPT's surface. Two intersect at the same line as three;

- *Equal-arcs (two great circles).* Based on equal arc reference points, two great circles are intersected. They intersect at a single line and define individual tessellation points on the PPT's surface. This procedure is computationally simple, but the resulting tessellation vertices are not rotationally symmetric and only mirror symmetric around one of the PPT's altitude axis;

- *Equal-arcs (three great circles).* Based on equal arc reference points. Three great circles are intersected, but they do not meet at a single line. Instead, they meet at three different lines, creating three points on the PPT. A small window triangle results and the window's centroid is projected to the surface of the sphere. The resulting grid is rotationally symmetric around the center of the PPT, and they are mirror symmetric around the PPT's three altitudes. Analysis will show that Equal-arcs, when developed with three great circles, produces a highly uniform grid, the best of the Class I schemas presented;

- *Mid-arcs.* Grids are generated by using the Mid-arc points on a triangle's three edges to define four others within. The technique is simple to calculate, rotationally and reflection symmetric, and applicable to all deltahedra. The hierarchical nature of triangles nested within triangles offers convenient referencing systems for individual triangles. The grid is quite uniform, but not as good as Equal-arcs (three great circles).

All four schemas result in a three-way grid over the PPT, where arcs and lines approximately parallel the PPT's edges. If the triangulation number is evenly divisible by three, selected grid points can define hexagons, pentagons, or diamonds. The 8^v examples in this chapter have a triangulation number of 64, which is *not* evenly divisible by three; thus, grid points will not define uniform patterns of these other polygons. However, a wide variety of surface treatments is possible. The examples shown here are simple surfaces and panels with incenter holes.

8.6   Class II: Triacon

Class II subdivisions create grids that are perpendicular to the edges of the spherical polyhedra, rather than parallel as with Class I grids. Its earliest use was in the Weatherbreak Dome in Montreal described in 2.8. Figure 8.27 shows examples of Class II tessellations. Notice that the tessellation grid is symmetrical to the polyhedron's edges and face medians. This means that Class II tessellations are always even frequencies. For this reason, Class II geodesic notation is easily recognized: b always equals c, and neither is equal to zero. Notice, too, that selective use of grid points produces triangles, diamonds, hexagons, or pentagons. In this section, we will continue to use a spherical icosahedron 8^v as our foundation polyhedron because we want to compare our results with the other schemas; its geodesic notation will be $\{3,5\}_{4,4}$.

There are many ways to create Class II tessellations. The schema developed here is called *triacon* and was developed by Duncan Stuart in 1952 to make it easier to prefabricate geodesic domes by reducing the number of different parts needed.[20] In the next chapter we will see just how innovative Stuart's schema is and why it is widely used today.

The triacon is unique. It achieves a remarkably uniform distribution of points over the sphere and does so with the fewest number of different chord/arc lengths, face shapes, and face areas of any subdivision method for a given frequency. These are key traits when spheres are manufactured or assembled from piece parts. This accounts for the triacon's popularity with dome builders and why most large geodesic structures still use the triacon method. The triacon is based on the subdivision of a single Schwarz triangle, rather than an equilateral triangle as with Class I. When the Schwarz triangle is subdivided and the resulting geometry mirrored two times, it forms a diamond shape, which is positioned symmetric to the polyhedron's edge.[21] The fact

Figure 8.27. Examples of class II subdivision schemas.

---

[20] (Stuart 1952).

[21] Triacon diamonds for an icosahedron resemble the faces of the rhombic triacontahedron polyhedron, the likely source for this nickname.

that triacon is based on a diamond shape that is edge-symmetric to the polyhedron gives us an advantage; triacon can subdivide *any* of the five spherical Platonic solids, not just deltahedra. Another advantage of triacon is its triangulation numbers are always evenly divisible by three. This means its triangular tessellation grid can produce regular layouts of triangles, diamonds, hexagons, and pentagons, as we see in Figure 8.27. From back to front: tetrahedron 16^v, dodecahedron 8^v, icosahedron 10^v, octahedron 22^v, and cube 8^v.[22] The triacon does not create grid members coincident to the polyhedron's edges like Class I grids do. Instead, the polyhedron's edges are simply geometric references used to compute the grid. In Figure 8.27, the polyhedron's edges are shown only to highlight the underlying polyhedron and point out that selected grid points along the edge and face medians also define hemispheric great circles. They can be useful in certain geodesic applications. Class II grids always result in a grid point in the middle of the polyhedron's face. This is not always the case with Class I schemas, and only occasionally happens in Class III schemas. This can be a useful feature in some spherical applications.

8.6.1 Schwarz LCD Triangles

A finite number of Schwarz triangles cover a sphere without overlap or gaps. On a spherical icosahedron, a single Schwarz triangle covers one-sixth of a PPT's face or 120th of the total spherical icosahedron surface; 60 of them are left-handed and 60 are right-handed. There are 24 Schwarz triangles on a spherical tetrahedron, 48 on a cube, 48 on an octahedron, and 120 on a dodecahedron. The Schwarz triangle is the Least Common Denominator (LCD) triangle we will subdivide.

We begin subdivision by orienting the spherical icosahedron edge-zenith and an edge coincident with the *xz*-plane. Figure 8.28(a) shows an oblique view of the

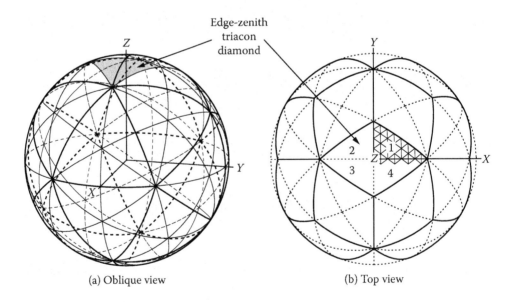

(a) Oblique view   (b) Top view

Figure 8.28. Edge-zenith triacon diamond.

---

[22] From back to front, the geodesic notations are: $\{3,3+\}_{12,12}$, $\{3,5+\}_{4,4}$, $\{3,5+\}_{5,5}$, and $\{3,4+\}_{11,11}$.

edge-zenith diamond. A top view of the same edge-zenith diamond is shown in (b) where its four quadrants are numbered 1 to 4. A partial triacon grid is developed in quadrant 1 of the edge-zenith diamond. Orienting the spherical Platonic edge-zenith with the diamond axis coincident to the xz-plane makes calculations easier. First, the LCD triangle is a right triangle and easily solved with spherical trigonometry. Second, the grid coordinates we calculate for quadrant 1 are all positive and can be mirrored to the other three diamond quadrants, simply by changing their x- and/or y-coordinate sign; their z values never change. For example, changing only the sign of the x-coordinate, a point in quadrant 1 is reflected to a symmetrical position in quadrant 2. Changing the y sign reflects points in quadrants 1 and 2 to quadrants 4 and 2, respectively. This is a pretty good return on our subdivision effort. Calculate one quadrant, and the others will fall into place if you simply change the signs of various x- and y-coordinates.

8.6.2   How Frequent

Class II schemas create grids that are perpendicular to the polyhedron's edges, and the grid is symmetrical around the face's medians. Because of this symmetry, Class II frequencies are always even and their geodesic notations b,c are always positive and equal integers. If a Class II frequency is greater than 2^v, it may contain harmonics or subgrids just like Class I grids. Figure 8.29(a) shows a sphere with a series of triacon diamonds for 2^v, 4^v, 8^v, 16^v, and 32^v in clockwise order. Each diamond is made of four LCDs and the icosahedron's edges are shown in the figure as dashed lines. Note that the 32^v tessellation contains all the vertices of the others grids; thus, they are all harmonics of the 32^v grid. Figure 8.29(b) shows a single 16^v triacon diamond with different line weights for the 2^v, 4^v, and 8^v harmonics it contains.

To determine if a tessellation has harmonics, start by listing its prime factors. The tessellation will have a harmonic for even products of prime factor pairs. Table 8.3 lists the harmonics for frequencies 2^v through 30^v. Notice that if you create geometric information for 16^v through 30^v tessellations, it will contain the same information for 2^v though 14^v tessellations, since these grids are also contained in the others.

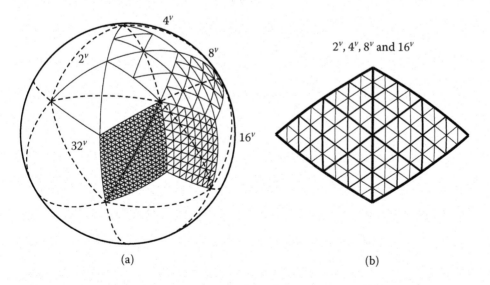

(a)   (b)

Figure 8.29. Triacon diamond subdivision harmonics.

v	Prime Factors	Harmonic Frequencies				
2	2 (Prime)					
4	2,2	2^v				
6	2,3	2^v				
8	2,2,2	2^v	4^v			
10	2,5	2^v				
12	2,2,3	2^v	4^v	6^v		
14	2,7	2^v				
16	2,2,2,2	2^v	4^v	8^v		
18	2,3,3	2^v	6^v			
20	2,2,5	2^v	4^v	10^v		
22	2,11	2^v				
24	2,2,2,3	2^v	4^v	6^v	8^v	12^v
26	2,13	2^v				
28	2,2,7	2^v	4^v	14^v		
30	2,3,5	2^v	6^v	10^v		

Table 8.3. Triacon frequency harmonics.

8.6.3   A Quick Overview

The triacon subdivision method is one of the easiest to implement. Figure 8.30 shows the sequence in a series of top views (from the positive z-axis) of the LCD and evolving triacon diamond. The basic procedure is as follows:

- Orient the spherical Platonic solid (an icosahedron in our examples) with an edge at the zenith and coincident to the xz-plane, as shown in Figure 8.30(a). Four LCDs, two left- and two right-handed, form a diamond that is symmetric to the zenith-edge;

- Subdivide the LCD in quadrant 1 of the edge-zenith diamond, Figure 8.30(b). The LCD is a spherical right triangle, and the bases of the smaller right triangles are multiples of Equal-arc subdivisions of the icosahedron's edge. Simple trigonometric relationships define every tessellation grid member; compute the coordinates of each of the tessellation vertices in the LCD, shown in Figure 8.30(b), and mirror the grid point coordinates to quadrants 2, 3, and 4 by changing the appropriate x or y signs of quadrant 1 vertices. This completes the definitions for the LCD and the edge-zenith diamond, Figure 8.30(c);

- Tessellate the remainder of the sphere by rotating copies of the edge-zenith diamond to all other polyhedral edge positions, Figure 8.30(d). Spherical icosahedra and dodecahedra will require 29 replications/rotations. Octahedra and cubes require 11 and tetrahedra require 5. Appendix B describes how this is done.

8.6.4   Establish Your Rights

The first step in triacon subdivision is to define the spherical polyhedron's LCD. Figure 8.31 shows the LCDs covering the spherical icosahedron and the way we label

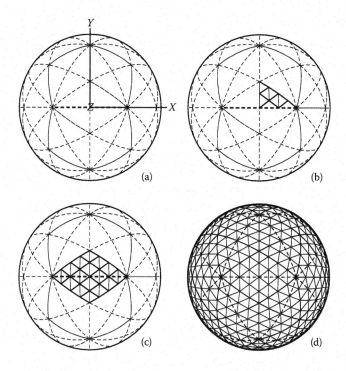

Figure 8.30. Triacon subdivision sequence.

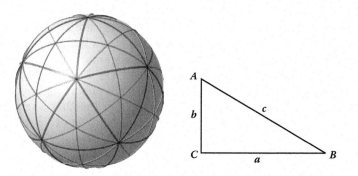

Figure 8.31. LCD right triangle.

the parts of a typical right triangle. Whether or not the LCD is left- or right-handed does not matter in our calculations. The first step is to find LCD sides a, b, and c, given its three surface angles. Referring to the LCD triangle labels in Figure 8.31, the basic relationships are:

$$\cos a = \cos A / \sin B,$$
$$\cos b = \cos B / \sin A, \quad \text{and}$$
$$\cos c = \cos a \times \cos b.$$

It's easy to establish the LCD's surface angles A, B, and C. Through symmetry and knowing that the sum of the surface angles around any point is always 360°, we know that $A = 60°$ and $B = 36°$, and, by definition, $C = 90°$. We now use the formulas in

Principal LCD Parts for Spherical Platonics				
Part	**Tetrahedron**	**Octahedron**	**Cube**	**Dodecahedron**	**Icosahedron**
Surface Angles					
A°	60.00000	60.00000	45.00000	36.00000	60.00000
B°	60.00000	45.00000	60.00000	60.00000	36.00000
C°	90.00000	90.00000	90.00000	90.00000	90.00000
Central Angles					
a°	54.7356	45.00000	35.26439	20.90517	31.71749
b°	54.7356	35.26439	45.00000	31.71749	20.90517
c°	70.52877	54.7351	54.73561	37.37738	37.37738
Arc Factors					
a	0.95531	0.7854	0.61548	0.36486	0.55257
b	0.95531	0.61548	0.7854	0.55357	0.36486
c	1.23096	0.95532	0.95531	0.65236	0.65236
Chord Factors					
a	0.9194	0.76537	0.60581	0.36284	0.54653
b	0.9194	0.60581	0.76537	0.54653	0.36284
c	1.1547	0.9194	0.9194	0.64089	0.64085
Full Edge	109.47118	90.00000	90.00000	41.81032	63.4335
Full Face Median	125.26434	90.00000	NA	69.09486	58.28254

Table 8.4. Principal LCD parts for spherical Platonic solids.

Figure 8.31 to find sides (central angles) *a*, *b*, and *c*. They are 31.7175°, 20.9052°, and 37.3774°, respectively. We only show a few decimal places here, but it is important to maintain the highest precision you can in your calculations.

We also need the arc and chord factors for the LCD's sides as well as the full PPT polyhedral edge and face meridian arc length. The arc factors for any of the LCD's sides are simply the radian value of their central angle, or $2\pi\theta/360$ where θ is the arc's central angle expressed in degrees. The chord factor between two points on the unit sphere is the straight-line distance between them and is simply $2r \sin(\theta/2)$ where r is the radius of the sphere and θ is the central angle between the two points. Given LCD sides, the polyhedron's edge and face medians are simply

$$\text{PPT polyhedron edge} = 2a, \quad \text{and}$$
$$\text{PPT face median} = b + c.$$

Table 8.4 summarizes the principal LCD parts for the icosahedron and the other four spherical Platonic solids. Angles are shown in degrees and arc/chord factors are for unit radius spheres.

8.6.5 Subdividing the LCD

The triacon schema subdivides the LCD by nesting a series of right spherical triangles within it. Figure 8.32 shows each step for an eight-frequency grid.

- *Step 1.* Define edge perpendiculars. Divide half the polyhedron's edge (arc) into $^{v}/2$ segments and drop perpendiculars to side a between every edge segment. The result is $^{v}/2$ nested right triangles, all sharing angle *B*. In this example,

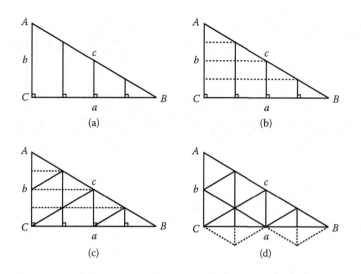

Figure 8.32. Triacon LCD tessellation sequence.

there are three nested right triangles within the overall LCD right triangle. We will label their angles and sides the same way we labeled the overall LCD triangle. Side a of each nested right triangle is an integral number of polyhedron edge segments. In the illustration, it appears that each of the nested triangles has the same angle A, but this is not the case. There are no similar triangles on a sphere. Each nested right triangle has a slightly different value for A, but angles B and C remain constant at 36° and 90°, respectively. They are shown in Figure 8.32(a). Given a, B, and C for each nested triangle, we find parts A, b, and c using these relationships:

$$\cos A = \cos a \times \sin B,$$
$$\cos b = \cos B / \sin A,$$
$$\cos c = \cos a \times \cos b.$$

- *Step 2.* Subdivide LCD's side b. With all nested triangle parts solved, divide LCD's side b into arc segments equal to the b sides of each of the nested right triangles found in step 1. Each of the LCD side b segments is slightly different; there will be $^\nu/2$ segments in all. See Figure 8.32(b);

- *Step 3.* Strike diagonals. Triangulate the grid with diagonals by alternating nested triangle c arcs across the LCD. Each row of c arcs is the same length. The result is a series of triangles, each with their own a, b, and c arcs, as shown in Figure 8.32(c);

- *Step 4.* Complete the triangulation. Diagonals are constructed with nested triangle c arcs to alternate points, as shown in Figure 8.32(d). When completed, all interior tessellation vertices have three-way intersections of two diagonals and a perpendicular to side a. Sides a and c of the overall LCD triangle will be subdivided $^\nu/2$ times, while b is subdivided $^\nu/4$ times.

Step 1 solves all six parts (three angles, three sides) of each nested right triangle within the LCD. Steps two through four replicate the parts across the LCD to create

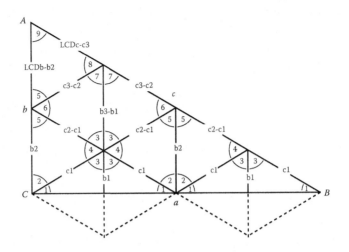

Figure 8.33. Triacon LCD tessellation 8^v summary.

the three-way grid. The elegance of the triacon schema lies in the fact that parts *A*, *b*, and *c* of the nested triangles are replicated across the LCD to define the angles and sides of the other triangles in their respective rows. Figure 8.33 shows how the replication works. Starting at the smallest nested right triangle, the one immediately next to LCD vertex *B*, notice how its angles 1 and 3 and side *b1* carry across the LCD to all other triangles in its row. Notice also that angle 2 is the complement of angle 1. Other angles are found this way, they are either *complements* or *supplements* of one or more nested triangle parts. For example, angle 4 is the supplement of two angle 3s. Some diagonal c arcs are found by factoring out c arcs from another nested triangle. The figure shows this easy step.

8.6.6 Grid Points

The triacon subdivision of the LCD has a special effect on the alignment of grid points within the LCD. Figure 8.34 shows that every grid point falls at the intersection of one great circle with one lesser circle, not two or three great circles shown in Class I schemas. Notice also that the planes of the lesser circles are parallel to the *xz*-plane of the zenith diamond (the great circle of the icosahedron's polyhedral edge), a result of replicating side *b* of each nested right triangle across the LCD during subdivision.

It is easy to find the Cartesian coordinates of grid points in the triacon diamond. Only two angles, *a* and *b,* are needed for each point and we already have their values when we solved the parts of each nested right triangle in the previous step:[23]

$$x = \sin a \times \cos b,$$
$$y = \sin b,$$
$$z = \cos a \times \cos b.$$

Angle *a* is measured along the *xy*-plane and is positive, if it is on the +*x*-axis side of the plane; otherwise, it is negative. Angle *b*, measured on the *yz*-plane, is positive, when it is on the +*y*-axis side and negative, if on the −*y* side. To find vertex coordinates for grid points in quadrants 2, 3, and 4, simply change the signs of the coordinates in

---

[23] These equations are similar to the general spherical coordinate system (ρ, φ, λ) described in Chapter 4.

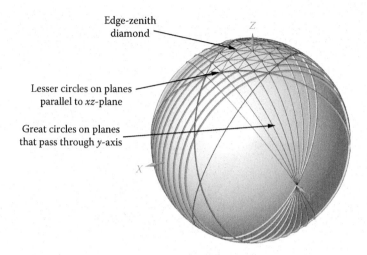

Figure 8.34. Triacon 8^v diamond great and lesser circle planes.

quadrant 1. The z-coordinate for every point in the edge-zenith diamond will always be positive. Appendix C shows how.

8.6.7 Completed Triacon

We have completely defined the triacon grid. Every triangle is solved and we have found the coordinates of every grid vertex. Figure 8.35 shows the triacon diamond and highlights the LCD in quadrant one (b). The overall diamond tessellation is quite uniform, and only four different triangles are needed (there are left- and right-handed versions of each).

Figure 8.36 shows a fully subdivided sphere and the part totals for an 8^v subdivision of an icosahedron. Notice that the total number of faces, edges, and vertices is far fewer than it is for any of the Class I schemas: 960, 1440, and 482, respectively, versus

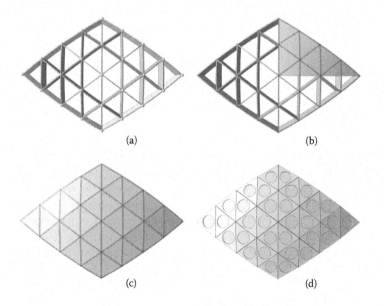

Figure 8.35. Triacon tessellation schema.

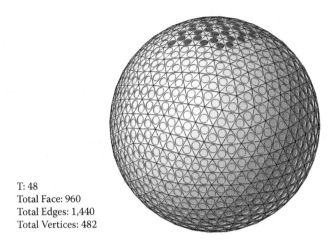

T: 48
Total Face: 960
Total Edges: 1,440
Total Vertices: 482

Figure 8.36. Triacon tessellated sphere.

1280, 1920, and 642. The completely subdivided sphere is the result of replicating the single tessellated diamond over the remaining 29 icosahedral edges. The same graphics conventions used before are used here. The initial triacon diamond in its edge-zenith position is highlighted.

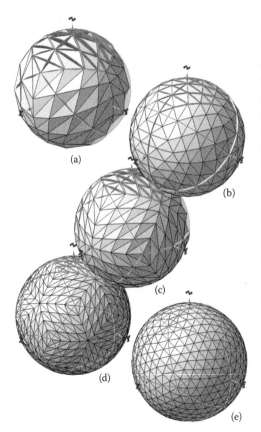

(a)

(b)

(c)

(d)

(e)

Figure 8.37. Triacon reference diamonds for Platonics.

8.6.8 Subdividing Other Polyhedra

The triacon subdivision is the only schema that can subdivide every Platonic solid. The triacon can do this because its basic subdivision unit is a diamond symmetrical with the polyhedron's edge. Figure 8.37 shows all five spherical Platonic solids with 8^v triacon tessellation. A single subdivided edge-zenith diamond and all polyhedral edges, shown in red, are superimposed on a fully tessellated sphere. From top down, left to right they are tetrahedron (a), octahedron (b), cube (c), dodecahedron (d), and icosahedron (e). The subdivision technique is identical in each case; only the initial LCD triangle characteristics differ.

Notice the difference in tessellation face orientations on each Platonic solid's grid. The octahedron and icosahedron, (b) and (e), have uniform orientations and appear almost like faceted spheres. By comparison, the tetrahedron's faces appear to be quite angular to one another. Also notice that, even at the small scale of these illustrations, it is clear that the icosahedron has the most uniform subdivision grid of the five solids. Since uniform face orientations and uniform grid are important in some geodesic applications, we will analyze them in the next chapter.

8.6.9   Summary

The Class II (triacon) subdivision method results in grids that are perpendicular to the edges of a spherical polyhedron's faces and symmetrical to face medians. As a result of this symmetry, frequencies are always even and tessellations will contain harmonics for the products of any of the frequency's prime factors that are not themselves prime frequencies. Triacon subdivisions have a distinctive geodesic notation: b,c specifications are always equal, and neither is ever zero.

The principal subdivision unit is a spherical right Schwarz triangle, also referred to as an LCD. Four LCD triangles, two left-handed and two right-handed, define a diamond that is symmetric to each edge of the polyhedron. Due to the edge symmetry of the triacon diamond, this subdivision schema can be applied to every spherical Platonic solid. The subdivision process is identical for any Platonic solid:

- Select the spherical Platonic solid and desired subdivision frequency;

- Orient one edge at the zenith and coincident with the xz-plane;

- Define the basic LCD triangle characteristics (angles and edges);

- Compute surface and interior angles for a series of right spherical triangles within the LCD;

- Compute arc and chord factors for the same;

- Replicate the b and c arcs across the LCD to complete the grid;

- Complete triangle angles and edges by finding complements and supplements;

- Establish the naming convention for tessellation vertices based on a naming hierarchy of the diamond reference, the quadrant within the diamond, and the row/column of the grid point;

- Compute xyz's of all vertices in the first diamond quadrant and mirror them to quadrants 2, 3, and 4 by changing the signs of their x- and y-components;

- Replicate the completed edge-zenith diamond to all other parts of the sphere.

It is worth mentioning that there are other Class II schemas. The triacon, developed by Duncan Stuart, divides the LCD's base (tangent to the polyhedral edge) into equal arc segments; see Figure 8.32(a). However, it is also possible to create Class II grids by evenly dividing side b or c of the LCD.[24] At first, you might think the results would be the same as triacon, but remember, there are no similar triangles on a sphere. Any tessellation based on Equal-arc subdivision of these other LCD sides will produce a different grid.

The triacon has been popular with geodesic dome builders because it minimizes the number of parts for any given frequency when compared to other class schemas. The triacon creates a very uniform distribution of vertices on the sphere, which results in the fewest number of different chords, triangles, areas, and shapes. A wide variety of face types—including triangles, diamonds, and hexagon-pentagons—is possible, because all Class II subdivisions have triangulation numbers, evenly divisible by three.

---

[24] (Clinton 1971, I-26–I-35).

8.7 Class III: Skew

Perhaps the most unusual spherical subdivisions are Class III subdivisions, or *skew* as they are sometimes called. Class III is quite different from the other classes. Tessellations are not parallel or perpendicular to the spherical polyhedron's edges; they cross at oblique angles, as shown in Figure 8.38. Because of this, their reflections are not symmetrical, either. Unlike Class I and II schemas, Class III schemas do not have convenient Equal-chord and Equal-arc reference points that can be used to start the subdivision process; thus, we will use a totally different technique to subdivide the PPT.

The method we will use here (there are others) is projection. We start by making a 2D triangular grid that meets the geodesic specification. This grid is transformed to three dimensions and is projected onto the surface of the PPT. This chapter describes how the initial 2D trigrid is defined, converted into a 3D grid, and then projected onto the surface of a PPT. Although triangular grids are developed here, the same technique can project a wide variety of nontriangular grids and designs onto the sphere. The projection technique can create Class I and II grids as well, though the results are not as optimal as the methods already presented for those classes. Optimization aside, projection is the most versatile technique we present.

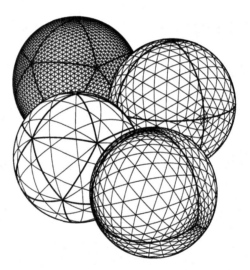

Figure 8.38. Examples of Class III subdivision schemas.

8.7.1 What Is Class III

Like the other class subdivisions, Class III subdivisions apply to spherical deltahedra—tetrahedra, octahedra, and icosahedra. Class III subdivisions are hybrids: their grids are oriented somewhere between Class I and II subdivisions. Tessellation frequencies can be even or odd, but their geodesic coefficients b,c are never equal to each other (Class II) or one of them equal to zero (Class I). Figure 8.38 shows some of the many Class III possibilities. From back to front they are icosahedron $\{3,5+\}_{12,8}$, octahedron $\{3,4+\}_{7,5}$, icosahedron $\{3,5+\}_{2,1}$, and tetrahedron $\{3,3+\}_{5,13}$. In the figure, the spherical polyhedron's edges are highlighted to show the tessellation orientation relative to the polyhedron's edges, but the polyhedron's edges are not part of the tessellation. Even without these accented edges, the polyhedral base can be identified by locating points where the grid valence changes. The change occurs at each polyhedral vertex. In triangular tessellations, the majority of grid points are *hexavalent*; that is, six grid members meet at a grid point. However, a different valence occurs when a tessellation point coincides with a polyhedron's vertex, and these grid points will be pentavalent (5), tetravalent (4), or trivalent (3) at the respective vertices of spherical icosahedra, octahedra, or tetrahedra. We can see the valence changes in Figure 8.38 if we ignore the polyhedral edges, which are not part of the Class III tessellation. Notice that most grid points on any sphere are hexavalent, but they change to a lower valence at a polyhedral vertex.

8.7.2   Snubbed Relatives

Class III subdivisions of the octahedron and icosahedron are related to the snub cube and snub dodecahedron.[25] Figure 8.39 shows them in symmetric positions where the octahedron's six vertices align with the snub cube's 6 square faces and the icosahedron's 12 vertices align with the 12 pentagonal faces of the snub dodecahedron. If the square and pentagonal snub faces were triangulated, the tessellations would have Class III geodesic notations $\{3,4+\}_{1,2}$ and $\{3,5+\}_{2,1}$. The snub cube's six square faces circumscribe and center over the octahedron's vertices in Figure 8.39(a). The snub dodecahedron's faces circumscribe the icosahedron in Figure 8.39(b), and its pentagonal faces are centered over the icosahedron's vertices.

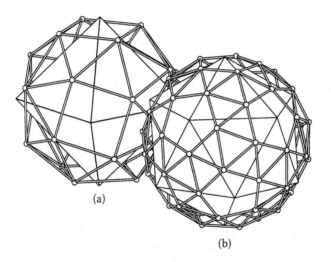

(a)

(b)

Figure 8.39. Snub cube and dodecahedron, octahedron, and icosahedron.

8.7.3   Enantiomorphs

One of the most distinguishing features of Class III subdivisions is their asymmetry. *Enantiomorphs* are geometric forms that are the mirror images of each other. Class III grids are rotationally symmetric, but their reflection symmetry is *enantiomorphic* or opposite-handed. Figure 8.40 compares the reflections of two icosahedron 8^v subdivisions. The top row [Figure 8.40(a) and (b)] is Class I, and their left-right reflections are

---

[25] Snub cube—24 vertices, 60 edges, 6 square faces, and 32 triangular faces. Snub dodecahedron—60 vertices, 150 edges, 80 triangular faces, and 12 pentagonal faces. Schläfli symbols (S3/4) and (S3/5), respectively. The tetrahedron has no snub relative. See Section 6.4.6.

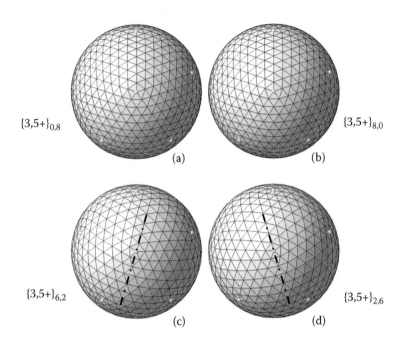

$\{3,5+\}_{0,8}$ (a)

$\{3,5+\}_{8,0}$ (b)

$\{3,5+\}_{6,2}$ (c)

$\{3,5+\}_{2,6}$ (d)

Figure 8.40. Reflection symmetry and enantiomers.

perfectly symmetric. However, left-right reflections for the Class III tessellation in the bottom row [Figure 4.40(c) and (d)] are opposite-handed. The $\{\ \}_{b,c}$ specifications, each Class I and Class III examples, are simply the reverse of each other. In the Class I example, reversing the $\{\ \}_{b,c}$ specification has no effect on the grid layout; however, this reverses the handedness of the Class III grid in Figure 8.40(b).

8.7.4 Harmonics

Just like the other class subdivisions, Class III subdivisions can contain harmonic frequencies. This can occur if b and c are not prime numbers. The number of evenly spaced points coincident with the PPT's edge chord is one less than the *greatest common factor (GCF)* of b,c.[26] The GCF of two whole numbers is the largest whole number that divides evenly into each of them. If the geodesic notation b,c coordinates of any coincident edge-point sum to a frequency that is a multiple of the overall subdivision frequency, it is a point on a harmonic grid.

A simple way to find the GCF of b and c is to list all the factors for both specifications. The GCF will be the largest factor common to both of them. Next, we show how this works. In Example 1, the overall PPT tessellation $b,c = 4,16$ or 20^v. The greatest common factor of 4 and 16 is 4; thus, three grid points (GCF-1) are coincident at each PPT edge. Their b,c coordinates and frequencies are listed in the figure for one PPT edge. A harmonic occurs at those points where $b + c$ frequencies evenly divide into the overall frequency. There are two harmonics, 5^v and 10^v, in Example 1. In Example 2, $b,c = 12,18$ or 30^v and five points are coincident at each PPT edge. Three of the edge points have frequencies that evenly divide into the original 30^v grid: 5^v, 10^v, and 15^v.

---

[26] Also called the greatest common denominator (GCD) or highest common factor (HCF).

> ### Example 1
>
> $b,c = 4,16$ or 20^v
> 4 factors 1,2,4
> 16 factors 1,2,4,8
> GCF = 4
> b,c points on PPT edge = GCF-1 = 4-1 = 3
> 1,4 5^v; 2,8 10^v; 3,12 15^v
>
> ### Example 2
>
> $b,c = 12,18$ or 30^v
> 12 factors 1,2,3,4,6
> 18 factors 1,2,3,6,9
> GCF = 6
> b,c points on PPT edge = GCF-1 = 6-1 = 5
> 2,3 5^v; 4,6 10^v; 6,9 15^v; 8,12 20^v; 10,15 25^v

Figure 8.41 shows the initial tessellation grid and its first harmonic for the two examples. In the figure, the original PPT tessellation is above and its first harmonic below. Grid points coincident to the PPT edge are highlighted, as are the edge points that are retained in their harmonic. A sample triangle within each tessellation grid is highlighted to make it easier to see how one tessellation relates to the other.

Class III tessellations tend to have fewer harmonics than the other classes because neither b nor c can be prime, even if their $b + c$ frequency is not prime. Harmonics also occur whenever b,c values double or are half, as long as the result is a whole number. Again, grid points coincident to the PPT's edges indicate a possible harmonic. Harmonic grids, like any other Class III grid, are rotationally symmetric but not reflection symmetric. All reflections are enantiomorphous or opposite-handed.

8.7.5   Developing Grids

Class III lacks the convenient 3D Equal-chord and Equal-arc references that were the starting points for Class I and II subdivisions. For Class III, we use a different approach and create a 2D grid, which we will eventually project onto the PPT's surface. Although we will develop a triangular grid in our example, it is possible to use this same technique to project any 2D geometry or pattern onto a sphere. Figure 8.42 graphically summarizes the entire sequence we will use; the remainder of this chapter details these steps.

The PPT's tessellation is first defined as a 2D trigrid with the required frequency and orientation relative to the PPT's edges. Figure 8.42(a) shows the grid and the triangular faces that are completely within the PPT. Trigrid point coordinates are converted to Cartesian values and positioned so that one of the grid's directions is perpendicular to the xz-plane and centered over the origin, where it is scaled to correspond to the final PPT's chord length, as shown in (b). The scaled grid is then translated up the $+z$-axis, a distance equal to the polyhedron's insphere radius. The center of the grid is tangent to the insphere and its apices are tangent to its circumsphere (c). Trigrid points are projected from the origin to the surface of the sphere by normalizing each grid point's coordinates. An arrow from the origin through one grid point shows its projection to the sphere (d). The result is a single face-zenith tessellated PPT. Copies are then replicated to cover the rest of the sphere.

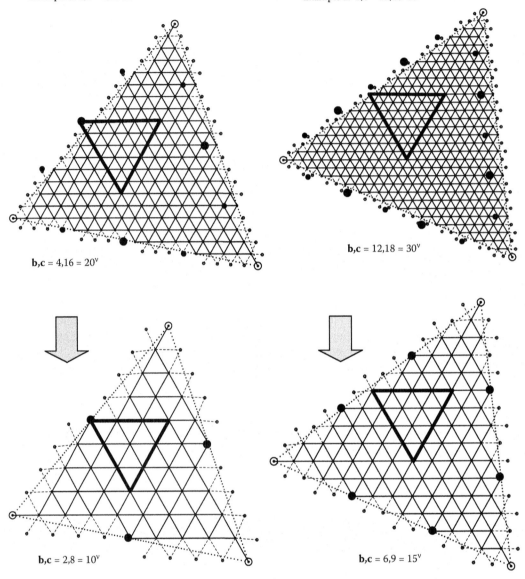

Example 1: b,c = 4,16 20ᵛ   Example 2: b,c = 12,18 30ᵛ

b,c = 4,16 = 20ᵛ

b,c = 12,18 = 30ᵛ

b,c = 2,8 = 10ᵛ   b,c = 6,9 = 15ᵛ

Figure 8.41. Examples of Class III tessellation harmonics.

8.7.6 The BC Grid

The first step is to define a 2D grid based on the geodesic specifications b, c. These two specifications determine the frequency and the grid's orientation relative to the PPT's face. The overall reference grid is called the BC grid. Figure 8.43(a) shows the basic system, its origin, b, c axis, and reference grid oriented at 0°, 60°, and 120°. The BC grid is also a fundamental reference in Chapter 10 Self-organizing Grids.

The standard grid unit is d, and the spacing between parallel grid lines is $d \sin(60°)$ or 0.866025 when $d = 1$. In Figure 8.43(b), a 2D 8ᵛ PPT *(b,c = 2,6)* is mapped out on the trigrid. The PPT's trigrid apex coordinates, in clockwise order from the origin, are 0,0; 2,6; and 8,–2.

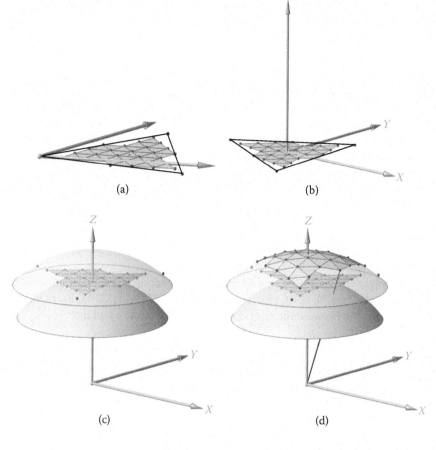

(a)

(b)

(c)

(d)

Figure 8.42. Class III grid development and PPT projection sequence.

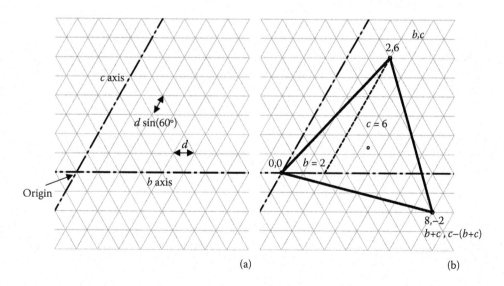

(a)

(b)

Figure 8.43. BC grid definition for 8^v tessellation $\{3,v+\}2,6$.

Trigrid points within the PPT or its edges are the points that will be projected onto the sphere. We are not concerned with any trigrid points that fall outside the trigrid PPT. Although this grid is not yet applied to a polyhedron's face, its geodesic specification is nearly complete. Lacking only the valence (v), the geodesic notation for the trigrid so far would be: $\{3, v+\}_{2,6}$.

8.7.7   From Two to Three Dimensions

The b,c coordinate system, shown in Figure 8.43, is convenient for specifying the tessellation grid, but it is not convenient for projection onto the sphere. Instead, we need 3D Cartesian coordinates. Converting trigrid coordinates to Cartesian coordinates is simple. If d is the length of one grid unit and b,c are trigrid coordinates, 3D Cartesian values for any trigrid point can be found by

$$x = d\left(\cos 60°c + b\right),$$
$$y = d\left(\sin 60°c\right), \quad \text{and}$$
$$z = 0.$$

In the example shown in Figure 8.44, the Cartesian coordinates for the trigrid PPT's apices A, B, and C are (0, 0, 0), (5.0, 5.19615, 0), and (7.0, −1.73205, 0). In each case, the z-coordinate is 0 because the trigrid system is still on the xy-plane.

Figure 8.44. PPT b,c coordinates.

8.7.8   Scale and Translate

The trigrid PPT is not the right size for projection; we need to scale it so that the distance between each of its PPT apices is the same as the chord distance of the polyhedron's face apices. To do this, all PPT points are scaled by the ratio of the polyhedron's chord length (inscribed in a unit radius circumsphere) and the chord of the PPT grid. The icosahedron's chord length is 1.05146[27] and the grid is 7.21110. The scale factor is the ratio of the polyhedron's chord length and the PPT's chord lengths. This is shown in the following formulas:

$$\text{scale factor} = \frac{\text{polyhedron's chord length}}{\text{PPT's chord length}},$$
$$\text{scale factor} = \frac{1.05146}{7.21110},$$
$$\text{scale factor} = 0.14581.$$

---

[27] Chord lengths for tetrahedra and octahedra with a unit radius circumsphere are approximately 1.63299... and 1.141421..., respectively.

By multiplying the PPT's point coordinates by the above scale factor, the vertex distance of the PPT grid becomes exactly the same as the vertex distance of the spherical icosahedron with a *unit circle* circumsphere.

8.7.9 PPT Standard Position

For Class III projections, the trigrid PPT must be repositioned with one of its edges perpendicular to the xz-plane and centered at the origin in a face-zenith orientation. The simplest way to reposition the edge is to rotate the PPT point set around the z-axis until the edge opposite the origin is perpendicular to the x-axis. One of the PPT's apices is already at the origin, so we simply rotate the PPT around this point. Earlier, we found the Cartesian value of apex $b,c = 2,6$ to be (5.0, 5.19615, 0). This point is 46.10122° with respect to the x-axis. If it were 30°, the grid's edge opposite the origin would be in the desired position. To reorient the PPTs, we simply rotate the grid points counterclockwise around the z-axis.

Figure 8.45. PPT centered, scaled, and edge normal position.

The amount of rotation is $46.10211° - 30° = 16.10211°$. To orient the PPT face-zenith, we only need to translate (slide it over) so its center is over the Cartesian origin. It is easy to do this. The PPT's centroid is simply the average of the sum of the x-, y-, and z-components of the PPT's apex points. Subtracting the centroid coordinates from the coordinates in the PPT's point set centers it perfectly. Figure 8.45 shows the PPT grid after the translation and rotation.

To review: a 2D trigrid on the xy-plane defines the vertices of the PPT based on the geodesic notation b,c specification. The trigrid ensures the correct frequency and grid orientation with respect to the PPT's edges. Only grid points, within or on, a PPT's edge are of interest. Grid points are converted from the 2D trigrid b,c system to 3D Cartesian coordinates. All points are still on the xy-plane; their z values are 0. The PPT is rotated around the z-axis to position the edge opposite the origin perpendicular to the positive x-axis and then centered over the origin of the Cartesian axis. The PPT's grid points are scaled to the size of the spherical polyhedron's face.

8.7.10 Projection

In Chapter 6 we saw that Platonic vertices are tangent to their circumsphere and their mid-face points are tangent to their insphere. We now translate the trigrid PPT up the $+z$-axis the distance of the insphere radius of the icosahedron by simply adding the radius value (0.7946544) to the z-component of every point (which was zero prior to this

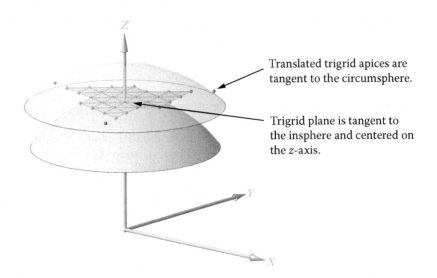

Figure 8.46. Tessellation grid preparation for projection.

addition). The effect is to move the grid up the positive z-axis where the grid's center is now tangent to the icosahedron's insphere and its three apices are tangent to the circumsphere, as shown in Figure 8.46. The grid is now in the proper position for projection.

Projecting the grid is easy; we normalize each grid coordinate. This makes their vector length equal to one, the same as the unit radius sphere. In effect, normalizing pushes each point onto the sphere's surface, projecting it along a line from the origin to the sphere's surface. Figure 8.47 shows the grid before and after projection. An arrow shows one point being projected from the origin to the surface of the circumscribing sphere. Normalizing the grid will not affect its three apex points; they are already on the surface of the unit sphere, a result of the translation in the previous step. This last step completes the tessellation of the PPT. The result of projection is a tessellated, face-zenith PPT with one edge "*parallel*" to the yz-plane.

Figure 8.47. Projecting a grid to the surface of the sphere.

8.7.11   Class III PPT

Figure 8.48 illustrates the PPT we just created using the same graphic standards as before. The schematic in Figure 8.48(a) shows the PPT's projected points as well as the points in its neighboring PPTs, so the triangle faces that straddle the edge are easier to read. In Figure 8.48(b), the resulting great circle arcs between vertices are shown along with the PPT's edge (red), which is not part of the final tessellation. Some tessellation surface treatments are suggested in Figure 8.48(c) and (d). PPT apices are marked with small black dots.

Note that in Class III subdivisions, tessellation triangles that straddle the edges of the polyhedron can only be created after the initial PPT is replicated and rotated because these triangles are created from points in PPTs on both sides of the edge. These edge triangles are included in (c) and (d); however, Figure 8.49 shows the complete spherical subdivision, in which the PPT has been replicated to cover the entire sphere. The total number of faces, edges, and vertices for the entire sphere are also listed. The process of covering the whole sphere is the same process used in earlier Class I schemas. It is detailed in Appendix B. The face-zenith orientation of the PPT is a consequence of developing the original trigrid on the xy-plane before translating the positive z-axis.

8.7.12   Other Polyhedra and Classes

The projection technique just described for Class III subdivisions applies equally well to the other two deltahedra—spherical tetrahedra and octahedra. Only three changes are required. First, the PPT trigrid must meet Class III b,c specifications. Second, the

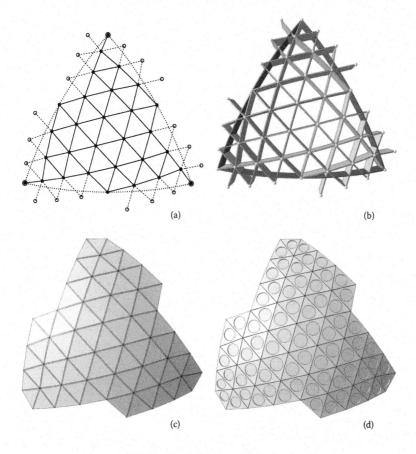

(a)

(b)

(c)

(d)

Figure 8.48. Class III tessellation schema.

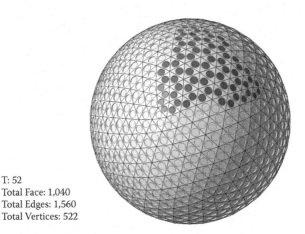

T: 52
Total Face: 1,040
Total Edges: 1,560
Total Vertices: 522

Figure 8.49. Class III tessellated sphere.

PPT grid must be scaled so that its vertices are the same distance apart as the polyhedron's face vertices, and third, the PPT is translated up the $+z$-axis a distance equal to the polyhedron's insphere radius. The scale factor and insphere radius are unique to each polyhedron. These are the only changes needed.

Figure 8.50 shows an inside view of the spherical PPT and a fully subdivided tetrahedron (a) and octahedron (b). The radius of each incenter circle is 80% of the distance from the incenter to the PPT's edges. You immediately notice the significant variation in face sizes, particularly with the tetrahedron.

Class I and II tessellations can be made by using the projection technique just used here for Class III.[28] Figure 8.51 shows Class I and II 8^v trigrids in (a) and (b) and their

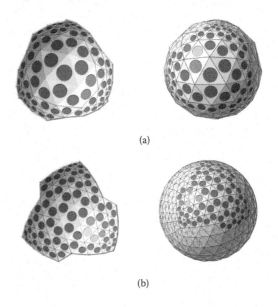

(a)

(b)

Figure 8.50. Class III 8^v tetrahedron and octahedron.

---

[28] In 1990, Christopher Kitrick developed a unified approach to generating the geometry of the Class I, II, and III grids by defining a set of mathematical algorithms for ten different geometric methods of generating geodesic triangulated grids. Several of these had not been previously found in the literature (Clinton 2002, 438).

resulting PPTs in (c) and (d) obtained by projecting the trigrids. The question arises, why not just use projection for all spherical subdivisions, since it can handle any class and any frequency?

The answer is the results are not as good as some of the techniques we have presented. For Class I Equal-chords, trigrid projection produces exactly the same result because Equal-chords is based on an equal subdivision of the PPT's planar edge and this is essentially the same as the planar trigrid before projection. Both subdivision surfaces are superimposed in Figure 8.51(e), and they are perfectly aligned and there is no difference. Trigrid projections for Class II grids (b) and (d) have similar results, as the triacon technique described earlier; however, they are not the same and give up many of the triacon's advantages.

In Figure 8.51 the trigrid projection results for Class II (d) are superimposed on a regular triacon subdivision for the same PPT and the same frequency. Unlike Figure 8.51(e), where Class I projection and Equal-chords methods produced identical results, it is obvious from (f) that the two Class II methods produce different results. The triacon's vertices and surface (in red) do not align with the trigrid's projection (in gold). Common vertices are both red and gold. Note that only the PPT's mid-face and mid-edge vertices are the same in both schemas; they are circled in the figure. The trigrid projection results in many more different chords as well as more tri-

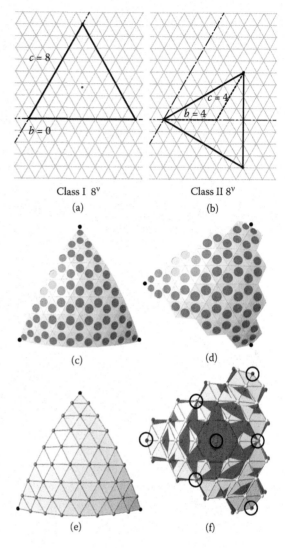

Class I 8ᵛ
(a)

Class II 8ᵛ
(b)

(c)

(d)

(e)

(f)

Figure 8.51. Trigrids for Class I and II tessellations.

angle area, as the two surface colors show. Thus, the big advantages of the triacon method are lost if projection is used. This result is not a total surprise. The triacon equally subdivides the spherical polyhedron's arc edge with a series of spherical right triangles. Equal-arc references along the polyhedron's edge, drive the grid development In contrast, projecting a b,c grid is more like a projection of an Equal-chord grid. It produces a Class II grid that is perpendicular to the edges of the polyhedral face, but it does not offer any of the benefits of the triacon, like a minimum number of different chords or face areas.

8.7.13 Summary

Class III skew applies to spherical deltahedra—tetrahedra, octahedra, and icosahedra. Tessellations grids are not parallel or perpendicular to the spherical polyhedron's edges, but cross at oblique angles. Their geodesic b,c must be positive, not equal to zero or equal to each other. Subdivisions of the octahedron and icosahedron are related to the snub cube and snub dodecahedron. Tessellations are rotationally symmetric, but their reflections are enantiomorphic, or opposite-handed. Like Class I and II tessellations, Class III tessellations may contain harmonics depending on b,c's common factors.

Class III tessellations lack convenient Equal-chord or Equal-arc references on the polyhedron's spherical polyhedron edge; thus, projection is used (other methods are possible too). Subdivision starts with a 2D trigrid on the xy-plane, where we define the final frequency, grid orientation, and included grid points. We convert points within the trigrid PPT or on its edge to Cartesian values and the points are then translated to the origin, scaled, and translated up the positive z-axis. The center of the grid (still planar) is tangent to the insphere, while the trigrid PPT's apices are tangent to the circumsphere. We then project grid points to the surface of the unit sphere by normalizing their coordinates.

We can also apply projection methods to other classes. When we apply these methods to Class I grids, the results are exactly the same as they are for the Equal-chords schema, but not the same as they are for the other Class I schemas we developed. For Class II schemas, the results are similar to but not the same as the triacon because Class III does not divide the polyhedron's edge into equal arc segments like the triacon does.

Class III projections involve a series of coordinate transformations: scaling, rotating, and translating. They can all be combined into a single transformation by combining their specifications. We give examples in Appendix B.

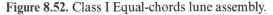

(a)

(b)

Figure 8.52. Class I Equal-chords lune assembly.

8.8 Covering the Whole Sphere

Each of the schemas developed in this chapter resulted in one tessellated PPT or diamond. But how does this describe the whole sphere? The answer is simple—rotation! Essentially, we copy the tessellated PPT or diamond geometry and use rotation to position the copy wherever we want it. We rotate copies of one PPT or diamond at a time or we can combine small groupings of PPTs or diamonds into a larger grouping and rotate the larger grouping to cover the whole sphere.

Rotation recomputes new coordinate values, in effect finding new values for coordinates, as they revolve around the x-, y-, or z-axis to their new position. Rotation does not alter the relationships of any grid point to one another within the PPT or diamond being rotated. All angles and distances between points and the origin are preserved during the rotation.

When making a complete sphere from a single PPT or diamond, it is often advantageous to create subassemblies and then copy-rotate them rather than handle each PPT or diamond separately. Figure 8.52 shows the exterior

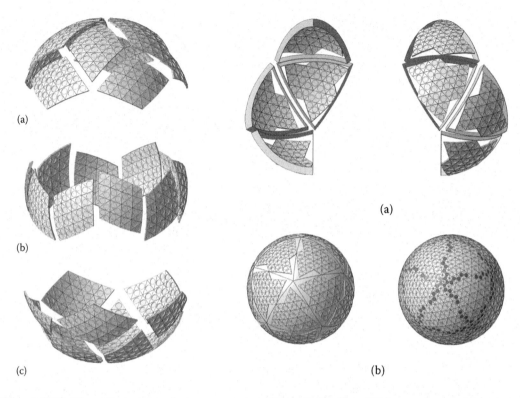

(a)

(b)

(c)

(a)

(b)

Figure 8.53. Class II pentagonal cap and ring assembly.

Figure 8.54. Class III lune subassembly.

and interior of a Class I subassembly of four PPTs. In (a), the subassembly resembles a spherical lune. It is created by rotating three instances of the original principal triangle, the topmost triangle in (a), and positioning them back-to-back, creating a lune-like segment that is one-fifth of the sphere. The assembly in (a) is not yet fully transformed in order to clearly show each PPT copy in the subassembly. In Figure 8.52(b) the sub-assembly in (a) has been copy-rotated four more times around the z-axis to complete the entire sphere.

Figure 8.53 shows how subassemblies might be used in Class II subdivisions. An upper pentagonal cap subassembly is created from the original subdivided diamond by copy-rotating it nine times. The cap, when completed, has ten diamonds. A lower cap is created by copy-rotating the entire upper cap subassembly. Assume the vertical axis of the sphere is the z-axis; the upper cap is copy-rotated 180° around either the x- or y-axis and then again 36° around the z-axis because the icosahedron's upper-lower pen-tagonal cap is offset by this much. The central ring of ten diamonds is accomplished by copy-rotating two diamonds from the upper (or lower) cap and positioning two adjacent to one another with their smaller diamond sides tangent. This subassembly of two diamonds is then copy-rotated four more times around the z-axis, each instance offset from the previous one by 72°. When all of the rotation transformations are fully applied, a completed sphere is the result.

Class III rotations can use the same strategy as Class I rotations. Figure 8.54(a) shows a lune subassembly. The triangles that straddle the PPT's edges can be com-pleted from tessellation vertices of PPTs that share the edge. The spherical icosahedron edges, shown in Figure 8.54(b), are not part of the final tessellation, but show where straddling triangles fill in to complete the sphere.

Additional Resources

Clinton, Joseph D. *Advanced Structural Geometry Studies Part I—Polyhedral Subdivision Concepts for Structural Applications*. National Aeronautics and Space Administration contractor report NASA CR-1734, Carbondale, IL: Southern Illinois University, 1971.

____. "Geodesic Math." *Domebook 2* ed. Lloyd Kahn et al. Bolinas, CA: Pacific Domes, 1971, 106–113.

____. "A Group of Spherical Tessellations Having Edges of Equal Length." 5th International Conference on Space Structures, London, 2002.

____. "A Limited and Biased View of Historical Insights for Tessellating a Sphere." *Space Structures 5* eds. G. A. R. Parke and P. Disney. London: Thomas Telford, 2002, 423431.

Gailiunas, Paul. "Twisted Domes." Newcastle, UK: Unpublished paper. web.ukonline.co.uk/polyhedra/Twist.pdf, 2008.

Kenner, Hugh. *Geodesic Math and How to Use It*. Berkeley: University of California Press, 1976.

Kitrick, Christopher J. *"A Unified Approach to Class I, II & III Geodesic Domes."* *International Journal of Space Structures* 5 (1990): 223–246.

Stuart, Duncan R. "The Orderly Subdivision of Spheres." *The Student Publications of the School of Design*, Raleigh: North Carolina State University, monograph, 1963.

Wenninger, Magnus J. *Spherical Models*. Cambridge [Eng.], New York: Cambridge University Press, 1979 (new appendix, 1999).

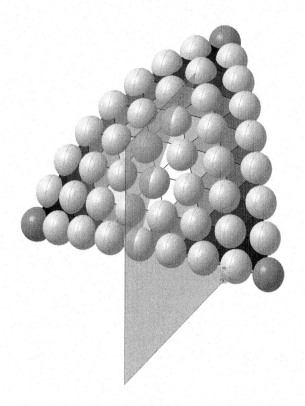

⑨ Comparing Results

We have developed six subdivision schemas, and it is natural to wonder which is best. The answer depends on the specific geodesic application and its design goals. No single schema fits every need. In this chapter, we offer a series of visual and statistical metrics to compare schemas. Some compare distances between vertices, triangle area, or volume of the subdivided sphere; others look at face shapes and their orientation. Although each metric highlights one characteristic and excludes others, there are common threads throughout and each provides a partial answer to the question of which is better.

What our analysis will show is that Class I Equal-arcs (three great circles) and Class II triacon produce the most uniform distribution of points on a sphere, and the most uniform face orientations of the schemas we developed. They also offer additional advantages. Triacon has the fewest different chord lengths, the fewest different face angles, and the largest number of equal-sized and equal-shaped faces by far. It is a superior performer. Equal-arcs (three great circles) is also a superior performer, with more variation than the triacon, but very uniform overall. It offers superior face orientations and a high percentage of symmetrical parts. However, these two schemas are not the last word. Mid-arcs and Equal-chords both offer unique features that make them preferred choices in certain applications. So once again, the answer to the question of *"which is best?"* depends on what you are trying to accomplish and your design requirements. We begin our analysis by discussing an intimate subject: kissing-touching.

9.1 Kissing-Touching

Chapter 4 introduced the concept of balls *"kissing"* and *"touching."* These terms, borrowed from the game of billiards, offer us a way to see how closely things pack together. Closely packed balls on a sphere measure the compactness and distribution of tessellation vertices. When the largest balls possible, all with the same diameter, are placed at each tessellation vertex, they give us a three-dimensional (3D) way to visualize the symmetry, distribution, and maximum packing a schema affords. Their pattern also makes it easy to see where the smallest and largest tessellation triangles can be found.

Some vertex balls will kiss and touch, but none will penetrate or overlap another. The spacing between balls and the maximum diameter possible for a schema indicate how uniformly the subdivision distributes points on the surface of the sphere. The best subdivision schema, in terms of point distribution, will have the closest packing of the largest diameter balls when compared to other schemas.

Kissing-touching does not measure the same parameter as spherical caps (discussed in Chapter 4). Both techniques are indicators of packing, but spherical caps measure how much of the sphere's surface area is covered. It is a density measurement, which will be used shortly. In the spherical cap technique, the cap's radius is based on the smallest *arc* distance between tessellation points. In contrast, kiss-touch ball diameters are based on the smallest chord length between them. Table 9.1 lists the kissing-touching distance between the closest balls in our schemas on the unit sphere.

Figure 9.1 shows kiss-touch vertex balls for the schemas developed earlier in the chapter for unit spheres. The viewpoint for each figure is from *inside* the sphere, where we get the least-obstructed view of their *principal polyhedral triangle* (PPT) or diamond packing. Contacts between balls, especially near the PPT's apices (red ball), are somewhat obstructed in the figure, but views from inside the sphere are still better than from outside.

In Figure 9.1, (a), (b), (c), and (d) are Class I Equal-chords, Equal-arcs (two great circles), Equal-arcs (three great circles), and Mid-arcs schemas, respectively. Class II triacon and a Class III skew are shown in (e) and (f). Each schema is represented by a PPT or diamond surface (pink) and its kiss-touch balls. PPT and diamond apex balls are indicated (red) and correspond to the vertices of the foundation polyhedra, the icosahedron in our examples. The figure points out typical kiss-touch pairs.

In Equal-chords, Figure 9.1(a), kissing-touching starts at each PPT apex (red ball) and the balls immediately around it. All other chords within the PPT are longer,

Kissing-Touching Ball Diameters	
Schema	**Smallest**
Class I	
Equal Chords	0.119459
Equal Arcs$_{(2\ GCs)}$	0.132679
Equal Arcs$_{(3\ GCs)}$	0.138283
Mid Arcs	0.138283
Class II	
Triacon	0.153316
Class III	
Full Face	0.128646

Table 9.1. Kissing-touching distances.

(a) Class I Equal-chords

(b) Class I Equal-arcs (two great circles)

(c) Class I Equal-arcs (three great circles)

(d) Class I Mid-arcs

(e) Class II triacon

(f) Class III skew

Figure 9.1. Class I, II, and III kissing-touching.

and space appears between the balls. For Class I subdivisions, the general pattern is that chord lengths increase from the PPT apices to the center of the PPT edge, and from the center of the edge towards the center of the PPT. This pattern is quite visible in the figure. The balls show that the size of the triangles increases as they progress from the PPT vertices and edges towards the center of the PPT face. The biggest gaps are in the middle of the PPT. Kissing-touching makes these relationships quite visible.

Figure 9.1(b) shows Equal-arcs (two great circles). Its kiss-touch ball diameter is larger than the diameter of Equal-chords, and the smallest chord length does not occur at a PPT apex, as might be expected. Instead, it occurs on the bottom PPT edge between the two tessellation vertices immediately inside to the left and right of the center edge. They are pointed out in the figure. Since these two balls set the maximum diameter possible for the others, none of the balls along the PPT edge kiss-touch, even though the schema is based on an Equal-arcs PPT reference point. All the chords along the PPT's edges are larger. Thus, there is some space between balls. Notice, also, all the interior balls are displaced towards the bottom PPT edge, a consequence of defining tessellation points with just two intersecting great circles. As a result, the PPT's tessellation points are not rotationally symmetrical.

Equal-arcs (three great circles), shown in Figure 9.1(c), demonstrates a highly uniform distribution throughout. All PPT edge balls kiss and touch, because the edge reference points are evenly spaced and all have the same shortest chord lengths. The distribution of tessellation vertices is symmetrical in terms of face, edge, and PPT apices. The uniform distribution across the PPT's face is a direct result of intersecting three great circles and handling the resulting window artifact. Earlier, we showed that the intersection of three great circle planes in this schema create three points, not one, on the surface of the PPT (we ignore the three on the opposite side of the sphere). The window's projected centroid is used as the tessellation grid point. The combined effect of using three great circle intersections and the extra step of processing the window points is well worth the extra computation to eliminate windows. Its close packing, shown in (c), demonstrates that it is the most uniform of the Class I schemas.

Figure 9.1(d) shows Mid-arcs. The nesting pattern of triangles within triangles, the basic subdivision method used, is evident in the ball pattern. Kissing-touching occurs along all three PPT edges because this schema recursively divides every arc at its midpoint and this includes the entire length of all three PPT edges. In this respect, it is similar to Equal-arcs (three great circles). All interior chords are larger than the edge chords; thus, all balls along the PPT's edges kiss and touch. Notice, also, the spacing between balls increases as they get closer to the center of the PPT. The three balls at the middle of the PPT mark the vertices of the largest tessellation triangle within the PPT.

Figure 9.1(e) shows the triacon diamond, which is symmetric to every edge of the foundation polyhedron. The method divides polyhedral edges into equal arc segments, which is reflected in the even spacing of kissing-touching along the icosahedral edge (dashed diamond axis between the red balls). Unlike the previous schemas, kissing-touching in a triacon diamond first occurs at mid-PPT vertices (brown) instead of at the PPT apices (red). The triacon is the only schema that consistently defines a vertex in the middle of the PPT's face. Figure 9.1(f)[1] shows a Class III skew. Of all the schemas developed, skew has the greatest difference in chord lengths, as evidenced by how quickly the kiss-touch balls spread out towards the center of the PPT. We will see just how much they differ when we compare chords in the next section.

---

[1] The Class III example is an icosahedron {3,5+}2,6. The view in the figure is from the *inside* of the sphere, and this creates the impression that the subdivision is {3,5+}6,2.

In summary, kissing-touching is an effective metric for evaluating the symmetry, distribution, and compactness of subdivision schemas. The larger the balls and the more closely they are packed, the better the points distribution. From the point of view of kissing-touching, Class I Equal-arcs (three great circles) and triacon perform best. Mid-arcs performs well, though not as well as the other two. Class III subdivisions perform the worst. Equal-arcs (two great circles) has good packing, but its asymmetry may preclude it from many applications.

9.2   Sameness or Nearly So

Statistics show how similar (or dissimilar) objects are, and we can use these statistics to quantify the variation of chord lengths and triangle areas within a schema. Schemas with low statistical variation tend to have more uniform grids, but there are important exceptions to this rule.

In statistics, *variance* measures how much a set of values spread out. These values could be the chord or arc lengths, triangle area, or other geometric measurements. Variance is the average squared deviation of these values from their mean. The *standard deviation*, another important statistic, is simply the square root of variance. The lower the standard deviation, the closer the values are to being the same or alike.

Is there a spherical baseline with no variation where every chord and face area is the same? Yes, the foundation polyhedron itself is a baseline. All chords are equal, all faces are congruent regular polygons and all have equal area, all central and surface angles are the same, and all vertices are evenly distributed on the sphere. For example, the icosahedron's 12 vertices are evenly distributed on its circumsphere, all faces are equilateral and congruent, and all 30 chords and central angles are the same. This means that a basic Platonic solid, planar, or spherical, has zero variation and zero standard deviation in each of these characteristics. However, when we subdivide a polyhedron, everything changes. Many chord lengths and triangle areas result; their variance tells us how different they are.

Intuitively, one would expect the best-performing schemas to have the lowest standard deviations and variances, but this is not always the case. We will find excellent schemas with somewhat high deviations, which eliminate trivial differences in chord length by making more pronounced differences in the few chords they create.

Tables 9.2–9.4 summarize the vertex counts, chord lengths, and triangle areas that result from the subdivision schemas developed. Table 9.2 shows the triangulation

Tessellation Vertices			
Schema	***T***	***PPT/Dia***	***Sphere***
Class I			
Equal Chords	64	45	642
Equal Arcs$_{(2 GCs)}$	64	45	642
Equal Arcs$_{(3 GCs)}$	64	45	642
Mid Arcs	64	45	642
Class II	48		
Triacon	48	34	482
Class III			
Midface	52	30	522
Straddling Edge			

Table 9.2. Schema vertex comparisons.

Tessellation Chords									
Schema	PPT/Dia	Sphere	Unique	Shortest (S)	Longest (L)	S/L	Variance	Std.Dev.	Rank
Class I									
Equal Chords	108	1920	20	0.119459	0.164647	72.60%	1.13E-04	1.06E-02	3
Equal Arcs$_{(2\ GCs)}$	108	1920	44	0.132679	0.176238	75.30%	1.29E-04	1.14E-02	4
Equal Arcs$_{(3\ GCs)}$	108	1920	17	0.138283	0.162173	85.30%	5.37E-05	7.33E-03	1
Mid Arcs	108	1920	15	0.138283	0.164647	84.00%	9.60E-05	9.80E-03	2
Class II									
Triacon	56	1440	8	0.011174	0.013765	81.20%	1.77E-04	1.33E-02	6
Class III									
Midface Straddling Edge	60	1560	20	0.128646	0.18247	70.50%	1.49E-04	1.22E-02	5

Table 9.3. Schema chord comparisons.

Tessellation Triangle Areas									
Schema	PPT/Dia	Sphere	Unique	Smallest (S)	Largest (L)	S/L	Variance	Std.Dev.	Rank
Class I									
Equal Chords	64	1280	15	0.00678	0.011738	57.80%	1.60E-06	1.27E-03	5
Equal Arcs$_{(2\ GCs)}$	64	1280	35	0.00807	0.011102	72.70%	6.94E-07	8.33E-04	2
Equal Arcs$_{(3\ GCs)}$	64	1280	15	0.009083	0.010308	88.10%	9.66E-08	3.11E-04	1
Mid Arcs	64	1280	15	0.009083	0.011738	77.40%	7.06E-07	8.40E-04	3
Class II									
Triacon	32	960	4	0.011174	0.013765	0.812	8.55E-07	0.000925	4
Class III									
Midface Straddling Edge	70	1040	16	0.007862	0.014417	54.50%	2.77E-06	1.67E-03	6

Table 9.4. Schema triangle area comparisons.

number, T, and the total number of vertices in a PPT/diamond and on full spheres. There is quite a range in total points distributed, even though the foundation polyhedra and subdivision frequency are the same for every schema developed. The Class I schemas all distribute the same number and the most points, while triacon distributes the fewest. The table lists the number of vertices falling within the PPT, diamond, or on any edge.

Table 9.3 compares the variation in chord lengths for each schema. The table tallies the number of chords for each PPT/diamond and for the entire sphere. The number of unique chords, the shortest and longest, and percentage difference is also listed. The table calculates the variance and standard deviation for the entire set of chords on the sphere, and ranks schemas from lowest to highest variation. It is remarkable that an 8^v triacon requires only eight unique chords, while Equal-arcs (two great circles) requires 44.

It is no surprise that Equal-arcs (three great circles) is the best performer. We have seen the pattern repeatedly in our previous analysis. It has few unique chords, and the difference between the largest and smallest length is small. The Equal-chords schema performs reasonably well with 20 unique chords and low variance in their lengths, but Equal-arcs (three great circles) and Mid-arcs perform far better, with much less variation in length and in the number of unique chords. Equal-arcs (two great circles) performs poorly because, compared to other schemas, it has more than twice the number of unique chords. The relatively high variance in its lengths tells us the difference between the members of this large set of unique chords is considerable, which is not a desirable characteristic.

Once again, the triacon stands alone. Its tessellation grid has only eight unique chords. The fact it develops a diamond symmetric to the polyhedral edge also helps. Unlike other schemas, in which small differences in length resulted from lots of unique chords, triacon eliminates trivial differences in length by increasing the differences between the ones it creates. Thus, its high variance is not a surprise, and the variance serves a useful purpose. Although we have not yet focused on geodesic applications, it is worth pointing out that from a manufacturing point of view, it is usually desirable to have fewer components used many times, rather than many different components used just a few times. The triacon offers many advantages, when manufacturing is a consideration. Class III performance is quite poor from a chord length perspective. Its chords have high variation in length. However, unlike the triacon, where variation led to a dramatic reduction in total unique lengths, variation here only led to lots of small differences in a high number of unique chords.

9.3 Triangle Area

Chord lengths directly affect a triangle's shape and area. Here we compare just the area of tessellation triangles. We will consider their shape in a moment. Our analysis is both statistical and graphic. Table 9.4 compares the variance in triangle area of each schema developed, and clearly, they are correlated with chord length. All triangles on the sphere are considered. This chord table, like the previous one, lists such information as the number of unique areas, the smallest areas, and the largest areas. Once again, Equal-arcs (three great circles) is a superior performer with the least variation in face areas. Class III subdivisions have the most variation. The contrast between triacon and the other schemas is dramatic. Triacon has only four unique triangles, while the other schemas have 15 or 35. This is an amazing reduction. Triacon achieves this by increasing the differences in the face areas it does create. Equal-arcs (two great circles) has the highest number of unique triangles, a consequence of having so many different chord lengths.

Figure 9.2 graphically shows the differences indicated in Table 9.4. The figure gives the triangles in each schema a gray value from black (smallest triangle) to white (largest triangle) to indicate their increasing area. The scale is relative to each figure and provides a quick way to see where similar faces are located and where symmetry occurs. In Figure 9.2(e) triacon is the easiest scale to read, with only four different-sized faces. The largest faces are along the diamond's long axes. Although they appear equilateral, they are slightly isosceles. We examine their shape more closely in a moment. The smallest faces are those at the diamond's short axis and fall near the midpoint of the principal triangle.

Triacon is unique in another way. Its largest triangles are adjacent to the PPT's apices and the smallest ones surround the PPT's center, just the opposite of Class I or III schemas.

All four Class I schemas have different patterns, but there are some common features. The smallest faces are usually located next to the PPT's apices and the largest in the middle of the PPT. Triangles along the PPT's edge increase in size as they approach midedge and as they move closer to the center of the PPT. However, Equal-arcs (two great circles) does not follow this pattern exactly. Its smallest triangles cluster along the bottom edge of the PPT, instead of around the PPT's apices; they are quite apparent in (c). There is also another exception to the pattern; its largest triangle is not in the middle of the PPT, it is above the PPT's center.

The triangle areas in Class I Mid-arcs, Figure 9.2(d), clearly reflect the recursive triangle-in-triangle subdivision method. The triangles within the PPT's central equi-

(a) Class I Equal-chords

(b) Class I Equal-arcs (two great circles)

(c) Class I Equal-arcs (three great circles)

(d) Class I Mid-arcs

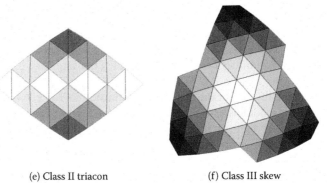

(e) Class II triacon

(f) Class III skew

Figure 9.2. Tessellation face area comparisons.

lateral area (outlined in the figure) are all larger than the triangles in the three outlying groups next to the PPT's apices. All triangles within the large central equilateral area are larger than those in the three surrounding groups next to the PPT's apices.

Equal-arcs (three great circles) has the least deviation in the area in 15 unique triangles, while triacon is ranked second, with four unique triangles. Class III skew shows the highest variance and ranks last of the schemas.

9.4   Face Acuteness

The schemas developed in this chapter show the influence the foundation polyhedron and subdivision schema have on the final tessellation grid and shape of the resulting triangles. Figure 9.3 shows just how much influence they have by comparing three

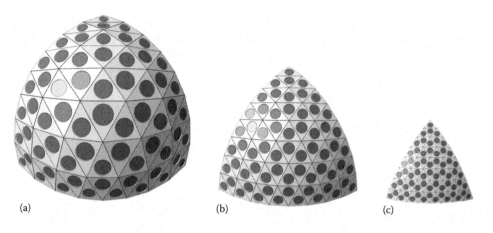

(a) (b) (c)

Figure 9.3. Face acuteness in subdivided deltahedra. Single tetrahedron
(a), octahedron (b), icosahedron PPT (c).

deltahedra 8^v subdivisions (Figure 9.3). What we see repeatedly is the icosahedron producing more uniform results than either the tetrahedron or octahedron. The basic reason is that the icosahedron's spherical excess (its area) is smaller than the spherical excess of the others. Thus, for the same frequency, the tessellation faces are smaller and have less variation in shape. Within a given polyhedron, and for the same frequency subdivision, there is still variation in the shape of triangles that result from the subdivision schema itself. A direct approach would compare a triangle's face angles. However, it is still hard to see overall patterns this way. Instead, we will use a simple graphic convention, which allows us to see the shape of a triangle very quickly—we will use the Euler line.

9.5 Euler Lines

In Section 4.17.2, we presented three important centers of a planar triangle—orthocenter, incenter, and circumcenter. Recall that in non-equilateral planar (not spherical) triangles, these points lie on a line called the Euler line. The length of the line segment between orthocenter and circumcenter is very sensitive to the ratio of the triangle's sides. A slight change in any of the triangle's chords causes a large change in the length of this segment and the angle it takes on the triangle's surface—herein lies the Euler line's analytic value. We can use this sensitivity and the Euler line segment to see how close or far away triangles are from being equilateral, and to analyze shape. We can also display the line segment itself and graphically tell if the triangle is equilateral (no segment, just a point), isosceles, or acute. Figure 9.4 displays the Euler line segments for all of the schemas developed. Illustrations (a) through (d) are Class I, while (e) and (f) are Class II and III, respectively. The orthocenter and circumcenter of each subdivision triangle is found, and the Euler line segment between them is shown. If the face is equilateral, both centers are coincident and there is no Euler line, only a point. Class I and III figures (a), (c), (d), and (f) show this. Their PPT's center triangle is equilateral, and there is no Euler line, only a point in the middle of this triangle. All other triangles in these schemas are either isosceles or acute, and their Euler line segments have measurable length and direction. There are no right triangles in these schemas, but if there were, the Euler line segment would touch the hypotenuse, or the apex opposite it, and be perpendicular to the hypotenuse.

Table 9.5 tabulates these lengths. For each schema, the total Euler line segment length is listed for the basic PPT or diamond and then extrapolated to the total length

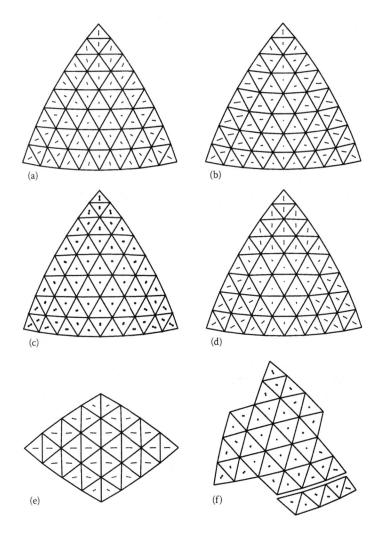

Figure 9.4. Subdivision schema Euler lines.

	Euler Line Segments				
	PPT / Diamond		**Full Sphere**		
Schema	**Total Triangles**	**Euler Line Length**	**Faces or Edges**	**Length Sub-Total**	**Total Length**
Class I					
Equal Chords	64	0.970213	20	–	19.40
Equal Arcs $_{(2\ GCs)}$	64	2.242257	20	–	44.85
Equal Arcs $_{(3\ GCs)}$	64	1.817358	20	–	36.35
Mid Arcs	64	1.235249	20	–	24.70
Class II					
Triacon	32	1.738888	30	–	52.17
Class III					
Midface	34	0.237650	20	4.75	
Straddling Edges	6 (per edge)	0.067179	30	4.03	8.7838

Table 9.5. Euler line segment comparisons.

of all segments on a full sphere. Classes I and II are straightforward, but the Class III schema require subtotals because some subdivision triangles lie totally within the PPT, while others straddle the edge. The last column lists the total length of the Euler line segments.

It is evident from the table that there is considerable variation in grid triangle shape. Clearly, Class I Equal-chords and Class III subdivisions (which are also based on a projection of Equal-chords) have the most uniform (near equilateral) faces. This result is consistent with what we saw in kissing-touching. The greatest deviation from equilateral faces is found in triacon. There is not a single equilateral face in the diamond, and those that straddle the polyhedra edge (long diamond axis for icosahedra) have the longest Euler line segments, as shown in Figure 9.4(e).

We have compared kissing-touching distances, chord lengths, triangle areas, and triangle shapes. We now need to see how subdivision grid components are affected by changes in frequency. We will find the triangulation number, T, very handy for this kind of analysis

9.6   Parts and T

The total number of parts in a fully tessellated spherical polyhedron depends on the polyhedron's basic characteristics: faces, edges, and vertex valence. It also depends on the schema's b,c frequency specification. We can use the triangulation number, T, and these characteristics to count the parts on the full sphere. Table 9.6 restates the faces, edges, and vertex valence for each deltahedron. The basic formulas for finding the total faces, edges, and vertices on the sphere are shown below:

$$T = b^2 + bc + c^2$$

$$\text{Total subdivision faces} = T \times \text{faces in base polyhedron},$$

$$\text{Total subdivision edges} = T \times \text{edges in base polyhedron},$$

$$\text{Total subdivision vertices} = \left[T \frac{(2 \times \text{valence})}{(6 - \text{valence})} \right] + 2.$$

The frequency of subdivision has a huge effect on part counts. Based on b,c and T values, Table 9.7 lists the face, edge, and vertex counts for 2^v, 4^v, 8^v, 16^v, and 32^v tessellations for the deltahedra. Shaded areas indicate the 8^v icosahedron schemas developed earlier in this chapter. Although it is not apparent from the table, the subdivision method within a class does not affect these counts. For example, Equal-arcs and Equal-chords schemas produce the same part counts; only their face areas, angles, and chords differ. Reversing coefficients b,c will not affect the counts. A 2,6 tessellation will have the same part counts as 6,2 for the same polyhedron.

Deltahedra Valences			
Deltahedron	Faces	Edges	Valence
Tetrahedron	4	6	3
Octahedron	8	12	4
Icosahedron	20	30	5

Table 9.6. Deltahedra valences.

v	Class	b	c	T	Tetrahedron			Octahedron			Icosahedron		
					Faces	Edges	Vertices	Faces	Edges	Vertices	Faces	Edges	Vertices
2	I	0	2	4	16	24	10	32	48	18	80	120	42
	II	1	1	3	12	18	8	24	36	14	60	90	32
	I	0	4	16	64	96	34	128	192	66	320	480	162
4	III	1	3	13	52	78	28	104	156	54	260	390	132
	II	2	2	12	48	72	26	96	144	50	240	360	122
	I	0	8	64	256	384	130	512	768	258	1280	1920	642
8	III	1	7	57	228	342	116	456	684	230	1140	1710	572
	III	2	6	52	208	312	106	416	624	210	1040	1560	522
	III	3	5	49	196	294	100	392	588	198	980	1470	492
	II	4	4	48	192	288	98	384	576	194	960	1440	482
	I	0	16	256	1024	1536	514	2048	3072	1026	5120	7680	2562
	III	1	15	241	964	1446	484	1928	2892	966	4820	7230	2412
	III	2	14	228	912	1368	458	1824	2736	914	4560	6840	2282
16	III	3	13	217	868	1302	436	1736	2604	870	4340	6510	2172
	III	4	12	208	832	1248	418	1664	2496	834	4160	6240	2082
	III	5	11	201	804	1206	404	1608	2412	806	4020	6030	2012
	III	6	10	196	784	1176	394	1568	2352	786	3920	5880	1962
	III	7	9	193	772	1158	388	1544	2316	774	3860	5790	1932
	II	8	8	192	768	1152	386	1536	2304	770	3840	5760	1922
	I	0	32	1024	4096	6144	2050	8192	12,288	4098	20,480	30,720	10,242
	III	1	31	993	3972	5958	1988	7944	11,916	3974	19,860	29,790	9932
	III	2	30	964	3856	5784	1930	7712	11,568	3858	19,280	28,920	9642
	III	3	29	937	3748	5622	1876	7496	11,244	3750	18,740	28,110	9372
	III	4	28	912	3648	5472	1826	7296	10,944	3650	18,240	27,360	9122
32	III	5	27	889	3556	5334	1780	7112	10,668	3558	17,780	26,670	8892
	III	6	26	868	3472	5208	1738	6944	10,416	3474	17,360	26,040	8682
	III	7	25	849	3396	5094	1700	6792	10,188	3398	16,980	25,470	8492
	III	8	24	832	3328	4992	1666	6656	9984	3330	16,640	24,960	8322
	III	9	23	817	3268	4902	1636	6536	9804	3270	16,340	24,510	8172
	III	10	22	804	3216	4824	1610	6432	9648	3218	16,080	24,120	8042
	III	11	21	793	3172	4758	1588	6344	9516	3174	15,860	23,790	7932
	III	12	20	784	3136	4704	1570	6272	9408	3138	15,680	23,520	7842
	III	13	19	777	3108	4662	1556	6216	9324	3110	15,540	23,310	7772
	III	14	18	772	3088	4632	1546	6176	9264	3090	15,440	23,160	7722
	III	15	17	769	3076	4614	1540	6152	9228	3078	15,380	23,070	7692
	II	16	16	768	3072	4608	1538	6144	9216	3074	15,360	23,040	7682

Table 9.7. Faces, edges, and vertices for 2^v, 4^v, 8^v, 16^v, and 32^v, Class I, II, and III tessellations.

For a given frequency and polyhedron, Class I tessellations create the highest number of parts, while Class II tessellations create about 25% fewer. This is a significant difference. Class III counts lie somewhere in between Classes I and II. For the same polyhedron and frequency, the greater the difference between Class III b,c values, the closer their part counts are to Class I, and vice versa. Likewise, the smaller the difference, the closer their counts are to Class II.

While part counts are critical factors in most geodesic applications, the way b,c combinations affect these totals is also critical. That's because the choice of polyhedra and frequency are made early in the design process, and the choice significantly

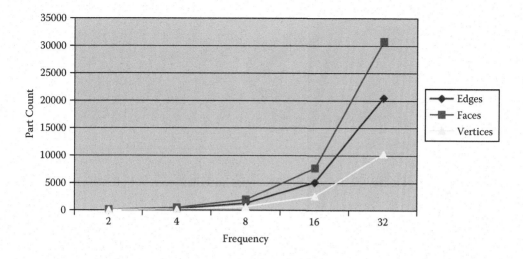

Figure 9.5. Icosahedron Class I part counts for 2^v, 4^v, 8^v, 16^v, and 32^v.

impacts all subsequent work and development. Figure 9.5 graphs part counts for the Class I icosahedral tessellations for various frequencies. The graphs for Class II tessellations are very similar; only the part counts are less. The curves show that part counts increase as a power, usually squared, of the frequency. This rapid buildup implies that most geodesic applications will tend to use lower frequencies rather than higher ones, especially if the spherical object is manufactured from separate components (as geodesic domes are).

9.7   Convex Hull

The *convex hull* is the smallest convex polyhedron that surrounds the points on the sphere. The convex hull of a subdivided sphere measures how uniformly points are distributed by "*shrink-wrapping*" the whole envelope around the sphere with a surface that defines a volume. In some cases, the convex hull resembles the planar surfaces of tessellated spheres. In Chapter 4 we compared the convex hull volumes of the five Platonic polyhedra, each with a unit radius circumsphere, and saw that as the number of vertices increased, the convex hull volume increased. The dodecahedron distributes the most vertices and has the largest convex hull volume of the five. As more and more points are distributed, the convex hull faces become smaller and the overall shrink-wrapping looks more "*spherical.*" Its volume also approaches, but never quite reaches, the unit sphere's volume. We also saw that when two spheres have the same radius and number of points on their surfaces, the one with the most uniform distribution or points will have the largest volume. Herein lies the use of convex hulls for comparing subdivisions. We can use convex hull volumes to compare subdivided spheres. The one with the largest volume has the most uniform distribution of points.

For the icosahedral schemas we developed earlier, we can use the surface of their triangular faces to define their convex hulls because all vertex chords connect the closest neighboring pairs of points. However, this is not always the case, as the 8^v triacon diamond in Figure 9.6 shows. For the same set of tessellation vertices, there is quite a difference between the schema faces for this octahedron (a) and its convex hull surface (b). The vertex points in both figures are the same.

(a)   (b)

Figure 9.6. Tessellation triangles and their convex hull surface.

Convex hull analysis demands precise surface definitions and volume calculations. The differences in our schemas are very small, and there is a computational trade-off here. We can find the convex hull of the whole sphere at once, or find the volume of a smaller representative part and then extrapolate it to the whole sphere. We choose the latter approach. Figure 9.7 shows examples of representative convex hulls for each class we developed. We calculate the volume of these sections and then extrapolate (multiply) it by the number of times required to define the total sphere's convex hull. The convex hull surface for each schema covers the faces of their subdivision triangles and their sides (to the origin of the sphere) to enclose a complete volume. The volume sides are created from a series of triangular surfaces defined by the end points of chords that run along the PPT's or diamond's edge and the origin of the sphere. For Class I and II convex hulls, a single 3D volume model is sufficient. For Class III convex hulls, two volume models are needed, one to define the volume for the triangles totally within the PPT and another to define the volume of triangles that straddle the PPT's edge. Each one has its own volume, and a particular number of times its volume is multiplied to describe the whole sphere.

Table 9.8 tabulates the convex hull volumes for each unit radius spherical schema we developed. The table lists the basic section, volume, and the number of times this volume is replicated to define the entire sphere's convex hull volume. Class III schemas require two volumes and the total number of times each appears in the total volume of the sphere; their subtotals are listed separately. The total convex hull volumes are then listed and ranked as a percentage of an undivided unit sphere's volume.

The analysis once again indicates that Class I Equal-arcs (three great circles) has the best tessellation point distribution within Class I; its convex hull volume is the greatest. The differences in convex hull volumes are due only to their point distributions because

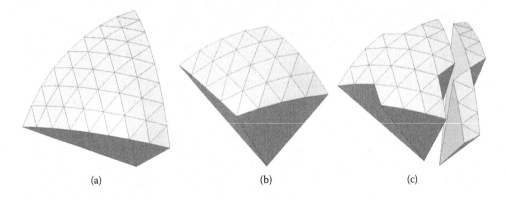

(a)   (b)   (c)

Figure 9.7. Example class I, II, and III convex hull models.

Subdivision Schema Convex Hulls						
Schema	Volume	Instances	Sub-Total	Total	%Sphere	Rank
Class I						
Equal Chords	0.207634	20	–	4.152685	99.138%	3
Equal Arcs$_{(2\ GCs)}$	0.207629	20	–	4.152571	99.135%	4
Equal Arcs$_{(3\ GCs)}$	0.207652	20	–	4.153034	99.146%	1
Mid Arcs	0.207637	20	–	4.15274	99.139%	2
Class II						
Triacon	0.138018	30	–	4.140535	98.848%	6
Class III						
Midface	0.144326	20	2.886529			
Straddling Edge	0.041928	30	1.257845	4.144374	98.940%	5

Table 9.8. Subdivision schema convex hulls.

each Class I schema placed the same number of points (642) on the sphere's surface and none are tangent. Convex hull comparisons can still be made if the number of points distributed by several schemas are not all the same. Class II and Class III schemas distribute 482 and 522 points, respectively. We are comparing volumes only, and neither of these two schemas are as good as Class I schemas.

Convex hull volumes are excellent ways to compare two or more tessellation point distributions. For comparisons to be meaningful, each sphere must have the same radius and it is best to have the same number of points on their surfaces. In all comparisons, high precision surface definitions are required because volume differences can be small.

9.8   Spherical Caps

A spherical cap is a slice of a sphere created by passing a plane completely through the sphere, without intersecting the origin. A cap looks somewhat like a contact lens and each one, centered over a tessellation vertex, has a spherical surface area and volume. Figure 9.8(a) shows caps for a Class II triacon icosahedron 10^v. All have the same radius, and they are as large as they can be without causing overlaps. We can use their total surface area to measure the efficiency of the subdivision schema. The ratio of the total cap area

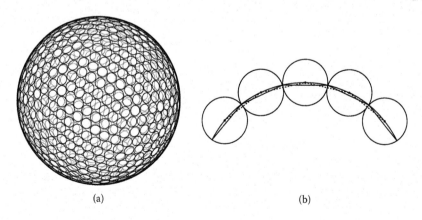

(a)   (b)

Figure 9.8. Spherical caps and kissing-touching balls.

to the sphere's surface area is a density factor that tells us the percentage of the sphere's surface they cover. The subdivision schema with the highest density factor for the same number of points distributed on the sphere has the better distribution of those schemas.

Caps kiss and touch, just as kissing-touching balls do, and they appear to be measuring the same thing. However, the cap's lesser circle diameter is based on the arc distance between the two closest tessellation vertices, as shown in Figure 9.8(b). The arc distance is measured along the surface of the sphere (accented arcs). The ball's diameter is also based on the two closest tessellation vertices, but it uses the chord distance between them instead (dashed lines). The result is the spherical cap's lesser circle diameter is a little larger than the kissing-touching ball radius for the same subdivided sphere.

The size and distribution of spherical caps is an important measure of how evenly distributed grid points are on the sphere's surface. We can compare one schema to another by comparing how much of the sphere's surface is covered by their caps. We can also make stereograms and visually see how spherical caps are distributed.

9.9   Stereograms

A *stereogram* is a 2D projection of the surface of a sphere onto a plane called the *primitive*. Every point on the surface of the sphere can be projected onto the primitive except one point: the projection point itself. In a sense, a stereogram is a 2D map, with some distortion, of the sphere's 3D surface (Appendix A describes the technique in detail).

Figure 9.9 shows the dimensions and formulas we use to find a cap's volume and area. Table 9.9 uses these formulas to find the area and density of the caps for the schemas we developed in the last chapter. Figure 9.10 through Figure 9.12 are the stereograms for each of the schemas. Each stereogram shows tessellation vertices and their caps in the northern hemisphere of the icosahedron. The ones in the southern hemisphere produce the same stereogram (but are rotated 32° around the center of the stereogram).[2] They do not add new information and, if superimposed, make stereograms difficult to read, so we omit them.

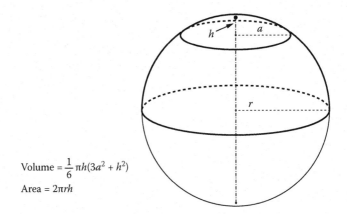

$$\text{Volume} = \frac{1}{6}\pi h(3a^2 + h^2)$$
$$\text{Area} = 2\pi rh$$

Figure 9.9. Spherical cap dimensions.

---

[2] For the vertices of an icosahedron, there are six axes with five-fold rotation symmetry. The icosahedron can be rotated around a vertex axis any multiple of 72°, and the appearance of the icosahedron is not changed. Faces in the southern hemisphere are rotated 32° with respect to those in the northern hemisphere. See Chapter 6 for details.

Spherical Cap Comparisons						
	Cap's					
Schema	Radius (a)	Height (h)	Area	Total Caps	Density	Rank
Class I						
Equal Chords	0.059729	0.001785	0.011218	642	57.31%	5
Equal Arcs$_{(2\ GCs)}$	0.066339	0.002203	0.013841	642	70.71%	4
Equal Arcs$_{(3\ GCs)}$	0.069142	0.002393	0.015037	642	76.82%	1/2
Mid Arcs	0.069142	0.002393	0.015037	642	76.82%	1/2
Class II						
Triacon	0.076658	0.002943	0.018489	482	70.92%	3
Class III						
Skew	0.06432	0.00207	0.01301	522	54.05%	6

Table 9.9. Spherical cap comparisons for unit radius sphere.

To facilitate comparing one schema with another and to highlight how caps relate to the base spherical polyhedra, the icosahedron is kept in the same vertex-zenith position in each figure. Our view is from the positive z-axis looking directly at the primitive on the xy-plane. The icosahedron's great circle edges (solid) and face medians (dashed) are projected as well to make it easier to interpret the pattern and see how points and their caps relate to the icosahedron's faces.

In each stereogram, one point (dot) surrounded by a circle appears on the primitive for every vertex-cap pair in the northern hemisphere or along the sphere's equator. Although the caps on the actual sphere all have the same radius, their projection on the primitive does not; their radii increase as they get closer to the circumference of the primitive. Projected circles are not centered on their corresponding vertex points, either. These subtleties are difficult to read at the scale of these stereograms, but the reasons are explained in Appendix A.

In Figure 9.10, (a) and (b) show spherical cap stereograms for Class I Equal-chords and Equal-arcs (two great circles), respectively. Caps are evenly distributed in Equal-chords, but few touch. The overall density (percentage of sphere's area covered) is relatively low—only 57.31% of the sphere is covered, as shown in Table 9.9. Caps appear around the circumference of the primitive, one of the icosahedron's six mid-edge great circles. The pattern of their spacing is regular, but the distances between points are not all the same. The cap size and vertex distribution of Equal-arcs (two great circles) are better than the cap size and vertex distribution of Equal-chords, despite the asymmetry of the points within each PPT. The largest cap possible is determined by the arc distance between points near one edge rather than at the PPT's apices. They are the same points that were highlighted in kiss-touch balls in Figure 9.8(b). If you look closely at the caps along the PPT's edges, you will notice they are not all the same distance from the edge. Compare the caps along the two edges circled. The ones along the edge labeled A in Figure 9.10(b) are closer to the edge than the ones along the edge labeled B. This is because this Equal-arcs schema defined tessellation points by intersecting only two great circles, rather than three.

In Figure 9.11, (a) and (b) show the spherical cap distributions of Equal-arcs (three great circles) and Mid-arcs, respectively. Equal-arcs' caps, along every PPT edge, are evenly spaced and touch each other. Caps also follow every PPT median, though they do not contact each other. From prior analysis, we expect this Equal-arcs schema to be a good performer and its stereogram confirms this. Caps are large, evenly distributed, and cover 76.82% of the sphere's surface, the highest percentage thus far.

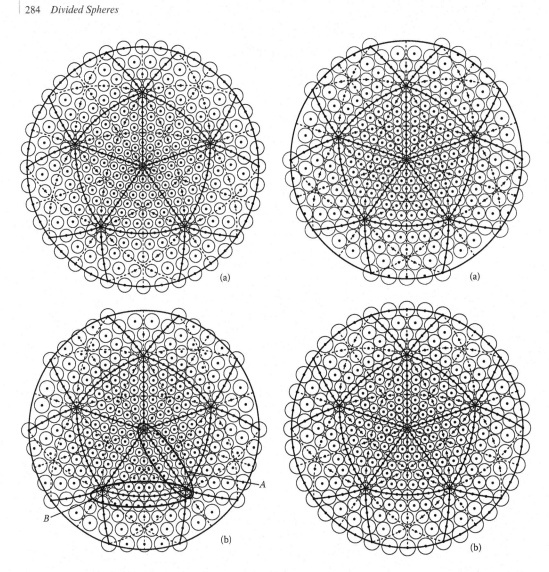

Figure 9.10. Equal-chords and Equal-arcs (two great circles) spherical caps.

Figure 9.11. Equal-arcs (three great circles) and Mid-arcs spherical caps.

Figure 9.11(b) shows Mid-arcs' caps, and they cover 76.82% of the sphere's surface. Surprisingly, their density is identical to Equal-arcs (three great circles); see Table 9.9. Like Equal-arcs (three great circles), its cap radius is determined by the closest two points in the grid and this occurs along the PPT's edge, where all of these caps touch. These two schemes perform identically and better than the others. They are tied for the first two rankings.

Mid-arcs' stereogram reveals one of its symmetry features. Recall from Section 2.5 that one of the icosahedron's many sets of great circles are six that span from midedge to midedge across each of its faces. These circles define equilateral triangles within every face of the icosahedron, exactly like Mid-arcs' initial subdivision does. As Mid-arcs continues to subdivide the PPT, the edges of this inner equilateral become increasingly subdivided. Its frequency is half the overall frequency. In our 8^v example, its edges will be 4^v. The equally spaced points along these inner-PPT edges project as equally spaced caps around the circumference of the primitive. We can verify that these caps belong to one of the six great circles. Each edge of the inner-PPT equilateral consists of four

segments with arc lengths of 9° for a total of 36° per edge. Each of these six great
circles passes through ten such edges as
they complete their 360° tour of the whole
icosahedral sphere. Unlike the caps along
the PPT's edges, these caps do not touch
because their tessellation vertex points are
farther apart than those along the PPT's
edges. Caps aside, the fact Mid-arcs subdivisions result in equally spaced points
on these six great circles may be exploited
in some design situations.

Triacon schemas result in a diamond
pattern (four LCDs back to back) symmetrical to each edge of the spherical polyhedra. Figure 9.12(a) highlights a typical
diamond. The pattern of caps within a
diamond is also distinct. They are evenly
spaced, but not tangent, along both of
the principal triangle edges. The triacon is
limited to even frequencies. Thus, they always create a vertex and a cap centered at
the intersection medians in the middle of
every PPT face. None of the Class I schemas do this. Unlike the other schemas, the
smallest arcs are found in the middle of
the PPT's face and not at its apices. The
triacon caps are the largest of the schemas,
but its grid distributes only 482 vertices
over the sphere, the lowest number, when
compared with the other schemas. Even
though its cap area is the largest, they cover only 70.92% of the sphere.

Figure 9.12(b) shows spherical caps
for a Class III schema. Inspection of the
spherical caps shows the vertex distribution is not very good. For starters, this 8ᵛ
subdivision distributes only 522 vertices
on the full sphere, compared to the 642
distributed by each of the Class I sche-

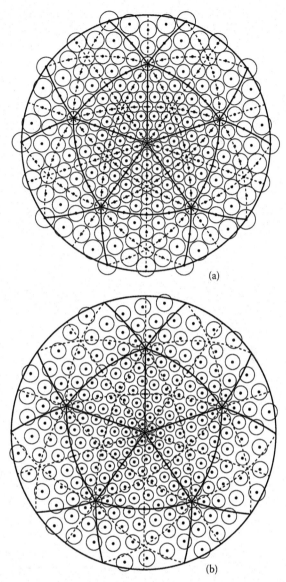

Figure 9.12. Class I triacon and class III
spherical caps.

mas. The smallest arc distance between grid points is one of the smallest in the six
schemas we review, and this determines the cap's radius. This schema also has the largest arc distances between vertices, which means spherical caps near the middle of each
polyhedral face are quite separate from their neighbors' caps. The stereogram clearly
shows this. Caps in this schema cover only 54% of the sphere, the lowest density of any
of the schemas we are comparing.

To summarize, spherical caps are slices of the sphere where the plane of the slide
does not pass through the origin of the sphere. The best schemas distribute the most
vertices with the shortest and least variation in arc lengths between vertices. The radius
of the cap's lesser circle is determined by the arc between the two closest vertices within

a subdivision schema. A spherical cap surrounds each tessellation vertex within the schema. All have the same radius and, while some caps kiss-touch, none will overlap. The ratio of the total cap surface area to the sphere's surface area is called the density factor and it tells us how much of the sphere is covered. The more area covered, the more uniform the tessellation vertices are distributed.

The larger the cap area, as a percentage of the sphere's area, the more uniformly the tessellation vertices are distributed. Stereograms provide a convenient way to see the cap distribution. Various symmetries within the schema are apparent, and they can be compared side by side or overlaid. Of the schemas reviewed, Equal-arcs (two and three great circles) and Mid-arcs schemas perform the best. Class II triacon was also a strong performer, but not as good as the other two. Triacon has larger caps than the others, but it also distributes the fewest tessellation vertices on the sphere of any schema.

9.10 Face Orientation

Thus far, we have compared kissing-touching distances, chord lengths, triangle areas, and shapes, convex hulls, and spherical caps. These metrics are very helpful when deciding which schema is best for an application. However, none of them tells us about the orientation of tessellation faces. Different shaped triangles can have the same kissing-touching distance, but very different orientations. We saw this earlier when we created triacon subdivisions for each of the Platonic solids. Here we measure the direction of faces. In some applications, face orientation may be the deciding factor when choosing one subdivision schema over another.

Every planar face we create in a subdivision is part of an infinite plane in 3D space, and each one has a normal. In Chapter 4 we described a normal as a mathematical entity that indicates both direction and magnitude. The direction of a normal to a plane is always perpendicular to the plane. Depending how the plane (or triangle) is defined, the normal's direction points towards the origin of the sphere, or away from it. In either case, however, the normal always points perpendicular to the plane of the triangle and everything on it.

Another way to visualize a normal is to think of its direction as a line that radiates from or towards the origin of the sphere and passes perpendicular to the plane of the triangle. The sphere's surface has an infinite number of planes tangent to its surface; thus, it also has an infinite number of normals. However, when a sphere is subdivided, triangular faces result and their planar versions each have a normal. Normal vectors have length, in addition to direction, and their length depends on the perpendicular distance between the triangle's plane and the origin of the sphere. We can use *face normals* as pointers to see which direction a face point is relative to the center of the sphere. Since the line of the normals pierces the sphere's surface at different points, we can use the normal to derive a Cartesian point that is on the surface of the sphere. Stereograms of derived points from the normals of all the subdivision faces tell us how uniformly subdivision faces are oriented. It is easy to define these points. All we have to do is "*normalize*" the normal, which means make its length equal to one, the radius of the sphere we are working with.

Figure 9.13 illustrates the relationship between subdivision faces, their face normals, and the projected point on the sphere derived from them. The figure shows a single Class II 8^v diamond subdivision for a spherical tetrahedron. Each arrow is perpendicular to the face at its base. The mixed orientation of triangles is evident. The arrows (blue) indicate the normals for their corresponding face. The arrow's base is

Figure 9.13. Projected face normals.

tangent to the sphere's surface, indicating where the normalized (unit) normal is on the surface of the sphere.

In Figure 9.13, the normal arrow's blue part is totally outside the sphere, and all arrows are the same length to make it easier to compare their directions. The base of each arrow (red) is extended inside the sphere's surface to be tangent to the point where the normal direction line intersects its triangle. The extended arrow lengths (red) indicate how far above the plane of the triangle the projected normal point is above the sphere. Clearly, they are not all the same length. Their lengths decrease to zero as their intersection points get closer to one of the triangle's vertices. Although there are no equilateral triangles in this example, the triangles that are nearly equilateral have normal arrows located closest to their centers.

In Figure 9.14, we see the normals for fully subdivided triacon spheres. Both are Class II 8^v subdivisions, and each triacon diamond in both spheres has 32 faces. Figure 9.14(a) shows a subdivided tetrahedron, while (b) shows an icosahedron. It is obvious that the icosahedron's face orientations are more uniform. The perpendicular

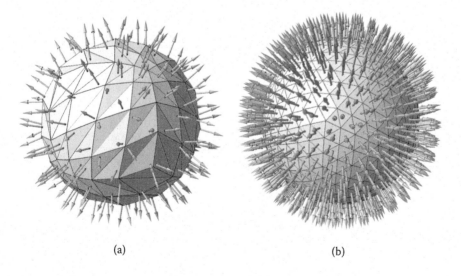

(a)   (b)

Figure 9.14. 8^v Class II spherical tetrahedron and icosahedron face normals.

face arrows point outward more consistently than the tetrahedron's. The main reason is that the icosahedron's triacon diamond is smaller than the tetrahedron's, which means more of them are needed to cover the sphere. More and smaller diamonds mean more vertices are distributed on the sphere and they are closer together, forming smaller triangles that are nearer to being equilateral than the tetrahedron's. The icosahedron has 30 characteristic diamonds (one diamond for each polyhedral edge), and this creates 960 triangles in the overall sphere, while the tetrahedron has only 6 characteristic diamonds, which create 192 overall faces. The improvement in face orientation is easy to understand. More diamonds and more points mean smaller faces with near-equilateral shapes. It all adds up to a better distribution of face orientations and its normals (arrows) show it.

To better understand face orientations, we use stereograms to compare the subdivision face normals from one schema to another. Normals to the planes of subdivision triangles cannot be used directly to make stereograms; however, we can derive a Cartesian point from the normal that is on the surface of the sphere and can be stereoprojected. The derived point will be located where the normal's ray pierces the surface of the sphere and it is this derived point that we stereoproject (see Appendix A for a complete explanation of how this is done.)

In Figure 9.15, (a) and (b) show the projected derived points for the face normals of the Class II tetrahedron and icosahedron subdivision just discussed. The fact that the projected points are more spread out in (a) than in (b) indicates that the tetrahedron's faces have a greater dihedral angle between them than the icosahedron's. The stereogram clearly shows this. In Figure 9.15(a), the tetrahedron's projected derived points are not only spread out more but they pair up in rows parallel to the diamond's long axis. This also happens in the case of the icosahedron (b), but to a far lesser extent. These stereograms confirm what we have already observed in the 3D models. For the same subdivision schema and frequency, the icosahedron has more uniform face orientations.

The differences, shown in Figure 9.15, are quite dramatic, largely due to the choice of foundation polyhedra (tetrahedron versus icosahedron). Here, stereographics merely verified something that was already obvious. However, when the foundation polyhedra are the same, the differences in face orientations between one schema and another for the same frequency can be subtle, and here stereographics prove to be an invaluable tool.

In Figure 9.16, (a) and (b) show stereograms for Equal-chords' and Equal-arcs' (two great circles) points derived from each subdivision face normal. Their stereograms are similar, but there are key differences. For example, it is apparent that derived

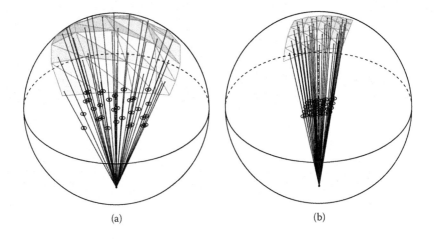

(a)　　　　　　　　　　　　　(b)

Figure 9.15. Class II 8^v tetrahedron and icosahedron stereograms.

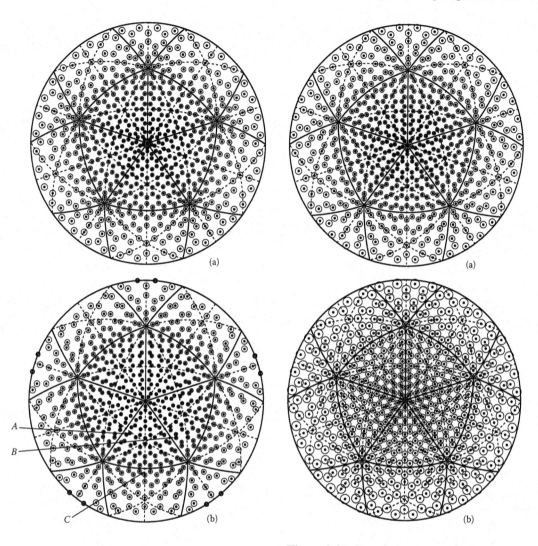

Figure 9.16. Equal-chords and Equal-arcs (two great circles) face orientations.

Figure 9.17. Equal-arcs (three great circles) and Mid-arcs face orientations.

points tend to pair up, more in Equal-arcs schema (b) along the icosahedron's edges marked *A* and *B*, whereas the points for Equal-chords (a) tend to be more uniformly distributed. Neither schema projects derived points on the icosahedron's edge, but both distribute points along face medians. Only trivial differences occur in the derived points within a principal triangle face.

Overall, face orientations for the Equal-arcs (two great circles) are not as uniform as Equal-chords. Equal-arcs' faces are only symmetric to one median per principal triangle, and the tessellation grid (and the faces it creates) is not rotationally symmetric around the center of the icosahedron's face. This can be seen in the projected derived points along PPT edge C. They are not the same as those along edges A and B, and this is a serious drawback for some applications. The stereogram clearly shows the problem.

Figure 9.17 shows the projected derived points from the face normals of both Equal-arcs (three great circles) and Mid-arcs. Both look similar and the differences are subtle. Mid-arcs' points are more closely paired up than Equal-arcs'. Of the two, Equal-arcs' faces are more uniformly oriented.

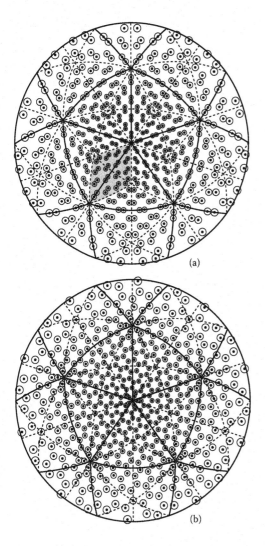

(a)

(b)

Figure 9.18. Class II triacon and Class III face orientations.

Triacon schemas are based on a diamond layout that is symmetrical to the polyhedron's edges. Figure 9.18(a) highlights a typical diamond. The most dominant features of the stereogram are the pairs of projected points in rows parallel to the icosahedron's edges. These rows are somewhat aligned to polyhedral edges on which the subdivision schema is based. The triacon schema also creates isosceles faces symmetric to every polyhedral edge; thus, it is no surprise to see projected derived points fall on the icosahedron's edge arcs. None of the other schemas routinely have derived points along the edges, though some frequencies of Class III will produce a few points there. Since the triacon does not produce a triangle in the middle of the PPT's face, there is no point at or near that point like there are in Equal-chords, Equal-arcs (three great circles), and Mid-arcs schemas. No triangle straddles the principal triangle's three medians. Thus, we will never find a projected derived point in the middle of the principal triangle, as we would in most Class I and some Class III schemas.

At first glance, Class III schemas seem almost impossible to compare to the others. The highly skewed nature of the overall tessellation grid, with respect to the icosahedron foundation, is immediately apparent in Figure 9.18(b). Unlike previous schemas, only one derived point per principal triangle falls on a face median. It is located at the midface point of the icosahedron. None of the other points fall on an icosahedra edge, median, or vertex. Despite all of this, the stereogram indicates a surprisingly uniform distribution of face orientations.

Uniform face orientations are a direct result of two factors: the number of points distributed on the sphere, and the degree to which the chord distances between them are equal. When the subdivision frequency is the same, subdivisions of spherical icosahedra produce better results than can be achieved with octahedra or tetrahedra because they start with more subdivided principal triangles and smaller chords. However, within the same polyhedral base, Class I Equal-arcs (three great circles) and Mid-arcs have the most uniform face orientations.

9.11 King Icosa

By almost any measure, the icosahedron is the best polyhedral foundation to use, when the goal is to create the most uniform point distribution possible, regardless of which subdivision schema is used. The main reason is its small face size: they have the least area (or least spherical excess) of any regular polyhedron for a unit sphere.

Figure 9.19. Polyhedra foundations compared.

The tessellation grid for a single icosahedral PPT defines only 5% of the total sphere's surface area, while the same frequency subdivision for an octahedron or tetrahedron must cover 12.5% and 25% of the sphere's area, respectively. We can illustrate this point graphically by subdividing all three polyhedra with the same frequency grid and simply noting the spacing between grid vertices. Figure 9.19 marks the vertices of an Equal-chords 8^v grid on the tetrahedron (a), octahedron (b), and icosahedron (c). In each case, kissing-touching occurs around the polyhedron's vertices, which are designated by red balls. The uniform distributions achieved with an icosahedral foundation (c) cannot be disputed. Its smaller and more numerous faces lead to a significantly more uniform distribution than the other two. The icosahedron is king!

9.12 Summary

In the previous chapter, we introduced three basic classes of spherical subdivision and a few of their variations. In Class I, the polyhedron's principal triangle was subdivided by defining edge-reference points (Equal-chords or Equal-arcs). These reference points were used to position great circle arcs intersecting where tessellation vertices occur. Class I grids run approximately parallel to the edges of the PPT, and the edge remains part of the final tessellation grid. Subtle differences in the way great circles are intersected lead to subtle differences in the way points on the PPT are created. These differences influence chord lengths and triangle area, shape, and orientation, and affect rotational and reflection symmetry.

Class II subdivisions produce grids that are parallel to the altitudes of the principal triangle. Instead of subdividing the entire principal triangle, they subdivide a diamond shaped area that is symmetrical to the polyhedron's spherical edge. This diamond has a major and minor axis, which further breakdown the diamond into four back-to-back right triangles. Subdividing just one of them defines the grid for the other three. Since Class II schemas are based on symmetrical areas around the polyhedron's edges (not faces), they can be applied to all five Platonic solids.

Class III is a variation of Class I, but its grid is skewed relative to the principal triangle's edges. Like Class I, Class III can be applied to tetrahedra, octahedra, and icosahedra, but not to spherical cubes and dodecahedra.

In this chapter, we analyzed the differences between each of the schemas for these three classes. The differences in the tessellation grids created from the various schemas are subtle, but significant. What we found is that there is no single *"perfect"* subdivision, each has characteristics that make one or two of them more appropriate for an application than others. Although we did not subdivide other polyhedra, it should be clear, the icosahedron generally provides the best spherical framework if uniform grids are the objective. Our hope is that the metrics developed here will provide a way of deciding which is best for your needs.

Additional Resources

Subdivision Schemas

Clinton, Joseph D. *Advanced Structural Geometry Studies Part I—Polyhedral Subdivision Concepts for Structural Applications.* National Aeronautics and Space Administration contractor report NASA CR-1734, Carbondale, IL: Southern Illinois University, 1971.

_____. "A Group of Spherical Tessellations Having Edges of Equal Length." West Chester, PA: Clinton International Design Consultants, undated.

Kitrick, Christopher J. "Plane Hexagonal Geodesic Tessellations." Proceedings of the International Association for Shell and Spatial Structures (IASS), 2015.

Wenninger, Magnus J. *Spherical Models.* Cambridge [Eng.], New York: Cambridge University Press, 1979 (new appendix, 1999).

Geodesic Math

Clinton, Joseph D. "Geodesic Math." *Domebook 2* ed. Lloyd Kahn et al. Bolinas, CA: Pacific Domes, 1971. 106–113.

Kenner, Hugh. *Geodesic Math and How to Use It.* Berkeley: University of California Press, 1976.

Kitrick, Christopher J. *"Geodesic Domes." Structural Topology* 11 (1985):15–20.

_____. "A Unified Approach to Class I, II and III Geodesic Domes." *International Journal of Space Structures* 5.3–4 (1990): 223–246.

⑩ Self-Organizing Grids

Christopher J. Kitrick

Many spherical division techniques and methods have been discussed throughout this book and it begs the question of whether or not new methods of dividing the spherical surface are still to be discovered that either provide a more efficient way to either handle traditional approaches or open up entirely new vistas. During the twentieth century we saw the development of numerous advances in spherical subdivision that focused on limiting differences in surface angles and face triangles, while minimizing production variations. With the rapid increase in computing power and ability of manufacturing to handle more complex designs some of the early limitations no longer pose the same restrictions. In addition, the entire scope of spherical division has expanded to address entirely different problems in global mapping, weather forecasting models, scientific instruments, communication, and even transportation.

This chapter will introduce the relatively new concept of spherical *self-organizing grids*. The simplest description of a self-organizing grid is an initial configuration of vertex elements on a spherical surface that self-modify their positions to converge to specific conditions or goals called organizing targets. Self-modification is a constrained and iterative activity that results in the vertices converging to a targeted final position. Why the self-organizing concept is so appealing is that it allows some very challenging groups of configurations to be precisely computed, across the full scope of topology, whereas traditional subdivision techniques, as outlined in Chapter 8, provide no workable solution framework. Some other equally important aspects to the approach are that it is relatively simple, easy to set up, computationally inexpensive, and converges to a result quickly. Since the approach was only recently developed it is still being investigated on how it can be applied to additional tessellations.

An important reason that self-organizing grids converge to a solution is because the target grid itself is not fully triangulated or constrained. An example of a fully constrained grid on the spherical surface is a triangle grid, where every vertex is directly linked to all its closest neighbor vertices. It does not matter how many triangles there are as long as the entire surface is composed of only triangles. A typical question of a spherical triangle grid is what is the minimum number of different lengths required. Of course, a spherical triangle grid that uses a single member edge would be wonderful but it is not possible, except in the simplest cases: the platonic polyhedron (icosahedron, octahedron, tetrahedron). All other triangular configurations on the sphere require more than one edge length.

The concept and examples of valid self-organizing spherical grids, covered in depth in this chapter, include hexagonal grids and grids that are a combination of hexagons and triangles. In each of these cases every vertex is not directly linked to all its closest neighbors. Lastly, throughout this chapter we will use the term edge to represent the distance between two vertices. In all the spherical examples in this chapter we will use a unit sphere, where the radius is one. On a unit sphere the term edge and chord factor (Section 4.4) have the same meaning.

10.1 Reduced Constraint Networks

A traditional geodesic arrangement is a triangular grid where every vertex is connected directly to its nearest neighbor vertex. In this type of arrangement there are too many constraints for the self-organizing method to be effectively utilized. But if we begin to remove elements, vertices and their connecting edges, the number of constraints is hereby reduced, thus producing a *reduced constraint network*. Rather than just randomly removing edges we'll consider symmetrical networks where the removal of edges is an orderly process. Figure 10.1(a) shows a fully triangulated network and three reduced networks (b, c, d) using different symmetric vertex and edge removals. In these particular reduced networks, every remaining vertex is only connected to three neighboring vertices. Figure 10.1(b) is the typical hexagonal arrangement, while (c) and (d) are different arrangements that have both hexagons and triangles, now referred to as *rotegrity* or *nexorade* arrangements when applied to a sphere. There are some excellent examples of geodesic structures that do not have a traditional fully triangulated frame, the majority of them being hexagonal-based configurations. An early example of a hexagonal configuration is the American Society for Metals International Headquarters dome[1], shown in Figure 10.2, that was completed in 1958. A more contemporary multiple shell implementation is the Eden project in the UK, completed in 2000.

10.1.1 Hexagonal Grids

Hexagonal configurations, shown in Figure 10.1(b), are the first example of tessellations that were proposed with the goal of finding a placement of vertices on the spherical surface such that the distance between every hexagonal corner was the same length. The first published work on *equal edge tessellations* was a 2002 conference paper[2] that presented a number of simple cases applied to the three platonic polyhedra. The paper

---

[1] Dome engineered and designed by T.C. Howard (Synergetics) located in Novelty, Ohio, added to the National Register of Historic Places in 2009.
[2] (Clinton 2002, A Group of Spherical Tessellations Having Edges of Equal Length).

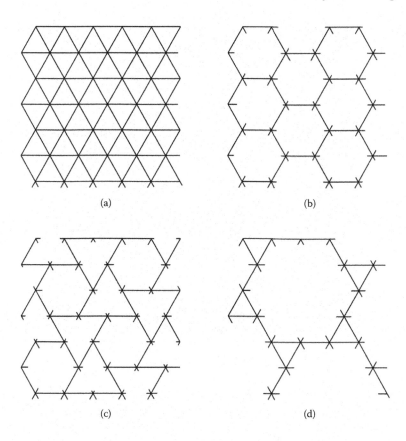

(a)   (b)

(c)   (d)

Figure 10.1. Regular and symmetric reduced constraint networks.

Figure 10.2. American Society for Metals International Headquarters.

Figure 10.3. Natural honeycomb.

posited the general solution as an open problem. Another paper[3] in 2015 extended the solution to cover a group of symmetric configurations but it was not until 2019 that a general solution[4] was published. The general solution for the hexagonal case, utilizing the self-organizing methodology, was also shown to be extensible to other tessellations. A benefit of a general solution is that it allows the full scope of topology to be explored not only for specific implementations but also to understand how well the self-organizing method serves to achieve an optimal solution.

A few words about hexagonal grids. They are a common occurring natural form. Honeycombs[5] (Figure 10.3) constructed by bees are a good analogy for the self-organizing method. Even though they are built by numerous independently and locally operating individual bees without a plan they produce a highly regular hexagonal form. For human managed bee hives the bees build nearly perfect hexagonal grids on a planar form. In these grids every cell is essentially the same and every edge of the hexagons are very nearly equal.

When attempting to construct a hexagonal grid on a non-planar surface such as a spherical surface, it is not possible to keep every hexagonal cell the same shape. Even when allowing for the cell shapes to deform, it is a challenge to adjust the geometry so that the edges of the cells remain a constant length. For grids, such as the hexagonal type, to be applied to a spherical surface we must also introduce some non-hexagonal cells to provide the grid the ability to wrap around the non-planar surface. If one of the Platonic solids is used to form the grid the shape of the non-hexagonal cells will be directly related to the base Platonic solid. Thus, a spherical hexagonal grid applied to an icosahedron will have exactly twelve pentagonal cells, and for an octahedral-based grid there will be exactly six square cells.

---

[3] (Kitrick 2015).
[4] (Kitrick 2019).
[5] (Image courtesy of David Hablützel).

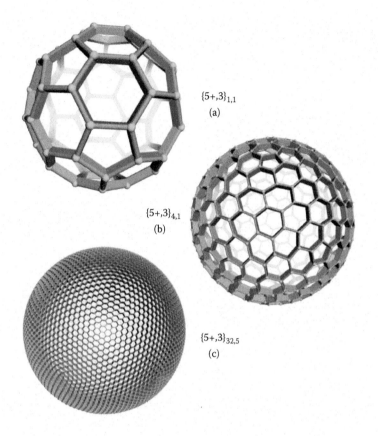

$\{5+,3\}_{1,1}$
(a)

$\{5+,3\}_{4,1}$
(b)

$\{5+,3\}_{32,5}$
(c)

Figure 10.4. Self-organized icosahedron-based honeycombs.

Figure 10.4 illustrates three self-organized spherical hexagonal grid configurations[6]. The first (a) at the top left is the familiar soccer ball pattern (truncated icosahedron, see Table 6.5), while the other two are increasingly more complex and chiral (Class III) hexagonal grids. In all three cases the cells internal shapes vary but cell edge dimensions are all constant. The methodology for finding the position of vertices on any complex arrangement that satisfies the invariant edge goal will be covered in depth in this chapter since the there is no regular tessellation method available that will arrive at these equal edge solutions.

10.1.2 Rotegrities, Nexorades, and Reciprocal Frames

The other reduced constraint networks shown in Figure 10.1 (c) and (d), which includes both hexagonal and triangular elements, when applied to a spherical surface, now identified as rotegrities, are also often called nexorades. Rotegrities were developed by Dick Boyt at Crowder College in the early 1970s borrowing from Fuller's geodesic tensegrity concept on how elements could be derived through rotation (Figure 10.5). Nexorades were developed by Olivier Baverel at the University of Surrey in 2000 as a general approach to free-form reciprocal frame structures.

---

[6] The three-dimensional images for all figures in Chapter 10 were created with *DisplaySphere*, the contributing author's software for displaying spherical geometry. The software and associated documentation are freely available through GitHub.

Figure 10.5. Fuller's central angle turbining.

Rotegrities share many geometric similarities to tensegrities but do not separate compression and tension elements. Boyt's rotegrities replaced each set of Fuller's tensegrity tension and compression elements with a single geodesic arc. Nexorades build upon the reciprocal frame concept that has been around for hundreds of years and has been used for roofs and other structures[7] where compression elements sit upon each other at some mid-point location (Figure 10.6). Baveral introduced the term *nexor* to represent a single reciprocal frame element and we will use that nomenclature for the elements in rotegrities.

In either rotegrities or nexorades the nexors can have multiple possible shapes. In rotegrities the nexors tend to be curved or bent arcs that represent a geodesic path as shown in Figure 10.5. In nexorades nexors are typically linear but in either a cylindrical or rectangular shape representing the straight line from one vertex on the sphere to the next. In either case the end of every nexor will sit somewhere along a neighboring nexor and thus every nexor will have two other nexors attached somewhere along its length. A fundamental aspect of the geometry of rotegrities and nexors is the positioning of two nexor ends across the shared nexor. As originally indicated in Figure 10.1 the triangular holes in the spherical reduced constraint types implementations (c) and (d) will either *underlap* or *overlap* giving us two basic configurations of rotegrities shown in Figure 10.7.

---

[7] An excellent overview of traditional reciprocal frames in architecture (Larson 2008).

Figure 10.6. Three nexors define simplest reciprocal frame.

Figure 10.8 highlights the geometric relationship between tensegrities, rotegrities, and nexorades. The left of the figure (a)–(c) shows the overlapping scenarios while (d)–(f) illustrate the underlapping scenarios. The rotegrities in (b) and (e) show the tensegrity strut and associated tension members replaced with arcs between the vertices. The nexorades in (c) and (f) shows the enlargement of the tensegrity strut's radius such that the struts are in direct contact with each other. In either the rotegrities or nexorades configurations some connection mechanism, such as straps, pins, bolts, etc., must be employed to maintain the position of the nexors to stabilize the structure.

In the development of the rotegrity, Boyt started with a normal triangular geodesic grid and then rotated every triangle edge arc around its midpoint by 36° (Figure 10.9), extending each edge till its end was coincident to its neighboring edges. The rotation of 36° resulted in a close approximation of the nexors being evenly divided. By varying the rotation angle and direction a rotegrity design could be either overlapped or underlapped. Conceptually straightforward and visually appealing the form has become quite popular. Boyt's original design methodology of relying on rotating edge arcs around their original center point did not allow the overlap proportions to be constant but provided a workable methodology for deriving the exact geometry

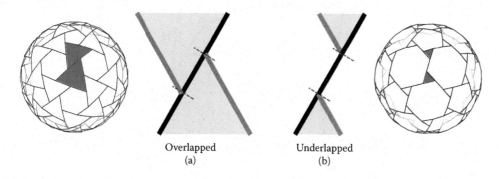

Overlapped
(a)

Underlapped
(b)

Figure 10.7. Basic nexor configurations of rotegrities.

(a)   (d)

(b)   (e)

(c)   (f)

Figure 10.8. Geometric relationships between tensegrities, rotegrities, and nexorades.

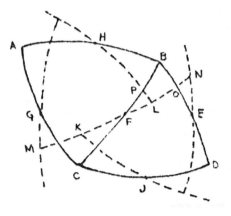

Figure 10.9. Boyt's edge rotation to derive rotegrities.

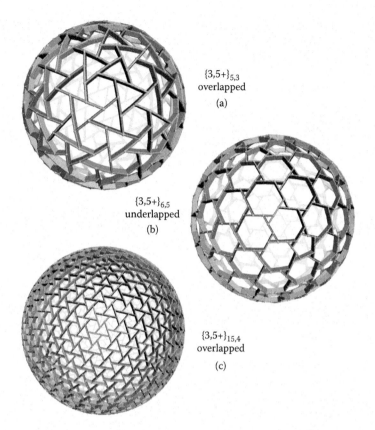

$\{3,5+\}_{5,3}$
overlapped
(a)

$\{3,5+\}_{6,5}$
underlapped
(b)

$\{3,5+\}_{15,4}$
overlapped
(c)

Figure 10.10. Self-organized icosahedron-based rotegrities.

using spherical trigonometry[8]. The self-organizing method, in contrast, allows us to set exact proportional ratios by freeing us from restraining the arcs' midpoint to their original locations.

Rotegrity tessellations are formally considered *non edge-to-edge tilings*. There are eight families of these types of tilings and the underlapping and overlapping rotegrities represent two of those families[9]. One characteristic that is common across most types of reciprocal frames, and always for rotegrities, is handedness, or enantiomorphism. Thus, an implementation will have either a right or left-handed topology, see Section 6.4.6. Figure 10.10 illustrates three self-organized rotegrity configurations. All three configurations show increasing complexity, with (a) and (c) being the overlapped type, while (b) highlights an underlapped type. In all three cases the nexors exactly follow a geodesic path that is divided into three equal sub arcs. Similar to the equal edge hexagonal configurations there is no regular tessellation method available that will arrive at these solutions.

10.1.3 Organizing Targets

For self-organizing grids to organize they require a specific design target. The target can be different based on the type of grid being developed but it is typically some form

---

[8] (Boyt 1991, unpublished).
[9] Planar non edge-to-edge tilings from regular polygons have been used since antiquity (Grünbaum and Shepard 1987).

of equivalence, the most common is equal edge lengths. In the case of hexagonal grids, the target is that all hexagonal cell edges are identical. As such the goal is not a specific value but a relationship between all the possible values that are being resolved. For the rotegrity form it is that every instance of member to member overlap is a constant ratio of a predetermined value. The constant ratio goal was not part of the original rotegrity design methodology; the ratios were a result of the amount of arc rotation based on the initial grid.

Organizing targets provide the end goal for the self-organizing method not the starting point. These targets allow the method to iteratively adjust its configuration as it progresses to a final solution by determining how far the current state of the grid is from the target as it self-organizes. In the cases that have been explored the initial starting point of the grid is not particularly relevant as the self-organization automatically adjusts vertices across the spherical surface to find an optimal solution.

The self-organizing grid method relies on the fact that there is a single solution to the desired configuration of the reduced constraint grid with respect to the target goal. This fact was not known apriori but is now understood from empirical evidence. The challenge has been to find a computational method that would find solutions in the general case.

10.2   Symmetry

Geodesic grids have always taken advantage of the inherent symmetry of polyhedra because only a small section of the spherical surface, represented by the Schwarz triangle, is unique (see Section 4.6.19). The rest of the geodesic grid can just be described as symmetrical replications of the unique Schwarz section through reflections and rotations. The self-organizing methodology does not require limiting ourselves to only symmetrically replicated grids but it is a great starting point to show how the process works and resolves the grid based on the organizing target using only a small subset of the full grid. First, we'll start by defining the unique part of the grid with respect to reduced constraint networks.

An important aspect of the self-organizing method is that the positions of vertices are inter-related and that relationship is iteratively adjusted as the grid transitions to an optimal arrangement of the vertices. Thus, instead of looking at vertices individually they need to be considered in small groups or neighborhoods and these neighborhoods near the boundary of the Schwarz triangle will extend to adjacent Schwarz triangles.

10.2.1   Uniqueness

Class III tessellations, though common is nature, particularly viruses, are much less common when considering standard subdivision techniques. Their rotational only symmetry makes it difficult to derive very highly symmetric sub-grids, such as those as seen in the highly repetitive Class II Triacon subdivision (Section 8.6), where all geometric uniqueness is contained in a single Schwarz triangle.

To be more robust when considering symmetrical uniqueness in this chapter we will focus on class III tessellations (see Section 8.7). As we have learned Class III tessellations do not have reflection symmetry only rotational symmetry and the smallest unique portion of the polyhedra is two Schwarz triangles. For the icosahedron that is 1/60th of the sphere rather than 1/120th. For the octahedron it is 1/24th of the sphere rather than 1/48th. As seen in Figure 10.11(a), the unique section is two adjacent Schwarz triangles and the entire face of a polyhedron can be constructed from two rotational replications of that unique section.

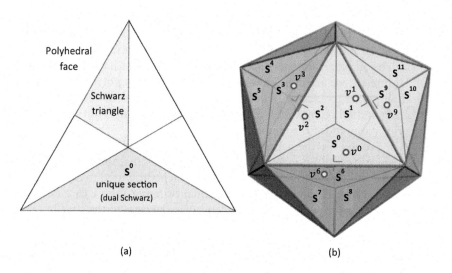

Figure 10.11. Symmetrical relationships.

When performing standard spherical subdivisions on Class I and II tessellations, focusing on a single Schwarz triangle, it is very efficient and straightforward to work directly with spherical trigonometry to divide the spherical right triangle. With Class III tessellations the grid will include an angular skew that is fixed and varies with the (b,c) values making the use of spherical trigonometry more complicated. Also, the self-organization methodology does not attempt to subdivide the surface of the sphere directly. The surface division is an indirect result of moving vertices along the surface of the sphere. In dealing with vertex positions and rotational symmetry it is more appropriate to utilize vectors to represent the relationships between vertices and matrices to encapsulate the rotational relationships between related vertices so we will utilize these mathematical constructs.

Considering the icosahedron, shown in Figure 10.11(b) we can see the replication of the reference unique section, labeled S^0, on all the adjacent polyhedral faces. Any vertex within or on the boundary of section S^0 can be replicated in any other section, S^N, by multiplying the vertex, represented as a vector, with the rotation matrix, M^S, associated with that section. The M^0 matrix associated with section S^0 is just an identity matrix. In the figure there is a labeled vertex, v^0, in section S^0. To find the coordinates of v^1 in section S^1 it would be defined as:

$$v^1 = v^0 M^1$$

And by using the transpose of the M^S matrix, M^{T_s}, we can do the reverse to find the vertex, v^0, from v^1 in section S^1:

$$v^0 = v^1 M^{T_1}$$

The rotation matrices associated with each section allows us to easily compute the coordinates of vertices that share a rotationally symmetric relationship. Details of how to derive the rotation matrices can be found in Appendix B while details of how to derive symmetric uniqueness can be found in Appendix C.

10.2.2   The BC Grid

In this section we'll introduce the **BC** grid used to define the triangular grids that will be the basis for all the vertices required for self-organized hexagonal and rotegrity grids. This will be an expansion on concepts first introduced in Chapter 8, Sections 8.1 and 8.7.6. Ultimately, we are interested in the subdivision of a polyhedron's face and the **BC** grid allows us to succinctly define any triangular class of tessellation where we can begin the orderly removal of vertices and edges that will result in the reduced constraint grids required for the methodology.

As noted in Section 8.1 the general form of a triangle grid's geodesic notation is $\{p,q+\}_{b,c}$ and since we will be focused on triangular grids the notation simplifies to $\{3,q+\}_{b,c}$, where q will depend on the base platonic polyhedron used to derive the grid. The standard used here is for **b** to always be equal or larger than **c** ($b \geq c$), and the three possible relationships between **b** and **c** determine the class of the tessellation. See Figure 10.12.

Now we will label all the vertices within an example **BC** framework ($b = 4, c = 1$), shown in Figure 10.13, where the polyhedron's triangle face is highlighted as well as the unique reference section, S^0. Notice that every vertex has two numbers. The top number is the vertex identifier going from left to right and bottom to top through the grid and it is associated with a unique Cartesian coordinate. The bottom number identifies the vertex's *symmetry group*. If the bottom numbers are the same between two vertices then the two vertices share a rotational relationship and can be transformed between each other. Notice that vertices that reside in or on the boundary of the base section, S^0, have a yellow background. In the ($b = 4, c = 1$) example only five vertices are symmetrically distinct. The coordinates of all the other vertices in the grid can be derived from rotational transformations of these distinct vertices in section S^0.

The **BC** grid is typically larger than the polyhedral triangle it defines as evident in Figure 10.12. The reason is that the self-organizing method operates on multiple vertices, in small groups, at a time and when that small group is near or at the edge of the unique section's boundary then some of the vertices will belong to another rotationally symmetric section.

Referring back to Figure 10.13 it specifically highlights all the vertices that share the symmetry group 2, the lower number at each vertex. The rotational relationship between these vertices is mathematically defined from the vertex transformations defined in Section 10.2.1. Now that we have defined symmetry groups we can proceed to the details of the self-organizing methodology.

Figure 10.12. BC grid.

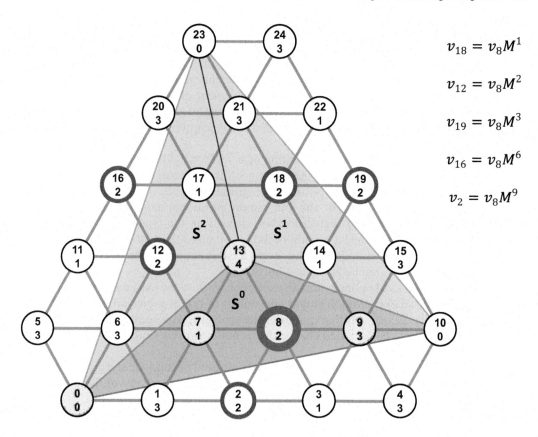

$$v_{18} = v_8 M^1$$

$$v_{12} = v_8 M^2$$

$$v_{19} = v_8 M^3$$

$$v_{16} = v_8 M^6$$

$$v_2 = v_8 M^9$$

Figure 10.13. BC grid notation for $\{3,q+\}_{4,1}$.

10.3   Self-Organizing—Key Concepts

In this section we will cover the key concepts of the self-organizing methodology. Essentially there are just a few so it is not difficult to fully comprehend the concept. Once you have a grasp of the methodology, implementing it to derive solutions for specific reduced constraint tessellations becomes just a matter of implementation.

10.3.1   Initial State

The self-organizing methodology is an iterative process that derives a grid solution that satisfies our target goal but it needs to start from some *initial state*. Fortunately, it turns out that the method is not sensitive to the initial state so we are free to choose an initial state that is easy to construct. The benefit is that we don't have to worry about how close the initial state is to the final state, whose exact configuration is unknown. One of the easiest methods to position vertices on a spherical surface is to sub-divide the flat triangles of the underlying polyhedron and project those planar positions onto the spherical surface. The projective method of creating a grid on the spherical surface is not used for actual designs since it generates a significant amount of variation in the components but as a simple method for generating an initial starting point for the grid it is entirely sufficient.

The projection technique was explained in Chapter 8, Section 8.7.10. Once we have the appropriate **BC** grid for the specific reduced constraint grid then projection of the vertex positions on the planar polyhedral face to the spherical surface is all that is required to obtain the initial coordinates of all the grid vertices. The **BC** grid is also used to provide the rotational relationship between sections and vertices.

10.3.2 Neighborhoods and Local Optimization

A *neighborhood* is a small group of vertices that can be both measured against the target goal and modified to bring us closer to the intended goal. In the hexagonal and rotegrity grids highlighted in this chapter a neighborhood consists of four vertices and the three associated edges, but the arrangement of the four vertices is different as illustrated in Figure 10.14(a)–(c). It is essential to understand that every vertex will belong to multiple neighborhoods and these neighborhoods overlap as shown in (d)–(f).

Local optimization is the modification of a vertex neighborhood so that their geometric arrangement is closer to the target goal. In the hexagonal neighborhood local optimization will only require positional modification of one of the four vertices, while for rotegrities two vertices of the four require positional modification. The specific method of re-positioning the vertices will depend on the type of grid that is being self-organized. Every time a neighborhood is optimized that optimization will impact overlapping neighborhoods and thus every local change has a broader impact to the overall grid. This is implied propagation.

During each local optimization we measure how close the local result is to the target goal. Each local optimization entails adjustment of one or more vertices in the neighborhood. The adjustment that optimizes the geometric arrangement of the neighborhood is a completely localized change that is both relatively simple and computationally inexpensive.

When operating on symmetrical grids, we only need to consider and optimize neighborhoods that reside in or overlap the base section, S^0, since the rest of the grid is symmetrically related and can be recreated from that section. Explicitly this means that neighborhoods that require optimization will have one or more of their re-positioned vertices on the edge or inside the base section.

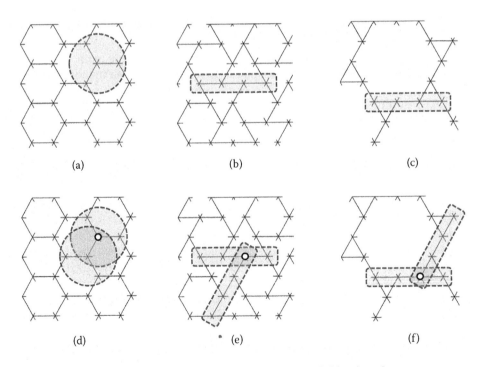

Figure 10.14. Hexagonal and rotegrity neighborhoods.

10.3.3 Global Propagation

Global propagation is the update of the non-unique sections of the **BC** grid after any neighborhood optimization that takes place in the unique section, S^0. Remember that each vertex in the unique section of the grid has one-to-one association with all its other symmetrically equivalent vertices in other sections, see Figure 10.13. Global propagation ensures that every local optimization is propagated across the symmetrically equivalent parts of the grid so the entire grid reflects the local optimization. This is an explicit propagation that keeps the vertices of the entire grid under development synchronized.

Once a vertex is re-positioned that change must be reflected in every other vertex that is symmetrically related. As shown in Figure 10.13 every vertex that shares the same symmetry group (lower identifier) must be updated by multiplying the modified vertex by the associated matrix. Figure 10.13 highlights vertex eight, which resides in the section S^0 of the **BC** grid. When vertex 8 is repositioned then every other vertex that shares the same symmetry group, 2, must be updated to replicate that positional change. Thus, when vertex 8 is moved then the entire group of symmetrically related vertices, (2,12,16,18,19), which belong to symmetry group 2, must be updated. In this way local changes are globally propagated and the entire set of related vertices remain synchronized.

10.3.4 Target Goal

The self-organization process is complete when the geometry of all neighborhoods is within the range of the numerical precision that was predetermined. For real world purposes the level of precision would be determined by what would be achievable in the manufacturing and construction process. From a purely geometric perspective we can set any level of precision and in the examples, we set the precision to be eight significant digits, which is more than sufficient for almost any purpose. If the geometry of any neighborhood is outside our target goal then another round of local optimizations is performed. Every time we optimize a neighborhood, we measure how close its geometry is to our target goal. When every neighborhood's geometry meets our target goal simultaneously then the self-organization process is complete. The method automatically converges without intervention as the movement of vertices is guided by the local optimizations and global propagation. There is no additional synchronization required between separate or overlapping neighborhoods. Neighborhood measurements are stand-alone but the effect of changing a neighborhood automatically reverberates to both the overlapping neighborhoods and also the symmetrical duplicates of that neighborhood across the larger grid. When a neighborhood is optimized it causes its overlapping neighborhoods' optimizations to either improve or worsen to some degree.

The number of optimizations required to achieve convergence of the self-organizing method is proportionally related to the required precision and as such when the precision requirement increases the number of iterations increases linearly.

10.3.5 Computation Sequence Algorithm

The computational sequence for the self-organizing method is a very simple iterative process with only a few basic steps. To start the desired reduced constraint tessellation geometry needs to be configured. With a desired tessellation, the initial vertex locations on the planar polyhedral triangles can be projected to derive the initial spherical coordinates. With a valid initial state only two layers of looping are required for the grid to self-organize to a solution that will meet the target goal. The inner loop includes optimization and global propagation of every neighborhood. The outer loop just checks if all neighborhoods meet their target goal simultaneously to decide that self-organization is complete.

Self-Organizing Algorithm
(*for symmetric grids*)

1. Initialize

    a. Create initial grid

    b. Designate unique and non-unique vertices

    c. Determine required transformation matrices

    d. Set target goal precision

2. Begin continuous loop

    a. Begin neighborhoods loop

        i. Measure how close neighborhood geometry is to target

        ii. Perform local optimization if geometry outside target

        iii. Perform global propagation on the re-positioned vertices

    b. If geometry of all neighborhoods meets target precision exit continuous loop

3. Complete—all vertices at final position that meets target goal

---

10.3.6 Summary

A self-organizing grid is an initial collection of vertex positions on the sphere that re-position themselves, through a guided iterative process, to achieve different optimal arrangements that cannot be found directly through any known standard divisions of the spherical surface. Unlike typical fully triangulated tessellations, self-organized grids are incomplete triangular tessellations that are referred to as reduced constraint arrangements where vertices no longer directly link to all of their original closest neighbor vertices.

The majority of this chapter focuses on the application of the self-organizing methodology to two reduced constraint tessellations, one hexagonal or honeycomb shaped, and the other known as rotegrities or nexorades that are non edge-to-edge tilings, but the methodology is extensible beyond these and will be discussed in Section 10.6. With rotegrities the term nexor is introduced to represent a set of vertices along a geodesic arc. A neighborhood, a key concept, represents a small group of connected vertices and every vertex is part of two overlapping neighborhoods. Each configuration has a very specific relationship between vertices in a neighborhood that needs to be achieved for the grid to be optimally self-organized.

Starting with a completely triangulated arrangement either a hexagonal or a rotegrity grid is created by a symmetrical removal of some vertices and edges to leave a partial reduced constraint grid. Depending on the geometry of the reduced constraint arrangement the specific relationship of vertices in neighborhoods determines how the iterative repositioning of vertices will proceed to reach the desired optimal relationship target.

By taking advantage of the symmetry of polyhedra and the small number of distinct vertices positions we minimize the number of vertices that need to be computationally manipulated for re-positioning, relying on simple matrix transformations to update symmetrically related vertices, allowing us to propagate our limited number of vertex changes across the entire grid. Through the use of the **BC** grid, symmetry identification, and vertex projection we derive the initial state for all vertex positions. A simple two-tier looping algorithm controls the entire self-organization process, with the inner loop dedicated to moving every neighborhood sequentially toward a local optimal arrangement, and the outer loop deciding the goal has been achieved when all local optimizations have attained their optimal arrangements simultaneously.

10.4   Hexagonal Grids

Now that the all the essential concepts are covered, we can show how to utilize these concepts to address the solution for any hexagonal grid such that every edge of the grid, the distance between its two end vertices, is the same length. Sometimes referred to as equal edge tessellations, they have posed a challenge to solve since they were proposed in 2002. The 27 example equal edge hexagonal tessellations (reduced constraint networks) that Clinton[10] derived and presented across three polyhedron (icosahedron, octahedron, and tetrahedron) and all three classes lead him to state that these hexagonal configurations could potentially be derived for any (b,c) configuration. Even though Clinton did not specify the details of his method for derivation these relatively simple scenarios can be derived from spherical trigonometry either directly or using iterative approaches. The transition from a spherical trigonometry approach to a vector manipulation approach was key to the development of the self-organization method that was both simpler and robust.

For hexagonal grids the general approach previously outlined will be followed. The only aspect of the approach that needs further clarification is the geometry of a hexagonal grid neighborhood and the specific method for performing a local optimization. The initial condition and target goal concepts are not fundamentally

---

[10] (Clinton 2002).

changed. Since hexagonal tessellations are incomplete triangulations, additional components, either tension or compression elements, such as double trussing as shown in Figure 10.2 or bicycle like spokes, would be required to create a fully stable physical structure.

10.4.1   Initial Condition

Prepare the appropriate BC grid on the planar faces of the polyhedron, with the hexagonal components delineated, and project these vertices outward to the spherical surface. Not every possible triangular BC grid will allow removal of vertices and edges that will result in a symmetric hexagonal layout. The valid subset of (b,c) grids that will provide a hexagonal grid is directly related to its triangulation number T $(b^2 + bc + c^2)$ (see Section 8.2 for details). Any combination of (b,c) whose triangulation, T, value is evenly divisible by 3 will yield a valid hexagonal grid. For example, in Figure 10.13 with $(b = 4, c = 1)$ the triangulation value T is 21, which is evenly divisible by 3.

10.4.2   Hexagonal Neighborhood and Local Optimization

A local neighborhood of a hexagonal grid is a group of four vertices on the spherical surface and the three edges between them. The four vertices and associated edges are shown inside the dotted circle in Figure 10.15. The center vertex is the common corner of three adjacent hexagons and the three other vertices are the nearest vertices. The center vertex resides in the base section, S^0. A local optimization is performed if the difference in length between the three edges is outside our desired precision for equivalence. The local optimization, if required, is to make the three edges (A,B,C) more equal in length. The local optimization will result in moving the center of the neighborhood so that three edges are closer to being the same length. The re-positioning of the center vertex is a first order approximation to achieve better equivalence.

For local optimization of a hexagonal neighborhood there are four steps. The measurement taken in the first step will determine if the other three steps are required. First, compute the average of the three edge lengths. If the difference between the average length and any of the three actual lengths is outside our precision target then perform the rest of the optimization steps. Second, extend or contract each length along its existing path to be the average length. Third, average the three resulting new endpoints to a new common endpoint. Fourth, normalize the new common endpoint so that it is returned to the spherical surface.

The example neighborhood in Figure 10.16 shows the four basic steps of the local optimization. For clarity the edge lengths vary significantly to make the process clear.

Figure 10.15. Hexagonal neighborhood.

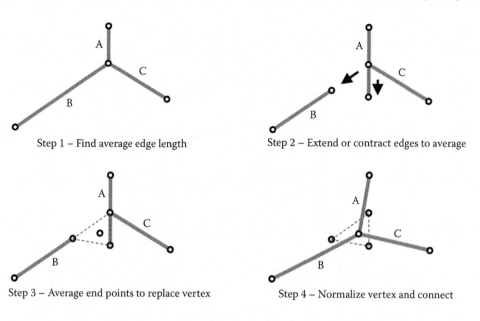

| Step 1 – Find average edge length | Step 2 – Extend or contract edges to average |
| Step 3 – Average end points to replace vertex | Step 4 – Normalize vertex and connect |

Figure 10.16. Local optimization of hexagonal neighborhood.

After the optimization only the hexagonal neighborhood's center vertex will have moved. The process only needs to track if the target goal was achieved or not, since the process only stops when every neighborhood optimization simultaneously reports that is has achieved the equivalence goal.

10.4.3   Global Propagation

If local optimization occurs to the hexagonal neighborhood the positional change of the neighborhood's center vertex must be globally propagated to all its symmetrical counterparts. The propagation follows the exact same approach outlined previously in Section 10.4.2. This is accomplished by multiplying the modified vertex of the neighborhood in section, S^0, by the transformation matrix, M^S, associated with each of the counterpart vertices. As indicated in Section 10.4.2 the center vertex will have originally been identified as being in the base section, S^0. After optimization the center vertex may move outside the Schwarz triangle but its symmetrical uniqueness will remain intact and the global propagation will ensure all its counterparts are equally transformed rotationally.

10.4.4   Hexagonal Grid Example One—Icosahedron

The example configuration in Figure 10.17 is a simple example of a Class III hexagonal grid $\{5+,3,\}_{2,1}$ that was derived by the symmetric removal of edges and vertices from the triangular grid $\{3,5+\}_{4,1}$. The initial coordinates of all the vertices are derived from the projection of the layout on the planar polyhedral faces. Vertices are assigned their appropriate symmetry group and the symmetrical matrices are initialized. The precision target for equivalence between edges is set to be greater than eight significant digits. Because a single face of the icosahedron covers only 1/20th or 5% of the entire spherical surface the variation in initial edge lengths from the grid projection is relatively small compared to other base polyhedron.

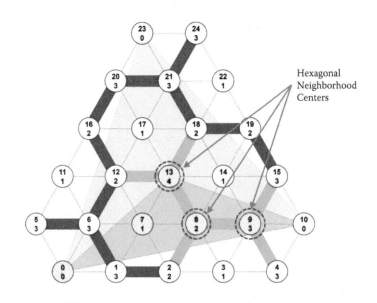

Figure 10.17. Hexagonal grid $\{5+,3\}_{2,1}$.

As shown in Figure 10.17 there are only three neighborhoods whose center vertices reside in section S^0. Since the neighborhood centered around vertex 13 is at the symmetrical center of the polygon face it can be excluded as a neighborhood since any optimization changes will result in no actual change in position. Remember that the three adjacent vertices around vertex 13 (8,12,18) belong to the same symmetry group, ensuring that the three lengths will always be the same thus negating the need to optimize. That leaves only two neighborhoods centered around vertices 8 and 9 that require optimization.

From here the continuous outer loop begins until all neighborhood precision requirements are satisfied simultaneously. The inner loop performs neighborhood optimizations on the two vertex neighborhoods centered around vertex 8 and 9. The neighborhoods are optimized in sequence. After the optimization of each neighborhood the related global propagation takes place. Following the optimization of vertex 8, global propagation will update vertices 2, 12, 16, 18, and 19 since all belong to the same symmetry group 2. After optimizing vertex 9 the propagation will update vertices 1, 4, 5, 6, 15, 20, 21, and 24. Figure 10.18 shows the propagation after each neighborhood optimization.

When the difference in edge lengths in both neighborhoods falls below the target precision for equivalence the outer loop exits and the final position of all the vertices is known. There is no explicit comparison between the edges of the two neighborhoods but since the neighborhoods overlap the edge equivalence of both separately results in a single edge length.

Figure 10.19 shows the initial and optimized grids extended to cover the entire sphere. On the initial grid there are four (4) different edge lengths, derived from the initial grid projection, indicated by different colors, while on the self-organized grid only one edge remains after self-organization. The difference is subtle visually but the methodology re-positioned the two unique vertices, 8 and 9, and reduced all edge differences into a single edge solution. Table 10.1 shows the initial four edge values of the tessellation from the projection of the grid to the spherical surface, and the final edge value after self-organization is complete.

Propagation of vertex 8
(a)

Propagation of vertex 9
(b)

Figure 10.18. Global propagation example.

(a)   (b)

Figure 10.19. Icosahedron—before and after self-organization.

Initial Edge Lengths	Self-Organized Edge Length
0.24867297	
0.25790173	0.26398687
0.26712491	
0.28016949	

Table 10.1. Hexagonal grid on the icosahedron.

10.4.5 Hexagonal Grid Example Two—Octahedron

The second example is essentially the same as the first except that the same grid will be applied to the octahedron. The topological layout shown in Figure 10.17 is identical. There are only three neighborhoods and one is at the center of the polygon face and can be excluded from optimization, leaving only two neighborhoods that require optimization. Since we are optimizing the grid as applied to an octahedron the symmetric matrices will be different and the initial vertex coordinates derived from projecting the vertices onto the spherical surface from the octahedral face will be different. Otherwise, the precision and algorithm are identical.

Figure 10.20 shows the initial and optimized grids extended to cover the entire sphere. On the initial grid there are four (4) different edge lengths indicated by the different colors, while on the self-organized grid only one edge is required. The difference is much more obvious visually as the methodology again reduces all edge differences into a single edge solution. Table 10.2 and 10.3 provides the edge lengths before and after self-organization.

10.4.6 Hexagonal Grid Example Three—Tetrahedron

The third example is a slightly more complex hexagonal tessellation applied to the tetrahedron and shown in Figure 10.21. The main reason to use a more complex tessellation for the tetrahedron is because with only four polyhedral faces the total number of hexagonal edges is quite small for the simpler tessellation we used on the

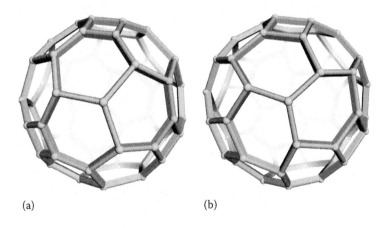

(a) (b)

Figure 10.20. Octahedron—before and after self-organization.

Initial Edge Lengths	Self-Organized Edge Length
0.35290527	
0.38858634	0.42177784
0.42413934	
0.48596892	

Table 10.2. Hexagonal grid on the octahedron.

Initial Edge Lengths	Self-Organized Edge Length
0.23766485	
0.24586865	
0.41456504	
0.45291082	0.47221275
0.45415855	
0.65299955	

Table 10.3. Hexagonal grid on the tetrahedron.

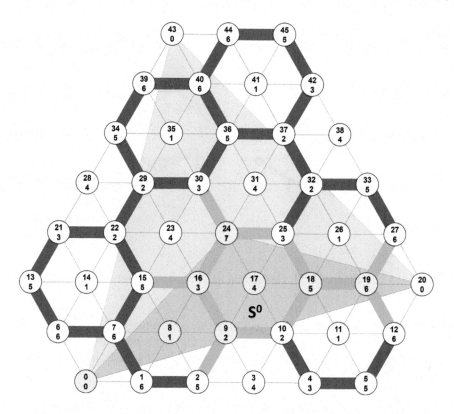

Figure 10.21. $\{5+,3\}_{3,1}$ Hexagonal grid example on a $\{3,5+\}_{5,2}$ **BC** grid.

icosahedron and octahedron examples and is harder to adequately visualize. A secondary reason is that it illustrates an increase in the number of neighborhoods that need to be optimized. For this tessellation there are five neighborhoods whose center vertices reside in section S^0 but the neighborhood at the center of the polygon face can be excluded thus leaving four neighborhoods to be optimized. Because a single face of the tetrahedron covers a quarter of the entire spherical surface the variation in initial edge lengths from the grid projection is quite significant. As expected, our symmetric transformation matrices will be different.

Figure 10.22 shows the initial and optimized grids extended to cover the entire sphere. On the initial grid there are six different edge lengths indicated by the different

(a)   (b)

Figure 10.22. Tetrahedron—before and after self-organization.

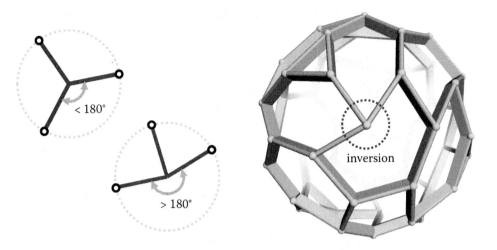

Figure 10.23. Hexagonal neighborhood inversion.

colors, while on the self-organized grid only one edge remains. The difference is very obvious visually since in the tetrahedron case the edges near the tetrahedron's vertices are initially quite short. Table 10.3 provides the edge lengths before and after self-organization.

When comparing the initial and self-organized location of vertices on the spherical surface not only have some vertices traveled significant distance, some of the hexagonal neighborhoods have required to be inverted during the self-organizing process to find a single edge. An inversion is when the angle between two connecting edges exceeds 180° as highlighted in Figure 10.23. The neighborhood inversion is a response to the method that is not seen with either the icosahedron or the octahedron but the inversion does not cause the hexagonal neighborhood to collapse.

10.4.7 Results

Before leaving this section on the hexagonal honeycomb configurations it is worthwhile to compare the single edge result from the self-organizing methodology with a theoretically optimal solution for the icosahedron. To find a theoretical optimal edge we will divide the entire spherical surface into the same number of hexagonal and pentagonal polygons as defined for the topology and make the assumption that every polygon is regular. The total number of polygon faces in a hexagonal tessellation is a function of (b,c). For the icosahedron the total number of faces, p, is given by the following equation.

$$p = 10T + 2$$

The value T is the triangulation number for the tessellation (Section 8.2):

$$T = b^2 + bc + c^2$$

Since we are using an icosahedral base, as we did in the first example, the total number of faces, p, will always include twelve (12) regular pentagons. The final formula for the optimal subtended edge angle and the corresponding edge length is given by the following two equations. In the first equation you will see a small constant value 0.248 that provides a correction when T values are small. Without the correction the error for the smallest value of T will be more than 3%. With larger T values the impact of the constant becomes negligible.

b	c	T	Self-Organizing Method		Ideal Approximation	
			Edge Length	Edge Angle	Edge Angle	Error
2	1	7	0.26398687	15.16960104°	15.17576850°	0.0406%
20	20	1200	0.02023021	1.15912537°	1.15035097°	0.7628%

Table 10.4. Optimality of hexagonal self-organized grids.

$$edge\_angle° = 2\cos^{-1}\left(\frac{\sqrt{3}}{2\sin\left(\frac{\pi}{3}\left(1+\frac{1}{10(t-0.248)}\right)\right)}\right)$$

The edge, is given by (Section 4.4)

$$edge = 2\sin\left(\frac{edge\_angle°}{2}\right)$$

Comparing edge equalization results with this ideal edge over a wide range of icosa-hedral T values, across all three classes, the maximum difference between actual results and the theoretically optimal does not exceed +0.77% even for very large values of T thus providing evidence that the self-organizing methodology converges to an optimal solution. Looking at the first icosahedral example which was $\{5+,3\}_{2,1}$ the triangula-tion number is seven. Using the theoretical formula provided above for the edge length we find the angle subtended by the hexagon edge is: 15.17576850° which is very close to the result of the self-organizing method. With a very highly tessellated icosahedron ($b = 20$, $c = 20$) the difference between the ideal edge angle and the edge derived by self-organizing is still well less than 1%.

10.4.8 Summary

The reduced constraint hexagonal grid is any triangular **BC** grid where the triangula-tion value, T, is evenly divisible by 3 and the symmetric removal of the relevant edges and vertices leaves only the hexagonal outline. The initial state of the vertex positions on the spherical surface is derived by projecting the remaining **BC** grid positions on the polyhedron's face to the spherical surface.

Each hexagonal neighborhood is four vertices and three edges arranged in a tri-valent pattern with one vertex at the center and three vertices outward from the center, each connected by an edge to the center vertex. An optimal arrangement of a neigh-borhood is when all three edges are identical in length. If any of the three edges are shorter or longer than the average of all three edges, an additional three steps are taken to re-position the neighborhood's center vertex in a coarse correction so that the edges are more equal in length. These additional steps lengthen and shorten the edges along their current vector path creating a small triangular hole at the center vertex, and the center of the hole becomes the new position of the center vertex, which is normalized in the final step to remain on the spherical surface.

If a neighborhood is optimized with the center being re-positioned then the change is globally propagated to all the symmetric counterparts of the center vertex to en-sure the entire grid stays symmetrically synchronized. Updates of a neighborhood also have an impact on overlapping neighborhoods that may improve or degrade their optimal arrangement thus providing a localized ripple effect or implied propagation.

The inner loop of the algorithm sequentially optimizes every neighborhood and reports if each neighborhood achieves the optimal arrangement of three edges being the same length. The outer loop completes when all neighborhoods report simultaneously that they are optimally arranged leaving us with all vertices in their final position on the spherical surface.

Multiple examples of the self-organization process show that the method successfully derives the difficult Class III equal edge configurations on any of three polyhedron and illustrates the amount of geometric movement has occurred from initial vertex position to their final optimal position to a high degree of precision. A theoretical model was presented to show how close the self-organization method comes to achieving an optimal solution.

10.5   Rotegrities

Now the self-organizing methodology will be applied to rotegrities. Once the rotegrity vertices have been self-organized the final form of the nexors can be arcs, straight members, or some hybrid combination since all require the same geometrical information as to where the vertices are positioned on the spherical surface. Both the underlapped and overlapped scenarios will be considered since the derivation is identical, only the proportions are different.

First let's look at the diagrammatic notation we will use for the rotegrity. Again, the **BC** grid will provide the overall grid of vertices to overlay the nexor arrangements. Each nexor is a linear set of four vertices that represent a local neighborhood. Also, each nexor will have its own unique identifier in the circle drawn in the middle of the nexor. Nexors that share the same unique identifier have the same sequence of four vertices indicated by their symmetry group numbers.

Figure 10.24 is a sample rotegrity grid overlayed on the polyhedron's triangle face. The figure only highlights the nexors that reside completely inside the **BC** grid. In this example there are six nexors completely inside the **BC** grid but only two different nexors, indicated with a dotted outline and by their different colors. Nexors, whose vertices share the same sequence of symmetry group identifiers, have the same color and the

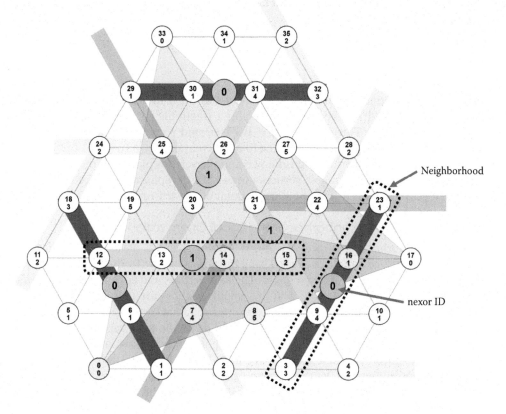

Figure 10.24. Rotegrity topology & nexors.

same identifier indicated at the center of the nexor. The two nexors we need to self-organize have at least one vertex in base section S^0. Looking again at Figure 10.24 all the blue nexors have the same sequence of vertex symmetry group identifiers (3,4,1,1) while the green nexors have the (4,2,3,2) sequence.

The rotegrity self-organizing methodology will produce valid results for any rotegrity grid applied to either the icosahedron or octahedron but unlike the hexagonal tessellations the self-organizing methodology will not converge to a solution for every rotegrity grid applied to the tetrahedron. Due to the requirement that tetrahedral nexors are aligned across a wide geodesic path, the convergence fails for the simpler overlapped tetrahedral cases.

10.5.1 Initial Condition

The rotegrity grid is a triangular grid reduced appropriately as discussed in Section 10.1. The topological description of rotegrities is not consistent in the literature. In this chapter rotegrities will be topologically described by the triangular grid used to determine its initial vertex positions. Using this method only a subset of triangular definitions (b,c) can be used for edge removal to yield a valid rotegrity. The valid subset of (b,c) grids that will provide a rotegrity are directly related to the triangulation number T. Any combination of (b,c) whose T value is evenly divisible by 7 (modulo 7) will yield a valid rotegrity. The simplest rotegrity $(b = 2, c = 1)$ with $T = 7$ only has one type of nexor. The number of different nexors, *n*, is directly related to a valid triangulation number (evenly divisible by 7).

$$n = \left\lceil \frac{T}{14} \right\rceil$$

The brackets around the division in the equation for *n* indicate a ceiling function, meaning if the result of the division contains a fractional part the result is the next highest integer value. Thus for (*b* = 6, *c* = 3) *T* is 63 and *T* divided by 14 is 3.5 so *n* will be 4. For this configuration there will be four different nexors.

Once the desired rotegrity grid is known and the vertices placed on the faces of the polyhedron, projection of these vertices outward to the spherical surface by normalization (Section 8.7.10) will be the starting point of the self-organization.

10.5.2 Rotegrity Neighborhood and Local Optimization

A local neighborhood of a rotegrity grid is a linear group of always four vertices on the spherical surface and three edges. Using the nomenclature from nexorades, the rotegrity neighborhood is a called a nexor. The four vertices and edges are shown inside the dotted area in Figure 10.25. The two middle vertices are the ends of two other neighborhoods or nexors. The initial position of the four vertices will not be aligned either in the proportions of the edges (A,B,C) or lie on the same geodesic arc.

The rotegrity's local optimization is two-fold; it moves the two middle vertices so they are (1) aligned in a single geodesic arc, and (2) they lie at the desired angular proportional ratio. Figure 10.26(a) shows an exaggerated initial position with the three initial edges clearly misaligned versus the target positions of the three edges aligned along a single geodesic arc. The figure's (b) and (c) parts illustrate the goal of equal angular proportion and arc alignment. The end points of the nexor are not modified.

There are multiple steps to accomplish the local optimization of a nexor. First determine the subtended angle of the geodesic arc between the end vertices of the nexor. Second, derive the desired proportional placement of the two middle vertices. Third, update the two middle vertex coordinates with the correctly proportioned positions derived in the second step.

Figure 10.26(a) illustrates a nexor whose two middle vertex positions are not close to the desired position on the geodesic arc between the nexor endpoints and both will need to be re-positioned. The local optimization of a nexor neighborhood will produce an exact correction, not an approximation, as we saw in the optimization of the hexagonal neighborhood. But since rotegrity neighborhoods overlap, the exact correction of one nexor will partially affect its overlapping neighborhood nexor. In addition, global propagation will also disturb the correction. None the less the corrections converge.

There are a couple of ways to decide when the rotegrity geometry has converged so that the continuous loop of neighborhood optimizations can be considered complete.

Figure 10.25. Rotegrity grid's local neighborhood—Nexor.

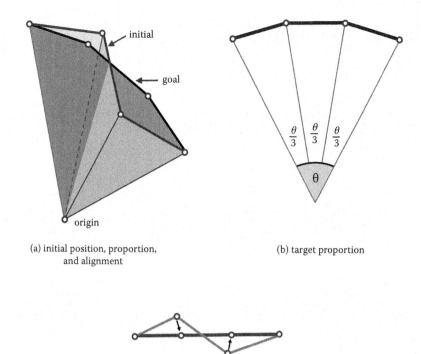

(a) initial position, proportion, and alignment

(b) target proportion

(c) target alignment

Figure 10.26. Rotegrity nexor optimization.

One method is to measure the distance between the middle nexor vertices, before and after the correction, and stop when the change in all distances become smaller than our precision requirement. A second method is to compare the distance between the nexor endpoints during the current neighborhood optimization with the distance derived from the previous optimization and stop when the change in distance falls under our precision requirement. Both methods will work but the second method is slightly cheaper computationally since we are always required to measure the distance between the nexor endpoints in step one of the optimization.

Now we will cover the specifics of how to perform the local optimization of each neighborhood or nexor. For simplicity we'll cover the overlapping case first since the two mechanisms are almost identical and then extend the idea to cover the underlapping case. The similarity in approach allows us also to start from the same initial condition for either case.

Each nexor of the rotegrity has four vertices, shown in Figure 10.27, which we will label p_0, p_1, p_2, and p_3. First determine the distance between the two endpoints. Since the endpoints are on the spherical surface, we know that the angle between the endpoints is directly related to that distance (Section 4.4). Second, find the position of the middle vertices (p_1, p_2) along the geodesic arc between the endpoints (p_0, p_3) that matches our proportional requirements. In our example case the desired proportion requires that the middle vertices be positioned equally spaced angularly. This requires them to be separate from each other and the endpoints by one third of the geodesic arc angle. To find the correct middle vertex coordinates we will first locate the positions on the straight line from p_0 to p_3 that corresponds to the angular division, then we will normalize the mid-line positions, which will in effect extend them to the correct position along

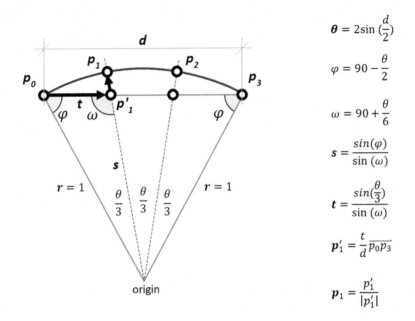

$$\theta = 2\sin\left(\frac{d}{2}\right)$$

$$\varphi = 90 - \frac{\theta}{2}$$

$$\omega = 90 + \frac{\theta}{6}$$

$$s = \frac{\sin(\varphi)}{\sin(\omega)}$$

$$t = \frac{\sin(\frac{\theta}{3})}{\sin(\omega)}$$

$$p'_1 = \frac{t}{d}\overrightarrow{p_0 p_3}$$

$$p_1 = \frac{p'_1}{|p'_1|}$$

Figure 10.27. Rotegrity proportional and alignment correction.

the geodesic arc. This process will accomplish the two optimization requirements: (1) the nexor's middle vertices will be at the correct angular locations, (2) all four vertices of the nexor will be aligned along a single geodesic arc. Figure 10.27 also shows the basic trigonometry for finding the position of the nexor's middle vertices along the geodesic arc.

Now let's consider the underlapping case. Rather than creating a separate reduced constraint tessellation that requires a different BC grid we can continue to use the same initial overlapping tessellation but change the application of the proportion ratios. In the overlapping scenario p_1 was 1/3rd between p_0 and p_3 and p_2 is either 1/3rd between p_3 and p_0 going in the opposite direction, Figure 10.28(a), or 2/3rd between going in the same direction along the $p_0 p_3$ path. For the underlapping scenario we can just flip the ratios so that p_1 and p_2 will cross over each other to correctly create the underlapping arrangement, Figure 10.28(b).

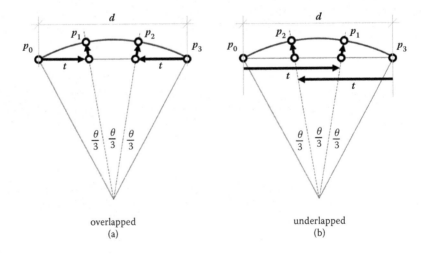

Figure 10.28. Overlapping and underlapping proportions.

10.5.3   Global Propagation

After each nexor optimization the two middle vertices will have been re-positioned and that change needs to be propagated to all of their symmetrical counterparts. This is accomplished again by multiplying the modified middle nexor vertices by the transformation matrices associated with their counterparts in the same symmetry group. The nexor's end vertices are not repositioned so no propagation is necessary.

Rotegrities have an interesting scenario with some nexors where a mid-vertex and an end vertex share the same symmetry group. Looking again at Figure 10.24, at either nexor 0 or 1, you will notice that re-positioning their mid-vertices will also change one of their own end points since both share a common symmetry group. Fortunately, this does not cause any issues with converging to a solution.

10.5.4   Icosahedral Rotegrity—Example One

The following is a simple rotegrity example derived from the Class III triangular grid $\{3,5+\}_{4,2}$ by removing edges and vertices. As mentioned previously the triangulation number T of the grid is 28 which is divisible by 7, making it suitable as a rotegrity. As shown in Figure 10.29 there are only three symmetrically distinct vertices and two different nexors. The configuration is the same as the one shown in Figure 10.24 except only nexors that reside fully in the **BC** grid are drawn. After deriving the initial coordinates of all the **BC** grid vertices associated with the nexors the optimization loop

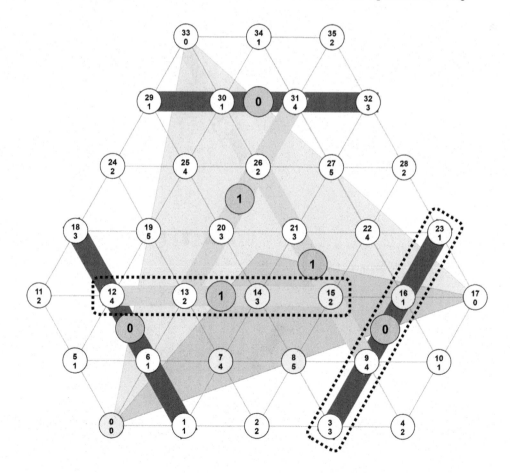

Figure 10.29. Rotegrity example.

commences with the two nexors. Each nexor neighborhood is optimized followed by the global propagation of the modified middle nexor vertices.

Figure 10.30 illustrates the required global propagation for each nexor. We will only optimize the two different nexors that have middle vertices in section S^0. Figure 10.30(a) focuses on nexor 0 and the propagation of the two middle vertices after the nexor's optimization. Since both middle vertices are modified during the optimization the changes in both vertex positions must be propagated throughout the **BC** grid. Different colors are used to differentiate the two vertices. After optimization of nexor 1 the global propagation of its modified middle vertices takes place and indicated similarly in Figure 10.30(b).

Figure 10.31 shows the initial and self-organized rotegrities replicated over the entire spherical surface. On the initial grid (a) there are five different edge lengths indicated by the different colors. Nexor 0 has three different edges and is not aligned, while nexor 1

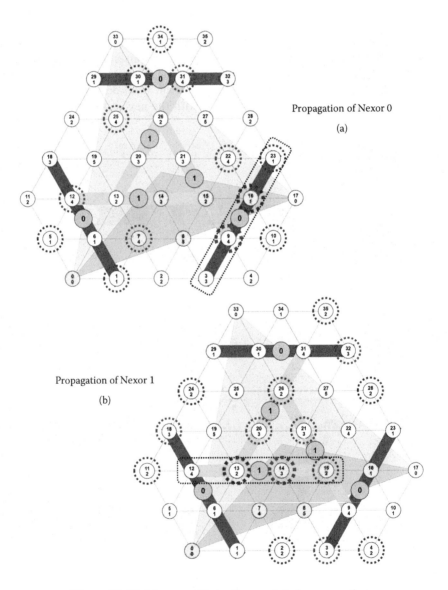

Figure 10.30. Nexor optimization—global propagation.

initial
(a)

self-organized overlapped
(b)

self-organized underlapped
(c)

Figure 10.31. Icosahedral rotegrity—before and after self-organization $\{3,5+\}_{4,2}$.

has two different edge lengths also not aligned. The initial configuration would not allow us to build a proper nexorade or compound arc rotegrity since the edge segments are not aligned. The nexors of the self-organized rotegrity are each composed of three equally spaced edges aligned along a single geodesic arc. Figure 10.31(b) shows the overlapped case while (c) shows the underlapped case that was derived from the same initial configuration but utilized the underlapping strategy of changing the direction and amount of proportion. Both the underlapping and overlapping rotegrities have the same number of different edges. Table 10.5 shows different edge values before and after self-organization.

Initial Edge Lengths	Self-Organized Edge Lengths (overlapped)	Self-Organized Edge Lengths (underlapped)
0.20935238		
0.21518030	0.21264461 0.24207636	0.15652627 0.17905402
0.22182768		
0.23339415		
0.24201994		

Table 10.5. Rotegrity grids based on the icosahedron $\{3,5+\}_{4,2}$.

10.5.5 Octahedral Rotegrity—Example Two

The second example is essentially the same as the first example except that the same rotegrity grid will be applied to the octahedron. The topological layout shown in Figure 10.29 is exactly the same. As in the icosahedral example there are only three symmetrically distinct vertices and two different nexors. Since we are optimizing the grid as applied to an octahedron the symmetric matrices will be different and the initial vertex coordinates derived from projecting the vertices onto the spherical surface from the octahedral face will be different. Otherwise, the precision and algorithm are identical.

Figure 10.32 shows the initial and optimized rotegrities extended over the complete sphere. On the initial grid there are five different edge lengths indicated by the different colors, while on the self-organized grids only two edges are required, one edge length per nexor. The difference is much more obvious visually because the octahedron's face is 12.5% of the total sphere compared to 5% for the icosahedron, as the methodology again aligns and proportionally distributes the nexor's vertices. Table 10.6 shows different edge values before and after self-organization.

initial
(a)

self-organized overlapped
(b)

self-organized underlapped
(c)

Figure 10.32. Octahedral rotegrity – before and after self-organization $\{3,4+\}_{4,2}$.

Initial Edge Lengths	Self-Organized Edge Lengths (overlapped)	Self-Organized Edge Lengths (underlapped)
0.28171809		
0.30361049		
0.32291350	0.30531576	0.22089772
0.37504752	0.41070164	0.30252139
0.41744238		

Table 10.6. Rotegrity grids based on the octahedron $\{3,4+\}_{4,2}$.

10.5.6   Tetrahedral Rotegrity—Example Three

The third example is a more complex rotegrity tessellation applied to the tetrahedron and shown in Figure 10.33. The reason to use a more complex tessellation is similar to the reason used for the hexagonal examples, because the tetrahedron only has four polyhedral faces the total number of nexors is quite small for the simpler tessellation

Figure 10.33. Rotegrity example for the tetrahedron $\{3,3+\}_{8,2}$.

we used on the icosahedron and octahedron examples and is harder to adequately visualize. A secondary reason is that it illustrates an increase in the number of nexor neighborhoods that need to be optimized. For this tessellation there are six nexor neighborhoods whose mid-vertices reside in section S^0 and all require optimization. Because a single face of the tetrahedron covers a quarter of the entire spherical surface and we are applying a more complex tessellation, the number and variation in initial edge lengths from the grid projection is significant. As expected, our symmetric transformation matrices will be different.

In this example we find that two different nexors (2 and 3), after optimization, result in identical angular dimensions. Thus, our original six neighborhoods self-organize into five unique nexors. Figure 10.34 shows the initial and optimized rotegrities extended over the complete sphere. On the initial grid there are 13 different edge lengths indicated by the different colors, while on the self-organized grids only five aligned edges are required, one edge length per unique nexor. The difference in the initial edge alignments is very obvious visually as the methodology again aligns and proportionally distributes the nexor's vertices. Table 10.7 shows the different edge values before and after self-organization.

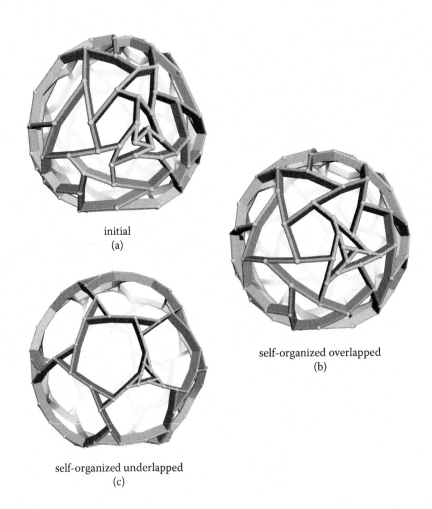

initial
(a)

self-organized overlapped
(b)

self-organized underlapped
(c)

Figure 10.34. Tetrahedral rotegrity—before and after self-organization $\{3,3+\}_{8,2}$.

Initial Edge Lengths	Self-Organized Edge Lengths (overlapped)	Self-Organized Edge Lengths (underlapped)
0.16587112		
0.16674134		
0.18136789		
0.19413511		
0.20086189	0.13142225	0.09582273
0.20344177	0.22647029	0.16552359
0.22434405	0.31670450	0.22958816
0.24809930	0.34278308	0.24839985
0.25153905	0.36975174	0.27158605
0.29037393		
0.37161987		
0.38858634		
0.47140452		

Table 10.7. Rotegrity grids based on the tetrahedron $\{3,3+\}_{8,2}$.

10.5.7 Summary

The reduced constraint rotegrity grid is any triangular **BC** grid where the triangulation value, T, is evenly divisible by 7 and the symmetric removal of the relevant edges and vertices leaves a non edge-to-edge tiling of hexagons and triangles. Depending on the removal selection the triangular holes may overlap or underlap. The initial state of the vertex positions on the spherical surface is derived by projecting the remaining **BC** grid positions on the polyhedron's face to the spherical surface.

Each rotegrity neighborhood, a nexor, is four vertices and the three edges arranged in a linear arc pattern with a vertex at each end and two vertices in the middle. An optimal arrangement of a nexor neighborhood is when all four vertices lie exactly on the same geodesic path and the placement of the two middle vertices meet the desired proportional division of the nexor's arc.

Optimization of a nexor involves re-positioning the two middle vertices so that they lie exactly on the arc defined by the current end vertices. This is accomplished in multiple steps that include determining the angle subtended by the two end vertices and then finding the positions on the vector between the end vertices that represent the proportional placement desired for the two middle vertices. Using the corrected placement of the middle vertices on the vector they are normalized to become the final correct position on the spherical surface.

When a nexor is optimized by repositioning the two center vertices the change is globally propagated to all the symmetric counterparts of the two middle vertices to ensure the entire grid stays symmetrically synchronized. Updates of a nexor have an impact on overlapping nexors that may improve or degrade their optimal arrangement thus providing a localized ripple effect or implied propagation.

The inner loop of the algorithm sequentially optimizes every nexor and reports if each nexor achieves the optimal arrangement of all four vertices being aligned and proportionally placed on a geodesic arc. The outer loop completes when all nexors report simultaneously that they are optimally arranged leaving us with all vertices in their final position on the spherical surface.

Multiple examples of the self-organization process show that the method successfully derives rotegrity configurations of both the underlap and overlap configurations on any of three polyhedra. The examples also illustrate the amount of geometric movement vertices have traveled from their initial positions to their final optimal positions.

10.6   Future Directions

The concept of the self-organizing grid as a design technique is very new and the potential scope of applying it to other tessellations is ongoing. This is an opportunity to mention some of the areas where new work has proceeded. Overall, there are three specific areas open for research that will be touched upon. One includes other reduced constraint configurations. Another includes reduced constraint configurations that have little or no symmetry. Lastly the methodology can be extended to reduced constraint configurations on non-spherical surfaces.

10.6.1   Additional Reduced Constraint Networks

In the previous examples in this chapter three reduced networks were discussed in detail to show how the application of the self-organizing technique could be used to derive solutions that are able to span the entire topological scope of spherical tessellations. There are other feasible non-fully triangulated tessellations that can be a basis for development. Diamond or rhombus tiling [3.6.3.6] is one case that self-organization can provide a limited but special range of solutions. The limitation is that valid reduced constraint grids can only be applied to BC grids that are Class III. If a Class I or II grid is used the self-organizing method will not converge since the grid has symmetrical constraints that cannot be met. The target goal for the rhombus-based tiling is similar to the hexagonal tessellation: every edge is the same length.

The family of diamond-based networks introduces two different neighborhoods that must be optimized rather than a single neighborhood as seen in the hexagonal and rotegrity tessellations. If the diamond network self-organizes successfully it produces a rather unique topological configuration where the boundary of every hexagonal or pentagonal neighborhood of five or six edges can be circumscribed by the same radius lesser circle. These special spherical circle coverings allow a single sized circle to be used to cover the entire spherical surface with minimal overlap and no gaps, like the one shown in Figure 10.35(f). These configurations are also a subset of *roundest polyhedra*, where the centroids of the polygon faces are tangent to a circumscribed sphere.[11]

---

[11] (Lengyel, Gáspár, and Tarnai 2017).

(a)

(b)

(c)

(d)

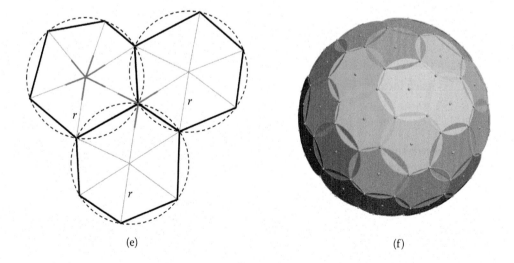

(e)

(f)

Figure 10.35. Alternate reduced constraint network.

Figure 10.35(a) shows our starting triangular network and (b) shows the reduced constraint network after symmetrical edge removal that leaves the rhombus configuration. Figure 10.35(c) highlights the two neighborhoods, one with seven vertices and six edges, the other with four vertices and three edges. Figure 10.35(d) highlights the relationship of the two types of neighborhoods. For the neighborhoods of four vertices and three edges, the center vertex is a trivalent corner of three polygons. When self-organization successfully completes every neighborhood of seven vertices will have six equal edges that represent the radii of a polygon that represents the spherical surface circle. The resulting boundary edges of the final polygons, shown in (e), are the edges we removed in the reduced constraint configuration and their dimensions are a result of the self-organization. Figure 10.35(f) shows a simple icosahedral case derived from a reduced $\{3,5+\}_{4,1}$ triangular configuration and concludes with a $\{5+,3\}_{2,1}$ configuration.

10.6.2   Non-Symmetrical Reduced Constraint Configurations

While we have focused on fully symmetric polyhedral examples to apply tessellations due to the advantage they provide with a high repetition of spherical components, identified previously as the unique section, that is not an essential requirement. The basic self-organizing technique can be applied to either semi-symmetric or completely non-repetitive patterns on a sphere or a section of the sphere. With limited or no symmetry in a reduced constraint grid the role of global propagation is either reduced or disregarded. The self-organizing methodology is adjusted to exclude global propagation after local optimizations. The rest of the methodology remains the same. The following shows the modified algorithm.

To illustrate let's look at a simple example of a small partial spherical hexagonal grid whose boundary vertices are fixed to a rectangular base. The grid is purposely not aligned to the base to ensure that there is limited symmetry in the final form. The grid is semi-symmetric but even that will be ignored during self-organization. The internal vertices need to be equally spaced on the spherical surface. The plan and section of the spherical cap are both delineated in Figure 10.36(a). There are 12 hexagonal neighborhoods. No symmetry is assumed so all hexagonal neighborhoods are optimized and there is no global propagation. After less than 250 optimizations of the 12 neighborhoods the distances between vertices on the spherical surface are within eight significant digits of each other. Figure 10.36(b) shows the original grid sitting on the ground plane in red, while the green grid represents the initial projection of the original flat grid onto the spherical surface resulting in 23 different

Self-Organizing Algorithm (*for non-symmetric grids*)

1. Initialize

   a. Create initial grid

   b. ~~Designate the unique and non-unique vertices.~~

   c. ~~Determine required transformation matrices, M^s~~

   d. Set desired precision

2. Begin continuous loop

   a. For each local neighborhood

      i. Perform local optimization and measure the result

      ii. ~~Perform global propagation on re-positioned vertices~~

   b. If all local optimization results meet target precision exit continuous loop

3. Complete—All vertices in final position that meets the target goal

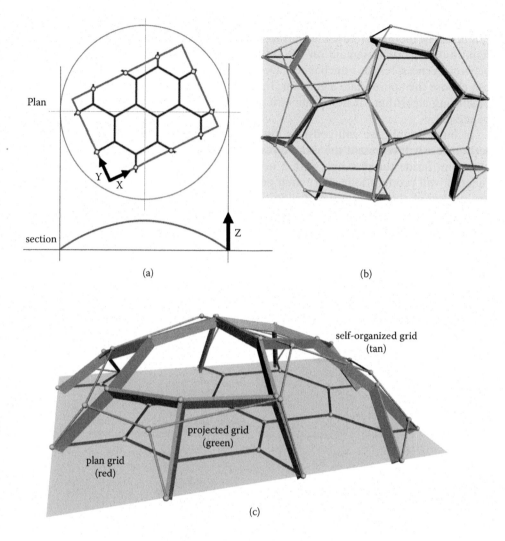

Plan

section

Z

Y
X

(a)

(b)

self-organized grid
(tan)

projected grid
(green)

plan grid
(red)

(c)

Figure 10.36. Semi-symmetric network.

edges. The tan-colored grid is the self-organized grid after complete optimization resulting in only one edge length. Figure 10.36(c) is a perspective view of the entire hexagonal grid with each phase color coded.

10.6.3   Reduced Constraint Configurations on Non-Spherical Surfaces

The self-organizing concept can be extended to non-spherical surfaces and that significantly widens it potential application. Tri-valent networks such as the hexagonal tessellation behave robustly when self-organized on any surface. The original self-organizing methodology controlled vertex movement so that the vertices would remain on the spherical surface as their positions were optimized. When working on a unit sphere centered at the origin the coordinates of any vertex on the surface will satisfy the simple equation $x^2 + y^2 + z^2 = 1$. With this relationship we can find any component (x, y, z) if we know any other two components. Also, any position that is no longer on the surface can be retracted back to the surface by normalization.

If the surface is not a sphere but can be represented by some function then optimization can be modified to use that surface function to ensure that during optimization re-positioned vertices remain on the surface. The measurement of edges, distances between vertices, remains unchanged. Depending on the symmetry of the surface we can choose the appropriate algorithm. In the following example we will use the self-organizing algorithm for non-symmetric grids, disregarding any symmetry that may be present.

In our example, we will consider another hexagon pattern applied to a square section that will represent the ground plane. Next, we will represent the z height of any point inside the rectangle with a non-spherical function. The example surface function will produce four alternating peaks and valleys. While keeping the boundary vertices of the grid fixed, we will initially position each of the interior vertices at the z height that results from the application of the height function at the respective initial (x, y) positions. Figure 10.37(a) is an orthographic view of the initial grid after vertex heights are adjusted, while (b) is an orthographic view of the same grid after it

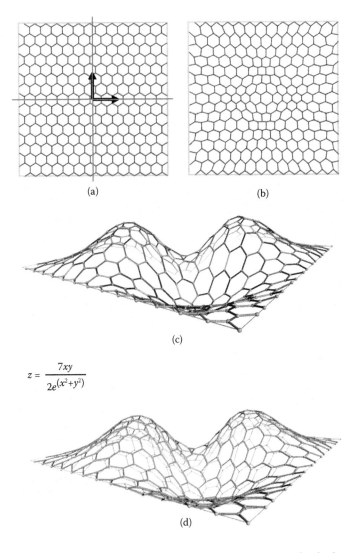

$$z = \frac{7xy}{2e^{(x^2+y^2)}}$$

Figure 10.37. Self-Organized hexagon network on a non-spherical surface.

has been self-organized. In the initial grid there are 390 neighborhoods that need to be optimized, and initially 333 different edge lengths, Figure 10.37 (c), but after self-organization there is only one edge length (d). All neighborhood center vertices move throughout the optimization process but always remain on the non-spherical surface given by the z function.

In this scenario the inner loop of the algorithm performs the typical hexagonal neighborhood optimization on all 390 neighborhoods sequentially, before continuously repeating until the distances between all neighborhood centers are equivalent to the desired level of precision. In the example for a precision of eight significant digits the optimization of 390 neighborhoods was repeated just over 3200 times.

10.7   Summary

In this chapter, we described the concepts around self-organizing grids with some specific groups of use cases that can be efficiently handled by this methodology; use cases that otherwise lack an alternative process for developing comprehensive solutions. Since this methodology is relatively new its application to spherical division problems is still being researched, while its application to non-spherical surfaces is almost untouched territory. Overall, the self-organizing methodology, illustrates that ideas in shell tessellations, that have experienced enormous explorations in the 20th century, still have secrets to share.

Additional Resources

Baverel, Olivier, Nexorades: A Family of Interwoven Space Structures, PhD thesis, University of Surrey, 2000.

Boyt, Richard D., *ROTEGRITY – An Alternate Method of Spanning Space with a Compound Curve*, unpublished, Crowder College, 1991.

Clinton, Joseph, A Group of Spherical Tessellations Having Edges of Equal Length, *5th International Conference on Space Structures* eds. G. A. R. Parke and P. Disney. Thomas Telford Limited, 2002.

Goldberg, Michael, "The Isoperimetric Problem for Polyhedra." *Tohoku Mathematical Journal* 40 (1935): 226–236.

Grünbaum, Branko and Shepard, G.C., *Tilings and Patterns* ed. W. H. Freeman. 1987.

Fejes Tóth, László, "Stable Packing of Circles on the Sphere." *Structural Topology*, University of Montreal, 11 (1985): 9–14.

Fuller, R. Buckminster, *Tensile-Integrity Structures*, Patent #3063521, 1962.

Larsen, Olga Popovic, *Reciprocal Frame Architecture*, Architectural Press, 2008.

Lengyel, András, Gáspár, Zsolt, and Tarnai, Tibor, *The Roundest Polyhedra with Symmetry Constraints*, Symmetry 2018, Polyhedral Structures, Symmetry, and Applications, 2017.

Kitrick, Christopher, Equal Edge Hexagonal Spherical Tessellations, *Proceedings of the IASS Annual Symposium*, 2015.

____. Icosahedral Roundest Polyhedra, *Proceedings of the IASS Annual Symposium*, 2017.

____. Non-Linear Solution to Spherical Coverings and Application to Forms, *Proceedings of the IASS Annual Symposium*, 2019.

____. *DisplaySphere*, Open source rendering tool for Spherical Geometry, GitHub, 2020.

Tarnai, Tibor and Wenninger, Magnus, "Spherical Circle Coverings and Geodesic Domes." *Structural Topology*, University of Montreal, 16 (1990): 5–22.

Ⓐ Stereographic Projection

In many spherical designs, it is helpful to visualize points and great and lesser circles in two dimensions or as a flat drawing. As the illustrations in this book demonstrate, there are many ways to project spherical geometry onto a 2D surface. However, there is one projection method that offers both visualization and direct analysis capabilities. *Stereographic projection* is a way to show the 3D relationship of points and arcs on a sphere in a 2D diagram called a *stereogram*. Geodesic subdivisions distribute many points and arcs on the surface of a sphere, and stereographic projection is the best way to represent them in a 2D way that preserves angular relationships.

Stereograms are like maps or graphs of the points and arcs on a sphere. We use stereographics to visualize how well a particular subdivision schema performs and to make comparisons among schemas. With experience, stereograms are easy to read. They immediately show symmetry, if it exists; they also show whether points are well distributed or if the subdivision faces are oriented uniformly. Often, you can identify the subdivision schema and the base spherical polyhedra being used simply by inspecting the stereogram.

Stereographics were studied by ancient Greek mathematicians and used to solve spherical trigonometry problems. Later, they were used to make astronomical instruments, such as the astrolabe. Today, the technique is used by cartographers, who make maps of the Earth or maps from satellite images of moons and planets. Geologists and civil engineers use stereographics to analyze large-scale rock formations, while crystallographers use the technique to study small crystals and their chemical composition.

A.1   Points on a Sphere

Stereographic projections depict points and arcs on the surface of the sphere. Points on the surface of the sphere are positions in 3D space, and we identify each of these points with the vector that points from the center of the sphere to this point. A vector has direction and magnitude. A line established by the point and the origin of the sphere define the direction. The magnitude is the length or distance between the point and the origin. For almost every case example in this book, the magnitude is one unit because we choose to work with a generalized sphere of one radius unit. That unit can be one millimeter or one mile; it does not matter.

In the geodesic applications developed here, points on the sphere can be the vertices of triangular grids or vectors that are normal to each face. Face normals are very useful vectors because their direction passes through the origin and is perpendicular to the face, or more precisely, to the plane of the face. The face normal has a magnitude as well. This magnitude is directly proportional to the distance of the face from the origin. However, we can project (or scale) the normal point and define a new point that *does* lie on the surface of the unit sphere. The new point is called a unit normal and maintains the original direction of the normal; thus, it remains on a line from the origin and remains perpendicular to the plane of the face. The stereographic projection of either geodesic tessellation vertices, or the unit normals of their faces, is the best way of representing the inter-relationship of their directions and orientations on a plane.

A.2   Stereographic Properties

Many techniques exist for projecting shapes on a sphere onto 2D surfaces. Cartographers have invented scores. But unlike other projection methods, stereographic projection is particularly suitable when analyzing and comparing spherical subdivisions.

First, stereographic projections are conformal. That is, the stereogram preserves the angular relationship between points or intersecting arcs on the surface of the sphere. If lines of latitude and longitude, which meet at right angles on a globe, are projected, they will meet at right angles on the stereoprojection. This also means that the angles between points or between crossing arcs can be directly measured on the stereogram. *Conformality* is an important property because it preserves the general shape of projected objects.

Second, circles on the sphere project as circles on the projection plane. If a great circle passes through the center of a projection, it projects as a straight line (a segment of a circle with an infinite radius). Other spherical projections can display angular relationships between poles on the sphere, but the stereographic projection is the only one that projects any circle on the surface of the sphere into a circle on the projection.[1] Geodesics distribute many great circle arcs and spherical caps, lesser circles around points on the sphere, and are often used to analyze how uniformly points are distributed. All of them project as circles on the stereogram.

Projections can be made from either the North or South Pole of the sphere, and the projection plane, called the primitive, can be the equator (any great circle) or tangent to the sphere at either pole (or any plane passing through the sphere). Each combination

---

[1] Some circles project as straight lines or circles with an infinite radius.

changes the appearance of the stereogram, but the information it contains and how it is used remain the same. These choices make stereographic projection flexible, although some combinations are more useful to geodesic work than others. We discuss some of these choices in this section, and we adopt some standard uses that we use to analyze subdivisions.

A.3 A History of Diverse Uses

The stereographic projection has its origins no later than the third century BC. Appollonius of Perga included theorems with direct application to proving circle preservation in his *Conics*, which dates from the late third century BC. Applications of the stereographic projection are certain to have been made by the Greeks no later than about 225 BC in the form of the anaphoric clock, which has a disk with stereographically projected stars rotating behind a grid including a stereographically projected horizon and tropics. Hipparchus certainly used the stereographic projection to make star maps and is incorrectly (in my opinion) credited by Ptolemy as being the originator of the projection method. Ptolemy's *Planisphaerium* demonstrates that a solid theoretical foundation for the projection was in place before 150 AD.

The Greeks called the process of representing a sphere on a plane *"unfolding the sphere."* The stereographic projection was named by François d'Aguilon (Franciscus Aguilonius) in *Opticorum libri sex* (published in 1613) from στερεοζ (solid). Hence, stereographic means *"drawing solids (on a plane)."* It is clear that the Greeks did not prove the conformal properties of the projection, but may have assumed it. It is equally clear that the Greeks did know and use the preservation of circles property. Preservation of angles (conformality) under the stereographic projection was first proved by Thomas Harriot (1560–1621).

Islamic astronomers used the projection and proved its basic properties no later than the ninth century. A notable entirely European contribution was *De plana spera,* an early thirteenth-century work by Jordanus de Nemore that presented the theoretical foundation for the stereographic projection. The stereographic projection was applied to astrolabes, quadrants, and other astronomical instruments in both Islamic countries and in Christian Europe in medieval times. Claudius Ptolemy (ca. 150 AD) wrote extensively on stereographics in his work entitled *Planisphaerium,* and in 1613, François d'Aguilon gave the technique its name, *"stereographic."*

A.4 The Astrolabe

Stereoprojection is a natural choice for star charts and reference grids for familiar objects or patterns in the night sky. Stereographics was the fundamental projection in the most important astronomy instrument ever devised: the *astrolabe*. An astrolabe is used to solve problems related to time and the positions of the sun and stars. The first astrolabes date to before 400 AD, and they remained very popular in Europe and the Islamic world into the seventeenth century. No other instrument enjoyed such a long life. The earliest astrolabesthat evolved in Greek antiquity were refined in the Islamic world and were virtually perfected by the tenth century AD. They were introduced into Europe in the eleventh century from Moslem Spain (al-Andalus). Figure A.1 shows the front and back sides of a classic astrolabe.[2]

The earliest known astrolabes, brass disks 6–12 inches in diameter, were made in Alexandria. Muslims first encountered the astrolabe in the eighth century in Harran,

---

[2] Astrolabe illustrations courtesy of James Morrison.

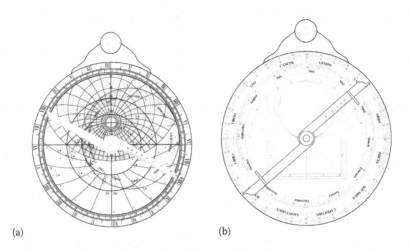

(a)   (b)

Figure A.1. Astrolabe front and back.

currently in Turkey near the Syrian border. Even the earliest designs were highly so-phisticated instruments. Every part was carefully designed and engraved with scales for specific functions. The main body, the *mater* (Latin for "*mother*"), was a round plate with a large, hollow circular pocket that accommodated a set of thin, circular disks called *climates*, *plates*, or *tympans* (Latin for "*plate*"). On the climate disk was engraved a stereoprojection celestial grid of the sky above the observer's local horizon, at the latitude from which the user would make his observations. The center of the cli-mate disk corresponds to the north celestial pole. Well-equipped astrolabes would also include a set of climate plates, stored in the mater, each engraved with a sky reference grid for a latitude where the instrument might be used.

Centered over the climate disk is a pierced disk called the *rete* (Latin for "*net*") which is free to rotate over the climate disk to simulate the daily motion of the stars. It is a frame with openings so the climate disk underneath can be seen. Pointers mark the positions of bright stars. The location of the sun is shown by the projection of the ecliptic as an offset circle on the rete. Scales on the back side of the astrolabe are used to determine the position of the sun on the ecliptic (i.e., the sun's longitude).

Early astronomers believed the Earth to be fixed in the heavens and that the stars revolved around it. Although we now know this is not true, the concept is still effec-tively used today by ocean voyagers who use sextants to navigate. The star's position over the stereographic reference lines indicates where in the sky (azimuth and altitude) the sun or a star can be found.

Many European astrolabes included a rotating rule over the rete. It served many purposes, acting like the hand of a clock, providing an altitude scale, and aiding in the interpretation of the rete's position over the climate. It also was an index to scales engraved on the outer rim, or *limb*, of the astrolabe.

Progressive design refinements and advanced craftsmanship, plus the widespread dissemination of astrolabes in Europe in the fifteenth and sixteenth centuries, led to instrument refinements and contributed to the astrolabe being applied to surveying, orienteering, and timekeeping along with astronomical uses. Many of these additional features required engraved scales to be added to the back of the instrument. The as-trolabe's popularity for casting horoscopes guaranteed its continued use even to the present day, and astrolabes were even available in paper and other affordable forms. The astrolabe was truly the most elegant and widely used scientific instrument ever made. Despite its many design variations, the fundamental use of stereoprojection remained constant.

A.5   Crystallography and Geology

New uses for stereographics have been found in the fields of crystallography, civil engineering, and structural geology. Crystallographers study the atomic structure of solids. Before the use of X-ray diffraction techniques, stereographic techniques were used to graph and analyze the geometry of crystals. Stereograms are made of the orientation of a crystal's planes and the angles between them. In order to make the stereogram, crystal is located inside an imaginary sphere. Points on the sphere are located directly above each face (on an axis normal to the face). These points are then stereoprojected from a pole in the hemisphere opposite the point. Figure A.2 shows the basic arrangement. The result is the dot-like pattern on the primitive, the equatorial plane where points are projected. Each dot is the projection of a point on the sphere directly above a crystal's face. The angles between dots indicate the angles between crystal faces. Other dot patterns indicate symmetrical relationships between vertices, edges, and crystal faces.

Stereographic projection is particularly useful in analyzing crystals. An ideal crystal is a perfect geometric shape. Many are microscopic, and crystal forms are found in four of five platonic shapes. For example, some forms of calcite galena resemble small cubic shapes, while pyrite can take cubic, octahedral, or pyritohedral (a dodecahedron with pentagonal faces) forms.

In nature, few crystals are perfectly formed. Neighboring bodies can interfere, and inconsistent flow of the crystalline liquid or multiple formations occurring in close proximity all distort a crystal's growth. The number of faces in a crystal and their orientations, however, are a function of the crystal's chemistry, and interference during their formation does not affect the angles between faces.[3] These angles are also independent of the face's size. Thus, samples from imperfect crystals, where only a small portion of a face has formed, can oftentimes be used to make accurate stereograms, as if the crystal were perfectly formed.

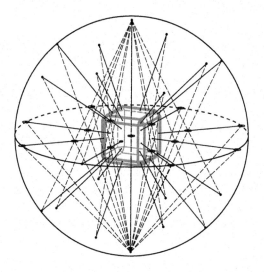

Figure A.2. Stereoprojection in crystallography.

---

[3] (Whittaker 1984, 1).

On a much larger scale than crystallography, structural geologists and civil engineers use stereographics to analyze the orientation of linear and planar geological features. Reference lines over geological features—such as the horizontal reference line of a sloping rock shelf or the dip slope angle of the shelf—are 3D references to the orientation of the geological feature. Stereographic projections of these features show how plates and strata are positioned with respect to one another. A stereogram can identify faults, folds, and other geological discontinuities where minerals form or oil and gas might be trapped. In structural geology, stereograms provide common visualization and allow comparisons. They also summarize large amounts of 3D data and serve as a predictive tool important in exploration.

A.6   Cartography

Azimuthal stereographic projection is a classic projection for mapmakers. The earliest known map of the world using stereographics was made by Walther Ludd (Galtier Lud) of St. Dié, Lorraine, in the early sixteenth century. The projection was made famous when Gerardus Mercator, a sixteenth-century Flemish cartographer, used the equatorial stereographic projection for the world maps of the atlas of 1595.

Today, stereographic maps of the Earth are still made and are sometimes used to display satellite images of celestial bodies, such as planets and moons. The projection plane and the point of projection can be placed at any location on a sphere, but two positions are commonly used. When these two positions are at the equator and projections are made from the North or South Pole, the map is called a *polar stereographic* or *polar aspect* projection. It is the most common projection of the polar areas of the Earth, moon, and planets since it is conformal. Directions measured from the center are true; the meridians are shown as straight lines radiating from the pole, while latitude circles appear as concentric circles around the pole.

Figure A.3(a) shows a polar stereographic map viewed from the North Pole. When the projection plane is through the middle of the Earth from the North to the South Pole, instead of at the equator, and projections are made from a point on the equator, it is called an *equatorial aspect* map. One is shown in Figure A.3(b), while (c) shows an oblique view. Later we will see a universal stereoprojection called the *Wulff net*, which is an equatorial stereographic projection.

Azimuthal stereoprojections are not distortion-free maps—no map projection is—but they preserve angular relationships, and circular features projected as circles. The scale of azimuthal stereographic maps is distorted, however, and becomes more pronounced as one moves away from the center of the map. These drawbacks are minimized, however, when the geographic area is smaller than a hemisphere.

From ancient astronomy and classic mapmaking to crystallography and structural geology, the stereographic projection continues to be relevant and a preferred projection in spherical work. In the next section, we look closer at this technique and examine the resulting stereograms in detail. In a later chapter, we apply stereographic projections in a novel way to evaluate spherical subdivisions. Stereograms are excellent ways to analyze how well a spherical subdivision distributes points on the sphere or how uniformly the resulting faces are oriented in three dimensions. We begin with stereoprojection basics.

A.7   Projection Methods

Stereographic projections are made by passing infinite rays (lines) from a projection point, usually the North or South Pole of a sphere, to other points on the sphere. Rays will pass through a plane that is either tangent to one of the projection poles or at the

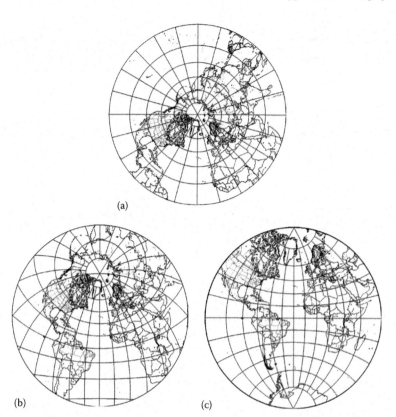

(a)

(b)   (c)

Figure A.3. (a) Polar, (b) equatorial, and (c) oblique stereographic maps.

equator of the sphere. The projection plane is called the *primitive*. Stereoprojection points are defined where rays intersect the primitive.

For geodesic applications, it is convenient to locate north and south projection poles at the positive and negative z-axis of the 3D coordinate space and to use the xy-plane as the primitive. This is also the arrangement used by many crystallographers.

Figure A.4 shows the basic stereoprojection arrangement and a single point (on the sphere) being projected from the South Pole. Depending on the type of stereogram desired, either pole can be used for projection. If one pole is used for all projections, points near the pole will project near infinity and the pole point itself cannot be projected at all. To avoid these problems, we adopt the practice of projecting from the pole opposite the point's hemisphere.

Figure A.4(a) represents the projected point with a small circle on the primitive. A dot in the middle of the circle indicates the exact projection point. Points projected from the pole of their hemisphere will project outside the circumference of the primitive. Cases such as this are common when celestial stereograms are made for astrolabes. Points, which are tangent to a pole, must be projected from the opposite pole. We will see some examples shortly.

Figure A.4(b) shows another fundamental principle of stereoprojection: antipodal points always project to points that are equidistant and opposite from the center of the primitive. Antipodal points are the circumference of the same great circle, and this is also shown in the figure.

We adopted the graphic convention to represent projected points from the North Pole as small circles with a dot in the middle. This choice is deliberate. If points from

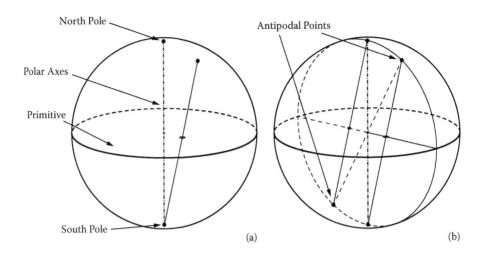

Figure A.4. Stereographic poles, primitive, and projection of points.

opposite hemispheres project to the same point on the primitive, the larger circle graphic for points projected from the South Pole will surround the smaller circle graphic from the points projected from the North Pole. In both cases, a dot will mark the exact projection point in the center. The resulting bull's-eye graphic is easy to spot on the stereogram, and the exact point of projection (the dot) is maintained.

A.8 Great Circles

The stereoprojection of a great circle is a circle. If the great circle is tangent to a projection pole (it will be tangent to both), it will project as a line on the primitive; or more correctly, it will project as an arc of a circle with infinite radius. Figure A.5 shows this case where a great circle is tangent to both projection poles. Its stereoprojection is a line through the center of the primitive across the entire diameter of the primitive. Two points on the great circle are projected separately, just to show the relationship

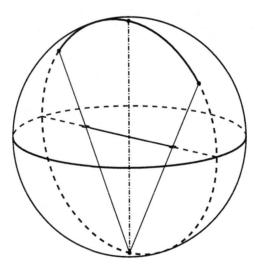

Figure A.5. Great circle tangent to projection axis.

Figure A.6. Great circle projection from the same pole.

Figure A.7. Great circle projection from poles opposite points.

between a great circle arc segment and the resulting line (infinite radius arc) segment on the primitive.

The equator of a sphere is also a great circle. Because its plane is coincident with the plane of the primitive, the equator is projected as a circle with its center at the projection of the pole.

In geodesics, most great circles or great circle arcs do not pass through the poles or are coincident to the equator (primitive). All of them, however, will project as arcs or segments of circles on the primitive. Depending on the choice of projection poles, the projection circle may be totally or only partially within the limits of the primitive.

Figures A.6 and A.7 show the difference the choice of projection pole can make. In Figure A.6 the entire great circle is projected from one pole—the South Pole. All the projected points intersect the plane of the primitive, but some projected points fall outside its circumference. The dihedral angle between the great circle and primitive in this figure is 30°. As the dihedral angle approaches 90°, the radius of the projected circle becomes infinite Great circles projected from a single projection pole are only practical if the projected circle is graphically clipped or the dihedral angles are relatively small.

In Figure A.7, the projection pole is always opposite the hemisphere of the point being projected. The resulting stereogram is compact and lies entirely on or within the circumference of the primitive. Rays from the North Pole are shown dashed, while rays from the South Pole are solid.

Both figures illustrate another fact about stereoprojection of great circles: the line of intersection of two great circles in three dimensions—the primitive great circle and example great circle projected—always passes through the origin of the sphere and intersects the surface of a sphere at two antipodal points. If one of these great circles is the equator, these two antipodal points are tangent to the primitive. A line joining them passes through the origin of the sphere and, likewise, through the center of the primitive.

A.9 Lesser Circles

Like great circles, lesser circles also project as circles on the primitive. Figure A.8 shows several orientations of lesser circles and their projections. A lesser circle, entirely

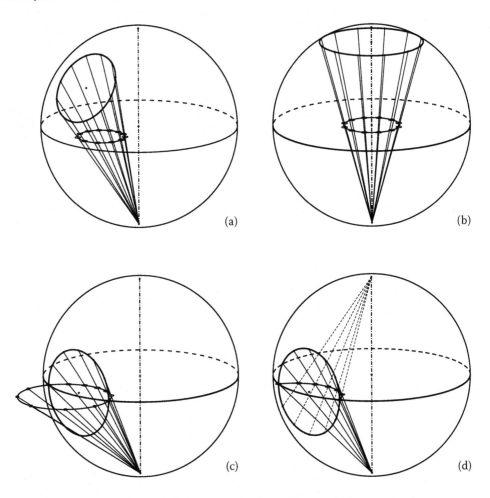

Figure A.8. Stereoprojection of lesser circles.

in one hemisphere, is projected from the pole in the opposite hemisphere, as shown in Figure A.8(a). Figure A.8(b) shows cases where the centers of lesser circles are tangent to the projection axis. These "*latitude*" circles project as concentric circles centered in the primitive. Some sample rays are shown as points on their circumference.

Figure A.8(c) and (d) illustrate projections where the lesser circle cuts through the primitive. If a single projection pole is used (c) for the entire lesser circle, the resulting projection points are on the plane of the primitive, but some points fall outside its circumference and the distance is dependent on how close the center of the lesser circle is to the origin of the sphere. It approaches infinity, as the lesser circle gets larger and approaches a great circle. If the lesser circle is projected from the pole opposite the point on its circumference (d), a compact projection results and all projection points are on or inside the primitive.

There are three additional subtleties about lesser circles. The first is how their position on the sphere affects their projected radius; the second is the projection of their centers; and the third is the projection of circles that cross but are not perpendicular to the primitive.

Figure A.9 shows the stereoprojection of four lesser circles with the same radius. The projected lesser circles, however, do not have the same radius. The smallest pro-

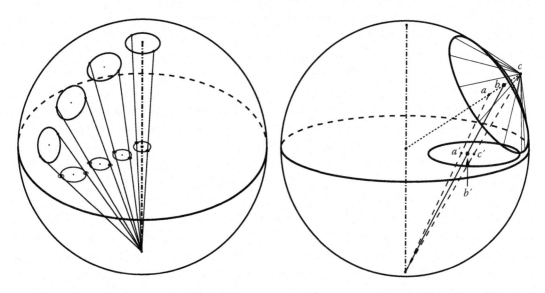

Figure A.9. Stereoprojection of equal lesser circles.

Figure A.10. Stereoprojection of lesser circle centers.

jected radius belongs to the lesser circle at the zenith centered on the projection axis. The projected circle radius increases as the lesser circle on the sphere approaches the primitive. The largest projected lesser circle radius will be the lesser circle whose center is tangent to the primitive.

Figure A.10 shows the projection ray (dashed) of a lesser circle's center, point a, from the South Pole. The ray's intersection with the primitive is marked by a small dot at a'. This point, however, does not qualify as a stereoprojection of point a because a is not on the surface of the sphere; it is on the plane of the lesser circle and inside the sphere. Point b is the projection of a from the origin to the sphere's surface. Mathematically, point (vector) a has been normalized to have a magnitude of one; thus, point b is on the surface of our unit radius sphere. Point b qualifies for stereoprojection and its projection ray (solid) to the South Pole intersects the primitive at b'. Both b and b' are indicated with larger dots.

When the points a' and b' are compared, neither one is in the center of the lesser circle on the primitive. It turns out that the true center of the projected lesser circle is point c'; it is the projection of point c, which is outside the sphere altogether and at the apex of a cone that is tangent to the sphere at the lesser circle. This was discovered by Michel Chasles, a nineteenth-century French mathematician, who developed theories of projective geometry.

Unlike the lesser circle examples, shown in Figure A.8(c) and (d), stereoprojection of lesser circles, which intersect the primitive but are not perpendicular to it, have somewhat unexpected representations. Figure A.11 shows this case. In (a), the whole lesser circle is projected from the South Pole, and part of its projection extends beyond the primitive. But when points on the lesser circle are projected from the pole opposite their hemisphere, as shown in Figure A.11(b), the projected lesser circle is totally within the primitive, and a part of the projection is "*folded back*" at the point where the lesser circle is tangent to the circumference of the primitive. This characteristic is not apparent in Figure A.8(d) because the plane of the lesser circle is perpendicular to the perfect. The "*folded*" halves of its projected circle (one from each pole) are perfectly superimposed and appear as just one arc on the primitive.

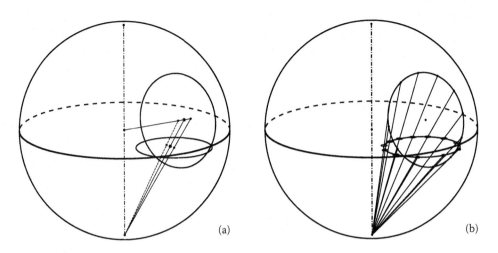

Figure A.11. Stereoprojection of lesser circles crossing the primitive.

A.10   Wulff Net

A Wulff net is a single, simultaneous stereoprojection of a large number of great and lesser circles. A classic Wulff net is shown in Figure A.12.[5] In a Wulff net, the orientation of the great and lesser circles to be projected is very specific—they all share the same axis, and their axis lies on the primitive. For an analogy, imagine that a sampling of the Earth's latitude and longitude circles are stereoprojected. Also assume the perfect bisects the Earth not at its equator,[4] but from the North to the South Pole. Now imagine that you stereoproject the set of longitude (great) circles and latitude (lesser) circles from one of the primitive's poles. Keep in mind that the primitive is cutting the Earth in half from north to south. Thus, the primitive's stereoprojection poles are on the Earth's equator, and these poles are on opposite sides of the Earth. Because of the symmetry in both sets of longitude and latitude circles, a stereoprojection from either the primitive's North or South Pole will produce the same stereogram.

The Wulff net, shown in Figure A.12, projects sample longitude and latitude circles every 2°, with emphasis every 10°. The number of longitude and latitude circles projected is arbitrary, limited only by application needs and the graphic standards employed. In this example, more than 1700 combinations of circles are shown in one stereogram. For this reason, the Wulff net is also called the universal stereogram.

In the Wulff net illustration, latitude circles are labeled in degrees from each of the poles, contrary to the convention of referencing the Earth's latitude and measuring north and south from the equator. The labeling on the Wulff net figure is for zenith distance and is used for setting the latitude of a place on a saphea astrolabe. When the rule on a saphea instrument is set to a number on the scale, the rule represents the horizon for that latitude.

So why is the Wulff net useful? It's because it is a totally generalized stereographic framework. In a sense, it is a general-purpose stereographic protractor. It can overlay other stereograms to measure angles or be used like graph paper when plotting points and arcs. It can measure surface and central angles, and within the accuracy

---

[4] A polar net is a companion to the Wulff net and does bisect the Earth at the equator.
[5] Illustration courtesy of James Morrison.

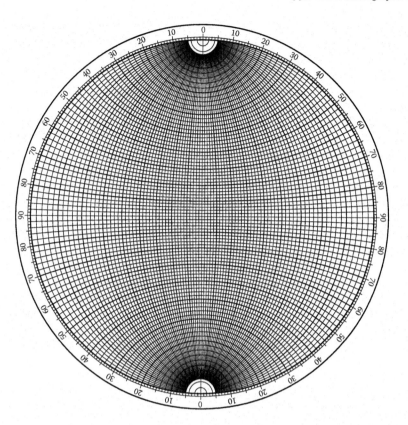

Figure A.12. Stereographic wulff net.

of its graphics, it can be used to solve many spherical trigonometry problems. Angles between great circles (dihedral angles) correspond to interior spherical angles. Lesser circle projections measure angles along great circles and can be used for rotating points or arcs.

A.11   Polyhedra Stereographics

We introduced stereographic projection in the context of astronomy, cartography, and crystallography. Now we will place it in a polyhedral context and use stereographics to map the distribution of subdivision grid points, geodesic arcs between them, and the orientation of polyhedra faces. We treat polyhedra as if they were crystals, and our approach is similar to the way crystallographers use stereographics.

The vertices of regular polyhedra and the grid points resulting from spherical subdivision are on the surface of the polyhedron's circumsphere and can be stereoprojected directly.

Less obvious is the fact that a vector describing the orientation of a polyhedron's face can be used to define a point on the surface of the sphere that can also be stereoprojected. This vector is called a *normal*, and its direction is perpendicular to the plane of the face. Depending on how the normal is calculated, the direction can point out from the origin or in toward it. We can use normals to polyhedral faces, or to the plane of triangles created by subdivision grids, to tell us something about the shape and orientation of the face. What's more, we can make stereograms of them.

A normal vector is perpendicular to all points on the face and the plane the face defines. The normal is independent of the size of the face (or triangle on a subdivided sphere) no matter how large or small the face is. The normal vector to the plane of a face, a triangular face for example, does not necessarily point to the face's center. When the face is a regular polygon (equilateral triangles, squares, and pentagons) and its vertices are on the surface of the sphere, its normal will point to the polygon's centroid. This can be seen in the normals to the faces of the five Platonic solids. But regardless of the shape of the polyhedron's face, a normal vector to that face will always point perpendicular to the plane of the face.

Stereographic projection adds little value for analyzing the regular polyhedra because the number of faces and vertices is small and their orientations are well known. However, when spherical subdivisions are developed from them, there is a dramatic increase in the number of vertices, faces, and diversity in the shape of faces. It is here, then, that stereographics become a powerful summary and analytical tool. Stereograms provide a visual, nonstatistical metric for comparing subdivisions, revealing symmetry, and identifying the fundamental polyhedra and subdivision techniques in use. As such, stereographics is a preferred method for determining which subdivision schemas are best when the application requires uniform or near-uniform face orientations. Given the simplicity of the technique and its useful results, it is difficult to understand why this technique is not exploited more often in geodesics. Building on the notion of polyhedra vertex and face vectors, we now make stereographic projections of them.

A.12   Polyhedra as Crystals

Figure A.13 shows the stereographic projection arrangement used by crystallographers. The primitive is at the equator, and its polar axis defines north and south polar projection points on the sphere. A single icosahedron face (equilateral triangle) is shown, and all three of its face vertices are tangent to its circumsphere. It is helpful to place the polyhedra vertex-zenith whenever possible. We will demonstrate later why this orientation makes it easier to measure the angles between face normals on the stereograms. In this figure, one vertex of the triangle is at the zenith and is coincident to the North Pole.

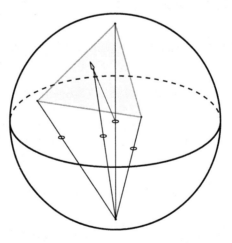

Figure A.13. Vertex and face stereographic projection.

In the figure, the triangle's normal direction is shown as an arrow from the origin through the middle of the triangle. The arrow indicates the normal's direction, and it points out and away from the origin of the sphere. A normal vector cannot be stereo-projected directly. Instead, it is used to define a Cartesian point that is on the surface of the sphere. This is accomplished by normalizing the normal vector; that is, scaling the vector's magnitude to one, the radius of the unit sphere. The resulting derived point is now on the surface of the sphere and on a ray from the origin that is perpendicular to the plane of the triangle. The triangle's vertices were already on the sphere; we can now stereoproject both the vertices and the derived point since they are all on the surface of the sphere.

To find the stereographic projection or a set of points, lines are drawn from each point to the pole in the opposite hemisphere. The stereographic projection of each point falls where its projection line intersects the primitive. In Figure A.13, all four projected points are shown as open circles on the primitive. Any point on the sphere coincident to a pole will project to the center of the primitive. All other points on the sphere project somewhere on the primitive, but not on its circumference or its center.

A.13   Metrics and Interpretation

Stereographic projections are *conformal projections*. This means the angle between any arcs or great circles on the sphere remains the same when projected onto the stereo-gram. Great circles passing through the projection poles project as straight lines and can be measured directly from the stereogram.

Earlier, we adopted a convention of orienting polyhedra vertex-zenith. This always places one vertex at the north projection pole. For all platonic polyhedra, except the tetrahedron, one vertex will always be at the South Pole if another is at the North Pole. By orienting vertex-zenith, we take advantage of the fact that any great circle passing through both the north and south projection poles project as straight lines onto the primitive and this makes it very easy to measure angles on the stereogram.

Figure A.14 adds great circles for two of the triangle's edges. Their projection lines are shown on the primitive as solid and dashed lines. A compass rose is added as a visual indication of the angle between lines; it measures 72°, the surface angle of the equilateral triangle face angle. Later, we will compare angles between projected face

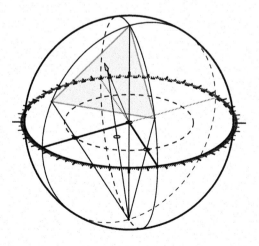

Figure A.14. Angle measurements from stereographic projections.

unit normals. This will alert us to the presence of gaps or unevenly distributed face orientations. Figure A.14 adds a concentric reference circle to the perfect. This illustrates the fact that projected points equidistant to the center of the primitive have the same north or south latitude.

A.14   Projecting Polyhedra

Let's examine the stereographic process for complete polyhedra. The icosahedron, tetrahedron, and octahedron form the basis of almost every subdivision developed in this book. Since their projected face normals will be found in a great many of the tessellations based on them, it is worth examining their stereograms.

Figure A.15(a) shows the stereographic arrangement for an icosahedron in a unit circumsphere. The icosahedron is oriented vertex-zenith, and two of its vertices are coincident to the North and South Poles of the primitive. Figure A.15(b) shows the normal directions to the planes of all 20 faces and their derived points on the surface of the circumsphere. Since the icosahedron's faces are equilateral and their vertices are tangent to the sphere's surface, the direction of face normals passes through the centroid (and incenter) of each face.

Figure A.15(c) shows the stereographic projection of the icosahedron's 20 points derived from each of its face normals. Points in the northern hemisphere are projected from the South Pole and indicated on the primitive with small open circles. Normals to each of the polyhedron's faces in (b) and the projection rays from derived points in

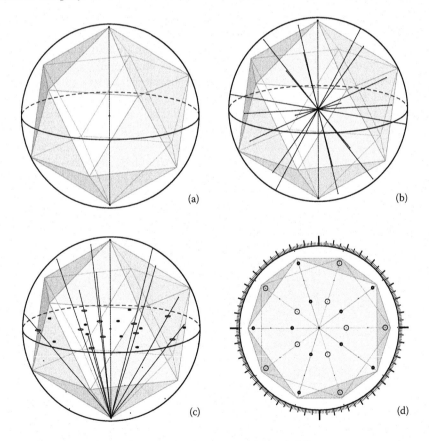

Figure A.15. Polyhedron stereographic projection sequence.

(c) are shown to better explain how the resulting stereoprojection in (d) is done. The derived points from face unit normals in the southern hemisphere are projected from the North Pole and these points are shown as solid-fill circles. Their projection rays are omitted to make the figure easier to read. The choice of projection symbols is deliberate and follows conventions used by crystallographers. If derived from opposite hemispheres projected to the same primitive point, they can still be distinguished. We will see cases such as this shortly, when we make a stereogram of an octahedron.

Interpreting the stereogram in Figure A.15(d) is easy. The shaded icosahedron superimposed over the stereogram associates projected points with their polyhedra faces. The icosahedron's rotational symmetry is immediately evident in the stereogram. As expected, there are 20 projected points (10 hollow circles from derived points in the northern hemisphere and 10 solid-fill ones from points in the southern hemisphere).

In this stereogram, every vertex of the icosahedron has a symmetrical one directly opposite the origin. This means that for the icosahedron's 12 vertices, there are six pairs of antipodal points, each creating a rotation axis. One pair of vertices is on the polar axis of the stereogram and they project to the center of the stereogram. In Figure A.15(d), a top view of the stereogram, we see that five equilateral triangles share one vertex coincident to the North Pole projection point, and their projections show the five-fold rotation symmetry. Rotation symmetry means that the polyhedron can be rotated about the stereographic axis any multiple of 72°, and the appearance of its stereogram is not changed. We see that the symmetry in the stereogram and the compass rose shows the projected unit normals for adjacent icosahedron faces are 72° apart. Rotation symmetry is a powerful concept in 3D geometry and is particularly useful in geodesics. The concept is developed in detail in Chapter 6.

The stereogram in Figure A.15(d) shows another relationship. Many of the points are arranged in patterns and are equidistant from the center of the primitive. When projection points are coincident with the same concentric circle on the primitive, it is an indication that derived face points have the same latitude, north or south, on the sphere.

Now, let's examine the stereoprojections of the derived points from the face normals of the tetrahedron and octahedron. Each example adds new projection cases and helps develop the skill required to read more complex ones quickly.

A.15 Octahedron

Figure A.16 develops a stereogram for an octahedron. The stereographic rays are shown for all eight derived points from the octahedron's face normals to demonstrate

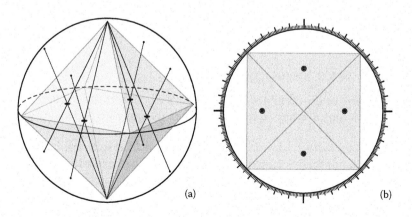

(a)   (b)

Figure A.16. Octahedron stereogram.

what happens when derived points from opposite hemispheres project to the same point on the primitive. In the case of the octahedron, there are four superimposed pairs. Due to the octahedron's rotation symmetry, the projected points are equidistant from the center because they all have the same north and south latitude on the sphere. Sets of them are 90° apart, as expected. The graphic convention of large and small symbols is particularly helpful here because four derived points in each of the hemispheres project to the same point on the primitive. The now-familiar patterns of equal latitude and angular spacing are also evident.

A.16　Tetrahedron

Figure A.17 shows a stereoprojection and stereogram of each derived point from the tetrahedron's face normals. The tetrahedron is arranged vertex-zenith. As in the previous examples, its vertices are on the unit circumsphere. Three of the four face derived points are in the northern hemisphere and are projected from the South Pole. Their projection points are indicated on the primitive with open circles. The bottom face's derived points projects from the North Pole and is shown as a solid fill circle in the middle of the primitive. Since the polyhedron is simple, all projection rays are shown as dotted lines. Figure A.17(b) shows the resulting stereogram with the image of the tetrahedron superimposed. As expected, the projections of the derived points in the northern hemisphere are 120° apart. The fact that these three projected points are equidistant from the center of the primitive also indicates that the unit normals all have the same or complementary latitude.

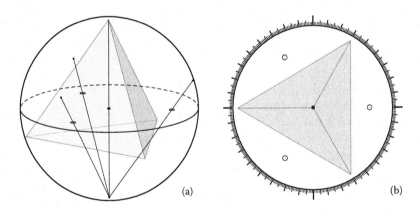

(a)　　　　　　　　(b)

Figure A.17. Tetrahedron stereogram.

A.17　Geodesic Stereographics

The previous polyhedra examples demonstrate the concepts of stereographics. Now let's take a quick look at what we can expect when we apply stereographics to geodesic subdivisions.

Figure A.18 shows the face orientations of a subdivided spherical octahedron. Arrows indicate the normal vector's direction, which is perpendicular to the plane of each triangular face. Although all subdivision vertices are tangent to the unit radius sphere, none of the subdivision triangles themselves are equilateral. Therefore, the perpendicular point (from the origin) is not the face's centroid or incenter, but somewhere to the side of it. It is possible, though not in this example, for the intersection of the

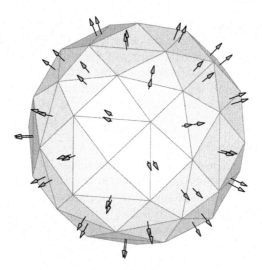

Figure A.18. Face orientations of a subdivided spherical octahedron.

normal's ray and the face's plane to be completely outside the triangle, even though the normal vector is perpendicular to the plane of the triangle.

Figure A.19(a) shows the projection arrangement for the spherical octahedron. The primitive, its polar axis, and its projection points are included. Derived points from face normals in both hemispheres are projected onto the primitive using the technique already established. The projection rays from the South Pole to derived points in the northern hemisphere are shown, but the rays from the North Pole are not depicted to make the figure easier to read. The graphic standards for projected points on the primitive are the same as before; points in the northern hemisphere are represented by the largest circles, while projections of southern hemisphere points use the smaller ones.

The stereogram's plan view is shown in Figure A.19(b). The high degree of symmetry in face orientation between the two hemispheres is apparent. With the exception of the derived points from normals tangent to the primitive's circumference, every point in both hemispheres has a symmetrical companion in the other hemisphere, and both project to the same point on the primitive. This example is the first to show derived points from normals coincident to the plane of the primitive. This means that their

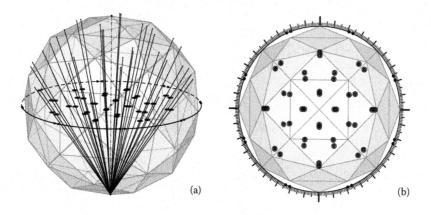

(a)   (b)

Figure A.19. Geodesic stereogram.

corresponding faces are perpendicular to the primitive. Figure A.19(b), a top view of the stereogram, shows that 16 faces are perpendicular to the normal. Since their derived points are already on the circumference of the primitive, no projection is needed. They are shown as small dark circles around the circumsphere of the compass rose. The fact that pairs of points are close to each other indicates that two adjacent faces are nearly coplanar; we see by inspection that this is the case.

A.18 Spherical Icosahedron

All of the subdivision schemas developed in earlier chapters use a spherical polyhedral framework. Although the stereograms shown in Chapter 9 were used to analyze the grid point distributions and subdivision face orientations of triangles, it is possible to analyze the icosahedron's symmetry as well.

Figure A.20 shows the stereoprojection of the 15 great circles of a vertex-zenith icosahedron. The zenith vertex is tangent to the north stereographic projection pole and our eye-point is above the north projection pole looking towards the primitive. A compass rose surrounds the circumference of the primitive and allows us to make direct measurements of the angles between the great circles.

Great circles, or arcs from great circles, project as arcs of circles on the stereogram primitive. In Figure A.20 icosahedron edges are shown as solid lines, while face medians are shown as dashed lines. It is easy to see the 15 great circles when they are combined.

Because the icosahedron's faces are equilateral, their median arcs intersect in the middle of each face. Great circle arcs (edge or median) that pass through the projection pole appear as straight lines, though they are actually arcs of circles with infinite radii. Due to the high degree of symmetry in an icosahedron, edge and median arcs intersect exactly at the equator and extend to the circumference of the primitive.

The compass rose has sufficient resolution to measure surface angles, the angle between two intersecting great circles. The surface angle of every icosahedral face is 72°,

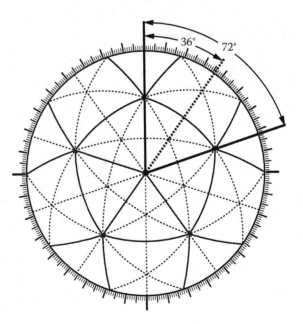

Figure A.20. Icosahedron great circle stereogram.

and this is confirmed by extending adjacent pairs of edges from the center of the primitive (the icosahedron's zenith point) to the primitive. Face medians are midway (dashed line) at 36°, as expected, (see Figure A.20). The stereogram of the icosahedron's great circles will become a familiar sight whenever the face orientations are analyzed.

A.19   Summary

Stereographic projection, one of the oldest techniques in descriptive geometry, is highly relevant to advanced geodesics and spherical subdivisions. The technique is simple to apply and results in a graphic that summarizes the distribution of points on a sphere. When those points are the vertices of polyhedra or points on the sphere derived from normal vectors to their faces, the resulting stereogram provides a complete map of how uniformly the vertices are distributed or the faces oriented.

A stereogram is the projection of one or more points, circles, or other features, such as arcs or derived points from face normals onto the primitive. A key characteristic of stereographic projections is that they are conformal; that is, they preserve the angular relationships between points and intersecting arcs. Angles can be measured directly from the stereogram; thus, stereograms can be used to graphically solve spherical trigonometry problems (within the precision of their graphics). All points, projection poles, and points to be projected must be on the surface of the sphere. Either the North or South Pole can be used to project a point. The stereogram application determines which is most suitable. A standard technique is to project points from the pole in the opposite hemisphere. This will ensure that projected points will lie inside the primitive or on its circumference. Points projected from the pole of their own hemisphere will project outside the circumference of the primitive. Points that are tangent to a pole cannot be projected from that pole because their projection is undefined. However, they can be projected from the pole in the opposite hemisphere. They project to the center of the primitive.

Antipodal points lie on the same great circle and project to points equidistant from and opposite to the center of the primitive. Different graphic symbols can be used to distinguish points projected from the North and South Poles. Crystallographers employ conventions, such as concentric circles, to identify which projection pole was used for a point and to distinguish points projecting to the same location on the primitive. All circles (greater or lesser) or arcs between points on the surface project as arcs of circles on the primitive. Any points coincident with the poles project to the center of the primitive. Antipodal points are on the same great circle and project to points opposite and equidistant from the center of the primitive. Great circles will intersect the circumference of the primitive at two opposite points. Great circles, passing through the projection poles, project as straight lines through the center of the primitive. These lines are arcs of infinite radii circles. The equator of a sphere is a great circle and coincident to the primitive. It is not projected. The line of intersection of two great circles passes through the origin of the sphere and intersects the surface of a sphere at two antipodal points. These two points are always on the primitive. A line joining them passes through the origin of the sphere and through the center of the primitive. Lesser circles project as circles on the primitive. The center of the lesser circle is not the center of the projected circle because it not a point on the sphere. The lesser circle's center is on a plane of the lesser circle and is actually inside the sphere. Lesser circles with centers tangent to the polar axis (latitude circles) project as circles centered in the primitive. A Wulff net is the stereographic projection of a series of greater and lesser circles.

Additional Resources

Lisle, Richard J. and Peter R. Leyson. *Stereographic Projection Techniques for Geologists and Civil Engineers*. Cambridge [Eng.], New York: Cambridge University Press, 2004.

Morrison, James E. *The Astrolabe*. Rehoboth Beach, DE: Janus, 2007.

Van Brummelen, Glen. *Heavenly Mathematics – The Forgotten Art of Spherical Trigonometry*. Princeton University Press, Popular Mathematics/History of Science Series, 2017.

Ⓑ Coordinate Rotations

Rotations are mathematical transformations that reposition one or more points on the surface of the sphere. Rotations were key transformations in the symmetry groups discussed in Chapter 6. Rotations are very useful when covering the entire surface of a sphere with a grid because we can subdivide a small section of the sphere, make a copy of the geometry, and then rotate the copy to cover another part of the sphere. By repeating this copy-rotate sequence, the entire sphere is covered without gaps or overlaps. Rotation does not add any new geometry to the original subdivided section; it only replicates what's there to another place on the sphere. Rotations are one of several isometry transformations that preserve distance between points; thus, the angles between points and the chords, and the arcs between them, remain the same as before.[1]

This appendix outlines general methods for rotating points on a sphere. We show simple and compound rotations around the x-, y-, and z-axes, as well as a more advanced technique to rotate points around an arbitrary axis. In this appendix, we show how the single principal polyhedral triangle (PPT) or diamond coordinates, such as those developed earlier in the book, are rotated to cover the entire sphere. All three class subdivision schemas for icosahedra, octahedra, and tetrahedral are illustrated,

---

[1] Reflections, translations, and glide reflections, like rotations, are isometry transformations; they do not change the distance or angles between points.

as well as the additional Class II dodecahedra and cube schemas. For each polyhedra and schema, a table lists the required angle-axis rotation sequences, and a companion figure shows step-by-step results. All information needed to rotate a single vertex- or face-zenith PPT and edge-zenith diamond to all other positions on the sphere is described here.

Mathematically, there are several ways to implement rotations. We use matrix techniques because they are easy to visualize and to program in computers or calculators. Complex rotation sequences involving more than one rotation can be performed in a series of single operations or by combining the matrices into a single matrix via matrix multiplication.

Rotations are not the only way to reposition geometry. Most subdivisions have highly symmetrical geometry; thus, PPTs and diamonds tend to have symmetrical counterparts elsewhere on the sphere. In these cases, geometry can be reflected to another side of a Cartesian plane, or vertex antipodes can place geometry on the opposite side of the sphere, simply by manipulating the signs of coordinates. Reflections and antipodal points are easy to figure out and can substitute for rotations in some cases. But both techniques have side effects, however, that you need to be aware of. We begin our discussion of coordinate rotations with general concepts.

B.1   Rotation Concepts

Rotation is a simple concept. One or more points are repositioned on the sphere by rotating them together around one of the Cartesian axes by some angle. Rotation does not affect the component of the points corresponding to the axes of rotation. That is, the x-component of points rotated around the x-axis do not change; only the y- or z-values change, depending on the amount of rotation. Rotating points, plus or minus any multiple of $360°$ around an axis, does not change the point's position.

Rotations can be simple or compound. A simple rotation involves only one axis and a positive or negative rotation angle around it. A compound one is several simple ones in sequence. When compound rotations are performed, the result depends entirely on the amount of each simple rotation and the sequence they are applied. Compound rotations are not *commutative*. A rotation by some amount around x followed by another around y usually does not result in the same position if the sequence of rotations is reversed, even if the angles specified are the same. This key point is discussed in detail in the following section.

B.2   Direction and Sequences

Mathematicians, physicists, and engineers use standard conventions to designate which direction is positive when rotating. We use the *right-hand rule (RHR)* convention because it is commonly found in mathematics texts and computer graphics literature. Use the following steps: Grasp the axis to be rotated around by your right hand. Point your thumb in the positive axis direction (away from the coordinate system origin). Your fingers curl around the axis, indicating the positive direction (angle) the object will be rotated. Figure B.1 shows the RHR positive angle direction around the z-axis, and Figure B.2 shows the before (a) and after (b) effect of rotating a cube $+15°$ around the z-axis. Note that the cube rotates; the xyz-reference axis does not.

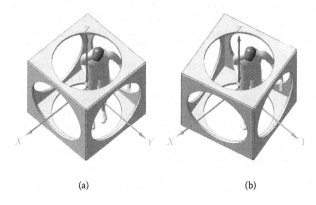

(a)   (b)

Figure B.1. The z-axis right-hand rule.

Figure B.2. Cube rotated +15° around the z-axis.

B.3   Simple Rotations

A simple rotation is a single axis-angle rotation. The choice of axis is important because the axis-angle combination usually affects the final result. Figure B.3 shows an example of simple rotation of an edge-zenith Class II triacon diamond (octahedron 8^v) rotated to the opposite side of the sphere. Two red circular surfaces have been removed—one from the long-axis and another from the short axis of the diamond—to make it easier to compare the effects of the two rotation sequences.

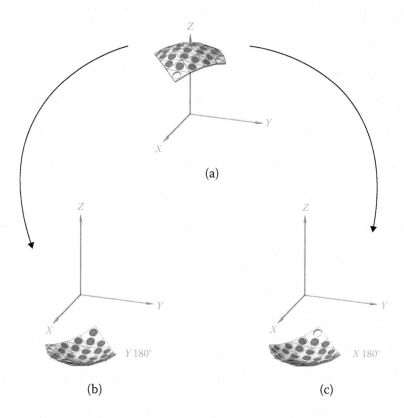

(a)

(b)   (c)

Figure B.3. 180° rotations around different axes produce different orientations.

The initial diamond position is shown in Figure B.3(a). Two different sequences are illustrated. In (b), the diamond is rotated 180° around the *y*-axis. In (c), the rotation is 180° around the *x*-axis. Both sequences placed the original edge-zenith diamond (a) on the opposite side of the sphere, as expected. Notice that the axes of the two diamonds are opposite one other in (b) and (c), as indicated by the missing red surfaces. For some geodesic applications, this difference might not be important, but at other times it may be. Next, let's look at how reflections can accomplish similar results without any rotation at all.

B.4   Reflections

Like rotations, reflections can reposition PPT and diamond coordinates elsewhere on the sphere. Reflections are isometry transformations and will not change the distance or angle between points within the set being reflected. In a reflection, all points are exactly the same distance from the plane they are reflected through; they are just on the opposite side of the plane. Polyhedra have multiple face-, edge-, and vertex-symmetries, as do their subdivisions (see Chapter 6). We can take advantage of symmetry where PPTs and diamonds have counterparts opposite one of the Cartesian planes. Points can be reflected to the other side of a Cartesian plane simply by reversing the sign of the component perpendicular to that plane.

Figure B.4 shows a number of these cases for an edge-zenith Class II icosahedron. In the figure, only the outline of the triacon diamond and their quadrant numbers are shown. The view is from the +*z*-axis looking towards the origin; the *x*- and *y*-axes are labeled. Notice that diamonds 6 to 11 on the positive *x*-side (right) of the *yz*-plane are reflected to the negative *x*-side, making diamonds 12–17 simply by reversing the sign of their *x*-components. We could continue this process and reflect all the diamonds we see in Figure B.4 to the opposite side of the sphere (*xy*-plane) by reversing the *z*-component of all the coordinates.

To be sure, reflection is the easiest way to cover the sphere where there are symmetric relationships around any of the three Cartesian planes.

Reflecting a PPT or diamond's coordinates has the effect of turning it inside out when it is placed on the opposite side of the reflection plane. This can have negative

Figure B.4. Reflecting diamonds.

Figure B.5. Reflecting coordinates.

effects on polygons if you are using left- or right-hand point sequences to know which face of the polygon faces away from the origin of the sphere.

Many 3D applications, such as hidden line removal or face-shading, depend on standard sequences of vertices when defining face boundaries. A sequence, in effect, "*walks around*" the perimeter of the polygon and references each vertex in the proper order necessary to define the polygon's perimeter.

Figure B.5 shows an example of a single edge-zenith diamond reflected through the xy-plane. The z-component of each vertex in the zenith diamond is reflected to the opposite side of the *xy*-plane by reversing the sign of the z-component of every vertex.

Two problems surface with the reflected diamond: its orientation is unique, and the handedness has changed. Notice that the three holes in the diamond remain on the same side of the *xz*-plane after the reflection. This is unique to reflections as no combination of rotations of the edge-zenith diamond around the surface of the sphere can produce this orientation. This can be proved by comparing this diamond to the orientations of the two in Figure B.3.

Another problem with the reflection concerns polygon vertices. We list polygon vertices using the right-hand rule, just as we do when specifying rotation angles. Imagine an axis perpendicular to the face and pointing away from the origin. You are outside the sphere. As before, you grasp the axis with your thumb pointing away from the origin of the sphere. Your fingers curl counterclockwise, defining the polygon. In Figure B.5, the three vertices of a single triangle in the upper diamond are numbered 1, 2, and 3 in RHR sequence. Their reflected points in the lower diamond are numbered 1′, 2′, and 3′. Notice that the reflected points are now left-hand ruled! If you were outside the sphere, their RHR order would be 1′, 3′, and 2′. These two triangles are symmetric but not congruent.

To see more examples of how handedness changes with reflection, take another look at Figure B.4. Notice how the quadrant numbers 1–4 change hand when diamonds are reflected across any of the Cartesian planes.

To summarize, reflections are isometry transformations: a kind of transformation that preserve point distances. They are a simple way to replicate geometry that is symmetrical to a Cartesian plane. However, if your application requires consistent handedness where point sequences define polygons, do not use reflection techniques; use rotations instead.

B.5   Antipodal Points

An antipode is a point diametrically opposite a point on the other side of the sphere; it is the point furthest away, yet still on the surface of the sphere. A line between any point and its antipode passes through the sphere's origin. Figure B.6 shows an example of two diamonds where every tessellation vertex in one is the antipode of the other.

Antipodal points have the same *xyz*-component magnitudes, but each has opposite signs. Unlike rotations or reflections, antipodes can only replicate geometry to the opposite side of the sphere. In this sense, antipodes are less flexible than reflections and rotations, but the technique can replicate geometry within certain limits and is easy to use.

Antipodal points have the same two drawbacks as reflections: the geometry is unique and polygon handedness is reversed. In Figure B.6, notice the position of the holes where selected red surfaces have been omitted. In the lower antipodal diamond, the holes are not in the same position as the reflected example, shown in Figure B.5, nor the simple rotation examples, shown in Figure B.3. The orientation of the antipode diamond is unique, demonstrates opposite reflection orientations, and cannot be replicated by any combination of rotations about the *x*-, *y*-, or *z*-axis. Notice, also, that the handedness of the polygons changes, as demonstrated by the vertex number sequences in the two triangles. The upper diamond is right-hand rule, while the lower one is *left-hand rule (LHR)*, when viewed from outside the sphere.

Antipodal points can be useful shortcuts in certain situations. But in general, we do not recommend using them to replicate geometry on the sphere if orientation or handedness is important to your application.

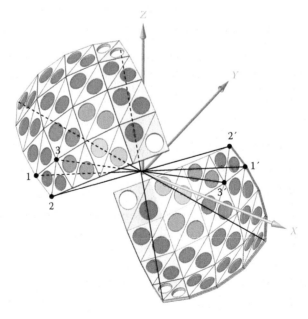

Figure B.6. Replicating geometry with antipodal points.

B.6 Compound Rotations

Most of the time, a PPT or a diamond's coordinates must be rotated two or three times to properly position it elsewhere on the surface of the sphere. The sequence of rotations is critical because compound rotations are *not* commutative. A rotation around one axis and then another does not always result in the same positioning, if the sequence is reversed. We will use an edge-zenith diamond in Figure B.7(a) with some missing red surfaces to illustrate this concept. Again, we use the right-hand rule to specify rotation angle directions.

In column one, the edge-zenith diamond is initially rotated 90° around the y-axis and then again 45° around the z-axis. In column two, the same axis-angles rotations are used, but the sequence is reversed (commuted). The z-axis rotation is performed first, then the y-axis rotation. It is clear that the final rotated diamond positions (b) and (c) are not the same.

To undo a compound sequence of rotations, you apply the rotation sequence that created them in reverse order and use opposite rotation angles

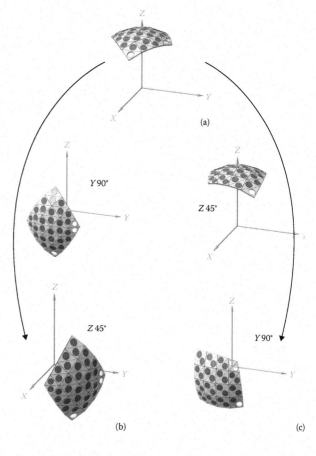

Figure B.7. Compound rotations are not commutative.

in each step. For example, to return the diamond in Figure B.7(c) to its original position at (a), rotate the diamond –90° around the y-axis, and then –45° round the z-axis.

B.7 Rotation Around an Arbitrary Axis

A PPT or LCD set of points can be repositioned anywhere on the surface of the sphere and maintain the handedness of any polygons within them with simple and compound rotations. A more intuitive and less error-prone approach for compound rotation cases might be to specify rotations around an arbitrary axis, not just the standard ones. This means you could define a new axis from the origin to any point on the sphere and then rotate any amount of geometry (PPT, diamond, etc.) around it. The procedure for doing this still relies on compound rotations, but excepting the arbitrary axis-angle specification, you do not have to know ahead of time what the other axis-angles are. This is because they are derived from the axis point you select.

Arbitrary axis rotation involves four steps. First, define the new axis by selecting a point on the surface of the sphere. The new axis is a line between the point and the sphere's origin. Second, the geometry you are rotating must be rotated twice in a way that makes the new axis coincident with the +z-axis. These rotation angles and axes will be described shortly. Third, the geometry is rotated around the z-axis, the specified positive or negative angle for the arbitrary axis. And last, the geometry is rotated twice more, opposite the order of the second step, to return the axis point back to its original position. This last sequence is the reverse of the rotations originally used to make the axis point coincident with the positive z-axis.

To repeat, a rotation around an arbitrary axis rotates the geometry to place the new axis coincident to the $+z$-axis. The geometry is rotated around the $+z$-axis the specified amount, and then rotated back so that the arbitrary axis is returned to its original position. This entire procedure sounds like a variant of compound rotations; it is. What is different is the simplicity of your original rotation specification. You define a convenient axis and rotation angle and let the computer/calculator handle the transformations.

Figure B.8 shows an example of arbitrary axis rotation. We want to copy-rotate the edge- zenith diamond to a new position immediately adjacent to one of its edges. The intended position is indicated by a half-tone diamond in (a). A convenient new axis is to use the edge- zenith diamond, point P, and to rotate the copy a negative angle around it. To do this, the new axis and a copy of the edge-zenith diamond is rotated the amount needed to bring the new axis coincident with the $+z$-axis, as in Figure B.8(b). The copied diamond is now rotated the specified negative angle around the z-axis (c). And finally, the diamond is rotated back the amount needed to restore the arbitrary axis to its original position. The new diamond is now where we want it (d).

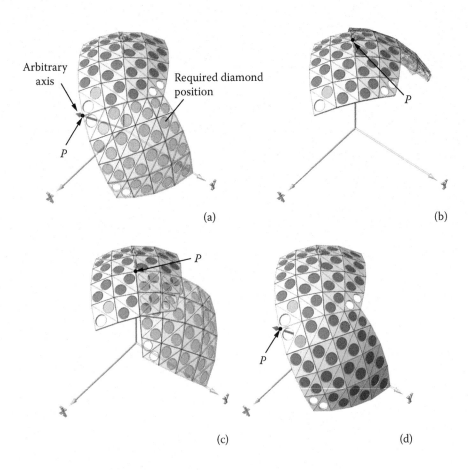

Figure B.8. Rotation around an arbitrary axis.

The idea behind arbitrary axis rotation is to use the axis point itself to provide the compound rotation specification needed to make the arbitrary axis coincident with the z-axis and then to return it to its original position.

But what angles are needed to make the arbitrary axis coincident to the positive z-axis and to reverse the sequence to return it to its original position? Only two angles are needed, and they are simply the spherical φ and θ angles of the selected axis point (see Section 4.14).

It couldn't easier. The compound rotation sequence is always the same. Find the spherical φ and θ angles of the axis point. Rotate the geometry $-\theta$ around the z-axis and then $-\varphi$ around the y-axis. The arbitrary axis is now coincident with the $+z$-axis, and the geometry moves with it. Yet the points maintain their positions relative to one another and to the axis point. Now rotate the geometry around the $+z$-axis by the angle specified for the arbitrary axis. Next, apply the reverse rotation sequences used to make the arbitrary axis coincident. This moves the axis point back to its original location and brings the rotated geometry along with it. The reverse rotation sequence is $+\varphi$ around the y-axis, followed by $+\theta$ around the z-axis.

The convenience of arbitrary axis rotations is obvious, but what about those five rotation steps? They can all be combined into one single matrix and applied only once to the geometry. Five transformation matrices are combined by multiplying each one together: two will move the arbitrary axis coincident with the $+z$-axis, one rotates geometry around the z-axis, and two more return the arbitrary axis and rotated geometry to where the arbitrary axis was originally defined. The algorithm for arbitrary axis rotation is at the end of this appendix. Its simplicity might surprise you.

B.8   Polyhedra and Class Rotation Sequences

Simple and compound rotation sequences give us all the tools necessary to rotate a single PPT or diamond's vertices to all the positions needed to cover the sphere without overlap or gaps. These sequences preserve point distance and handedness of polygon definitions.

We will not use reflection techniques here for the reasons discussed earlier. The following section details the sequences for all the Classes and polyhedra covered in Chapter 8. We use 4^v subdivisions to make it easier to visualize rotation specifications (axis and angles) and their effects. Most sequences involve two rotations; a few require three. The matrix math techniques used are outlined at the end of this appendix. We begin with rotations for icosahedron Classes I and III schemas, the most frequently used in spherical subdivision.

B.9   Icosahedron Classes I and III

Our strategy for covering the entire sphere with a grid is simple. We make copies of the single PPT's (or LCD's) point set we developed in Chapter 8 and then rotate each copy to a unique place on the sphere's surface. We repeat the copy-rotate process until the entire sphere is covered and there are no gaps or overlaps. In effect, we derive all new PPT point sets by copying a previous point set and rotating it to cover a new area of the sphere's surface. For Class I and III subdivisions of the spherical icosahedron, this means we need 19 copy-rotation sequences of the original PPT. Recall that in every Class I schema presented, we oriented the PPT vertex-zenith and symmetric to the yz-plane. The orientation of the original subdivided PPT is important

PPT	Ref PPT	Rotation 1			Rotation 2		
		X°	Y°	Z°	X°	Y°	Z°
6	1	0	0	180	0	116.5651	0
7	1	0	63.43495	0	0	0	36
16	1	0	−180	0	0	0	−144
2	1	0	0	72	0	0	0
3	6	0	0	72	0	0	0
4	7	0	0	72	0	0	0
5	16	0	0	72	0	0	0
8	1	0	0	144	0	0	0
9	6	0	0	144	0	0	0
10	7	0	0	144	0	0	0
11	16	0	0	144	0	0	0
12	1	0	0	216	0	0	0
13	6	0	0	216	0	0	0
14	7	0	0	216	0	0	0
15	16	0	0	216	0	0	0
17	1	0	0	288	0	0	0
18	6	0	0	288	0	0	0
19	7	0	0	288	0	0	0
20	16	0	0	288	0	0	0

Table B.1. Icosahedron Class I vertex-zenith PPT rotation constants.

because it determines the angles of rotation required for copies to cover the rest of the sphere. Although our schema examples in Chapter 8 were 8^v grids with 64 grid points each, the copy-rotate sequence described here is the same for any number of points in the PPT.

Table B.1 summarizes the copy-rotations necessary. We start with one PPT point set, the one we computed in Chapter 8, and refer to it as PPT 1. Each table row defines how a new PPT is derived from a previously computed PPT. The new PPT's number is listed in column one and shown in Figure B.9. The new PPT is derived from a previously computed PPT whose reference number is listed in column two. The reference PPT must be copied and then rotated one or two times to establish the position of the new PPT. The angles (listed in decimal degrees) and direction of rotation (right-hand rule) around the x-, y-, or z-axis for one or two rotations are listed in the same table row. Keep in mind that in most computer applications, rotation angles must be specified in radians, thus the rotation angles listed in the table must be converted.

Let's take an example. The first row in Table B.1 specifies that PPT 6's point set (column one) refers to a copy of PPT 1's point set (column two) that is rotated once around the z-axis 180° in a positive direction and then around the y-axis 116.5651° in a positive direction. Figure B.9 shows where these PPT references are located on the sphere and the images show rotation progress clockwise around the mid-figure diagram. When the copy-rotate sequence for PPT 6 is completed, two PPT point sets, our original PPT 1 and a new PPT 6 derived from it, cover the sphere.

The process for the remaining PPTs follows the same sequence. Each new PPT is a copy-rotate of a previous PPT, thus each new PPT must be found in the order specified in the table. Notice, too, that several PPTs—for example, PPTs 6, 7, 16, and 2—are all derived by copy-rotations of our original PPT 1's point set. Notice, too, that some PPTs further down the table only require a single rotation.

(a)

(b)

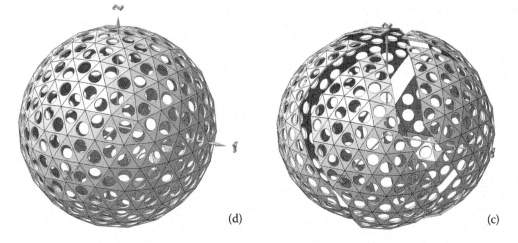

(d)

(c)

Figure B.9. Icosahedron Class I vertex-zenith PPT rotation sequences.

B.10   Icosahedron Class II

Icosahedron Class II rotations are a little more complex than their Class I and II counterparts, simply because more diamonds than PPTs are needed to cover the entire sphere—30 in all. The strategy we will use is to define five diamonds at each of the orthogonal positions. Table B.2 shows the angle-axis sequences for 2–5. Diamond 6 copy-rotates the original edge-zenith diamond to the opposite side of the sphere. Figure B.10(b) shows the positions of the first six diamonds.

With the orthogonal diamonds in position, an upper semi-hemisphere of 12 diamonds (7–18) is formed [see Figure B.10(c)]. These diamonds require three rotation sequences, each based on the original edge-zenith diamond 1. The icosahedral symmetry helps us significantly, which is why the specifications in Table B.2 have a symmetrical angle-axis pattern.

With the upper hemisphere diamonds in position, the entire set is simply copy-rotated 180° around the y-axis to complete the lower hemisphere diamonds 19–30.

When all rotations are completed, diamond 1 will have been copy-rotated a total of 29 times to cover the entire sphere.

Dia	Ref Dia	Rotation 1			Rotation 2			Rotation 3		
		X°	Y°	Z°	X°	Y°	Z°	X°	Y°	Z°
2	1	0	0	−90	0	90	0	0	0	0
3	1	0	0	−90	−90	0	0	0	0	0
4	1	0	0	−90	0	−90	0	0	0	0
5	1	0	0	90	90	0	0	0	0	0
6	1	0	180	0	0	0	0	0	0	0
7	1	0	31.7175	0	0	0	108	0	31.7175	0
8	1	0	−31.71749	0	0	0	−108	0	−31.71749	0
9	1	0	−31.71749	0	0	0	108	0	−31.71749	0
10	1	0	31.7175	0	0	0	−108	0	31.7175	0
11	1	0	−31.71749	0	0	0	198	−58.28255	0	0
12	1	0	−31.71749	0	0	0	−144	0	31.7175	0
13	1	0	−31.71749	0	0	0	144	0	31.7175	0
14	1	0	−31.71749	0	0	0	−198	58.2826	0	0
15	1	0	31.7175	0	0	0	198	58.2826	0	0
16	1	0	31.7175	0	0	0	−144	0	−31.71749	0
17	1	0	31.7175	0	0	0	144	0	−31.71749	0
18	1	0	31.7175	0	0	0	−198	−58.28255	0	0
19	7	0	180	0	0	0	0	0	0	0
20	8	0	180	0	0	0	0	0	0	0
21	9	0	180	0	0	0	0	0	0	0
22	10	0	180	0	0	0	0	0	0	0
23	11	0	180	0	0	0	0	0	0	0
24	12	0	180	0	0	0	0	0	0	0
25	13	0	180	0	0	0	0	0	0	0
26	14	0	180	0	0	0	0	0	0	0
27	15	0	180	0	0	0	0	0	0	0
28	16	0	180	0	0	0	0	0	0	0
29	17	0	180	0	0	0	0	0	0	0
30	18	0	180	0	0	0	0	0	0	0

Table B.2. Icosahedron Class II edge-zenith diamond rotation constants.

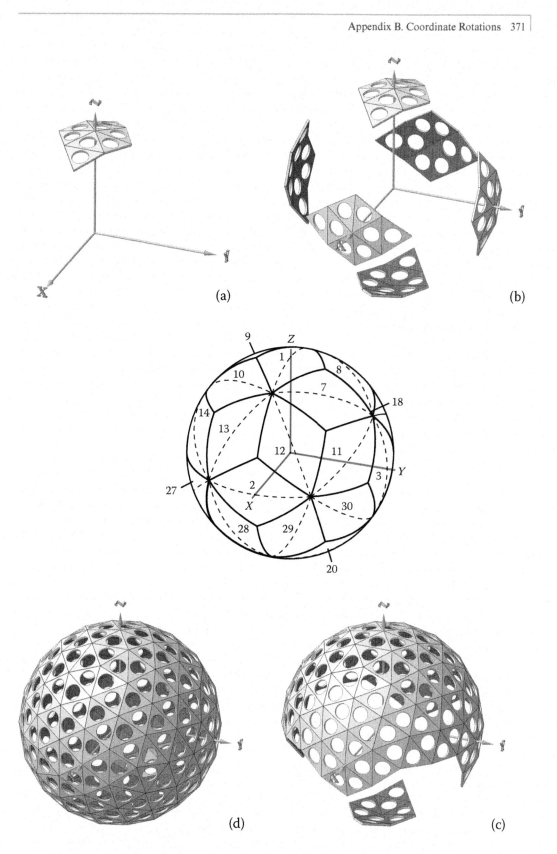

Figure B.10. Icosahedron Class II edge-zenith diamond rotation sequences.

B.11   Octahedron Classes I and III

Class I and II octahedra are the easiest rotation sequences of any polyhedra because each PPT is one-eighth of a sphere and all rotations are some multiple of 90°. Once again, we will use the gore strategy and define two PPTs (1 and 5) that span from the north to the South pole of the sphere. We will then copy-rotate the gore around the z-axis 90°, 180°, and 270° times to define 1 and 6, 3 and 7, and finally 4 and 8. Table B.3 and Figure B.11 show the sequence.

The gore strategy is quite useful and is used often in geodesics, especially in CAD assemblies.

PPT	Ref PPT	Rotation 1			Rotation 2		
		X°	**Y°**	**Z°**	**X°**	**Y°**	**Z°**
5	1	0	0	180	0	180	0
2	1	0	0	90	0	0	0
6	5	0	0	90	0	0	0
3	1	0	0	180	0	0	0
7	5	0	0	180	0	0	0
4	1	0	0	270	0	0	0
8	5	0	0	270	0	0	0

Table B.3. Octahedron Class I vertex-zenith PPT rotation constants.

B.12   Octahedron Class II

An octahedron has 12 edges; thus, we will need 11 rotation specifications to copy-rotate the original edge-zenith diamond to cover the entire Class II sphere. The strategy we use here is to form a *"ring"* of diamonds around the sphere and then fill in the areas on either side. Table B.4 shows that the angles are increments of 90° and the rotations for the first three instances are around the y-axis. Figure B.12(a) and (b) show the ring being formed.

Dia	Ref Dia	Rotation 1			Rotation 2			Rotation 3		
		X°	**Y°**	**Z°**	**X°**	**Y°**	**Z°**	**X°**	**Y°**	**Z°**
c	1	0	90	0	0	0	0	0	0	0
3	1	0	180	0	0	0	0	0	0	0
4	1	0	270	0	0	0	0	0	0	0
5	1	0	45	0	0	0	90	0	-45	0
6	1	0	-45	0	0	0	-90	0	45	0
7	5	0	0	180	0	0	0	0	0	0
8	6	0	0	180	0	0	0	0	0	0
9	5	180	0	0	0	0	0	0	0	0
10	6	180	0	0	0	0	0	0	0	0
11	7	180	0	0	0	0	0	0	0	0
12	8	180	0	0	0	0	0	0	0	0

Table B.4. Octahedron Class II diamond-zenith rotation constants.

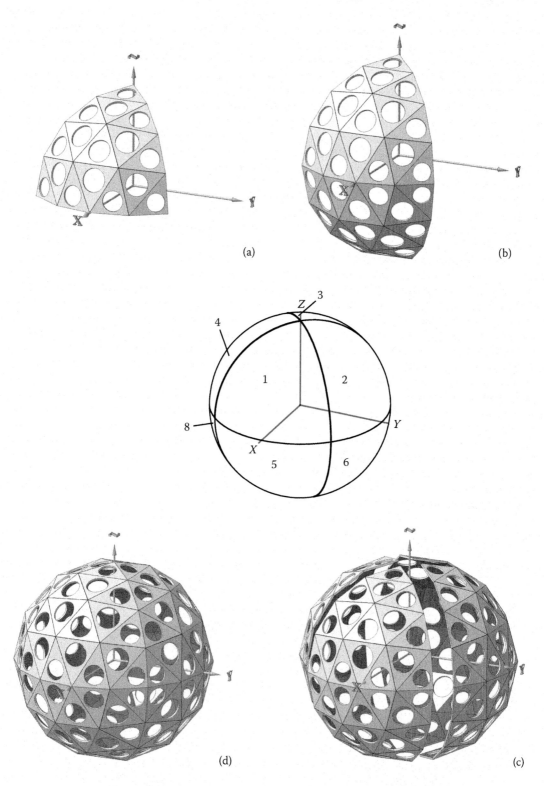

Figure B.11. Octahedron Class I vertex-zenith PPT rotation sequences.

(a)

(b)

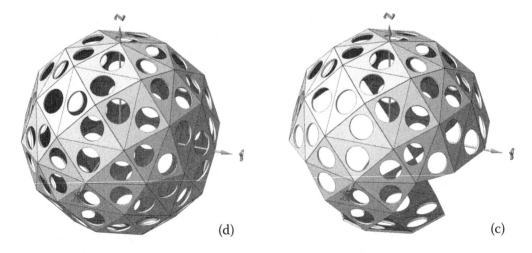

(d)

(c)

Figure B.12. Octahedron Class II diamond-zenith rotation sequences.

To fill in the areas on either side of the ring, diamonds 5 and 6 are rotated three times to complete two quadrants of the sphere. Diamonds 7 and 8 reference diamonds 5 and 6 and copy-rotate them 180° around the x-axis. This places diamonds 7 and 8 on the opposite side of the sphere and completes the upper hemisphere, as shown in (c).

The remaining quadrants in the lower hemisphere, diamonds 9 through 12, are copy rotations of diamonds 5 through 8 from the upper hemisphere. This fills in the remaining gaps in the lower hemisphere. All rotations in this set are 180° degrees around the x -axis. When the entire sphere is covered in Figure B.12(d), diamond 1 will have been copy-rotated 11 times.

B.13   Tetrahedron Classes I and III

Class I and III tetrahedron rotations are the easiest and require the fewest rotation sequences of any subdivided polyhedra. Table B.5 lists the rotation sequences for PPTs 2, 3, and 4; each is a copy-rotation of the original PPT 1.

The vertex-zenith PPT 1, shown in Figure B.13(a), is copy-rotated around the z-axis 120° and 240° (b) to define PPTs 2 and 3. PPT 4 on the −z-axis is created by copying PPT 1 and rotating it 180° around the z-axis and then 250.5°. This angle is the spherical height of the PPT around the y-axis. Figure B.13(c) shows the relationship of the PPTs in the completed sphere (d).

PPT	Ref PPT	Rotation 1			Rotation 2		
		X°	Y°	Z°	X°	Y°	Z°
2	1	0	0	120	0	0	0
3	1	0	0	240	0	0	0
4	1	0	0	180	0	250.5288	0

Table B.5. Tetrahedron Class I and III vertex-zenith PPT rotation constants.

B.14   Tetrahedron Class II

Class II tetrahedron rotations are quite simple. Only five rotations of the edge-zenith diamond are needed to cover the sphere. Except for the first one, they are all compound sequences with a similar pattern. Table B.6 shows the sequence and this repetitive pattern of angle-axis specifications. Notice that all diamonds are based on copy-rotations of diamond 1.

In this example, the initial edge-zenith diamond 1, shown in Figure B.14(a), is rotated 90° around the z-axis and then 180° around the x-axis, defining the bottom diamond PPT 2. It is shown in Figure B.14(b). The four diamonds that make up the

Dia	Ref Dia	Rotation 1			Rotation 2			Rotation 3		
		X°	Y°	Z°	X°	Y°	Z°	X°	Y°	Z°
2	1	0	0	90	180	0	0	0	0	0
3	1	−90	0	0	0	−45	0	0	0	−45
4	1	−90	0	0	0	45	0	0	0	45
5	1	90	0	0	0	45	0	0	0	−45
6	1	90	0	0	0	−45	0	0	0	45

Table B.6. Tetrahedron Class II diamond-zenith rotation constants.

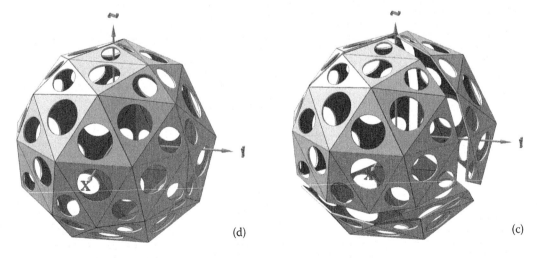

Figure B.13. Tetrahedron Class I and III vertex-zenith PPT rotation sequences.

(a)

(b)

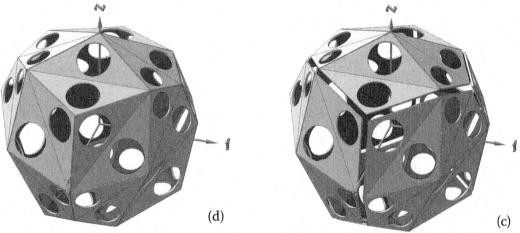

(d)

(c)

Figure B.14. Tetrahedron Class II diamond-zenith rotation sequences.

middle of the sphere, diamonds 3–6, all have the same general rotation sequences, where diamond 1 is rotated 90° plus or minus around the *x*-axis, followed by plus or minus 45° around the *y*-axis, and then 45° around the *z*-axis.

B.15 Dodecahedron Class II

Class II dodecahedra require more rotation sequences than any other subdivided spherical polyhedra. There are 30 edges in a dodecahedron; thus, it will take 29 copy-rotation specifications of the initial edge-zenith diamond to cover the rest of the sphere. The strategy we will use here is to define an upper hemisphere of diamonds and then copy-rotate the entire set to complete the lower hemisphere. A few diamonds are needed to span the equator, and we treat these separately.

Table B.7 shows the angle-axis sequences we need. Notice the repetitive pattern of rotation angles and axes for diamonds 2 through 13. Diamonds 14–17 span the equator and have 90° rotation angles. Notice, too, that all rotated diamonds thus far reference the original edge-zenith diamond.

Figure B.15 shows this sequence graphically. Notice that we build the upper hemisphere first and then copy it to the lower hemisphere. The lower hemisphere's 13 diamonds is easily specified by simply copy-rotating every upper hemisphere diamond that does not span the equator. The entire set is rotated 180° around the *x*-axis to complete the sphere.

Dia	Ref Dia	Rotation 1			Rotation 2			Rotation 3		
		X°	Y°	Z°	X°	Y°	Z°	X°	Y°	Z°
2	1	31.717	0	0	0	0	72	−31.717	0	0
3	1	−31.717	0	0	0	0	−36	−31.717	0	0
4	1	−31.717	0	0	0	0	36	−31.717	0	0
5	1	31.717	0	0	0	0	−72	−31.717	0	0
6	1	−31.717	0	0	0	0	72	31.717	0	0
7	1	31.717	0	0	0	0	−36	31.717	0	0
8	1	31.717	0	0	0	0	36	31.717	0	0
9	1	−31.717	0	0	0	0	−72	31.717	0	0
10	1	−31.717	0	0	0	0	−18	0	58.235	0
11	1	−31.717	0	0	0	0	18	0	−58.235	0
12	1	31.717	0	0	0	0	−18	0	−58.235	0
13	1	31.717	0	0	0	0	18	0	58.235	0
14	1	0	0	90	0	90	0	0	0	0
15	1	0	0	90	−90	0	0	0	0	0
16	1	0	0	90	0	−90	0	0	0	0
17	1	0	0	90	90	0	0	0	0	0
18	1	180	0	0	0	0	0	0	0	0
19	2	180	0	0	0	0	0	0	0	0
20	3	180	0	0	0	0	0	0	0	0
21	4	180	0	0	0	0	0	0	0	0
22	5	180	0	0	0	0	0	0	0	0
23	6	180	0	0	0	0	0	0	0	0
24	7	180	0	0	0	0	0	0	0	0
25	8	180	0	0	0	0	0	0	0	0
26	9	180	0	0	0	0	0	0	0	0
27	10	180	0	0	0	0	0	0	0	0
28	11	180	0	0	0	0	0	0	0	0
29	12	180	0	0	0	0	0	0	0	0
30	13	180	0	0	0	0	0	0	0	0

Table B.7. Dodecahedron Class II diamond-zenith rotation constants.

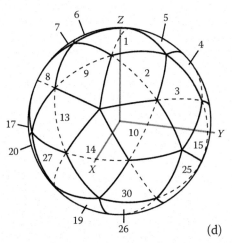

Figure B.15. Dodecahedron Class II diamond-zenith rotation sequences.

B.16 Cube Class II

The approach used to cover a Class II cube is similar to that used with the dodecahedron. The rotation sequences are shown in Table B.8 and illustrated in Figure B.16.

Dia	Ref Dia	Rotation 1			Rotation 2			Rotation 3		
		X°	Y°	Z°	X°	Y°	Z°	X°	Y°	Z°
2	1	0	0	35	0	60	0	0	0	35
3	1	−90	0	0	0	0	0	0	0	0
4	1	0	0	−35	0	−60	0	0	0	−35
5	1	0	0	35	0	−60	0	0	0	35
6	1	90	0	0	0	0	0	0	0	0
7	1	0	0	−35	0	60	0	0	0	−35
8	7	180	0	0	0	0	0	0	0	0
9	5	180	0	0	0	0	0	0	0	0
10	4	180	0	0	0	0	0	0	0	0
11	2	180	0	0	0	0	0	0	0	0
12	1	180	0	0	0	0	0	0	0	0

Table B.8. Cube Class II diamond-zenith rotation constants.

(a)

(b)

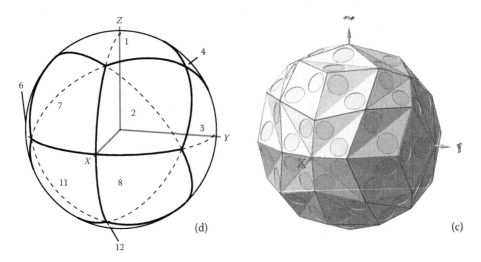

(d)

(c)

Figure B.16. Cube Class II diamond-zenith rotation sequences.

The sequence begins by copy-rotating the basic edge-zenith diamond 1 six times to create diamonds 2–7. This forms an upper hemisphere. Two diamonds, 3 and 6, span the equator. Once this is completed, the five diamonds that do not span the equator (1, 2, 4, 5, and 7) are copy-rotated 180° around the *x*-axis to complete the lower hemisphere's diamonds 8 through 12. All 12 diamonds are in place and cover the sphere perfectly.

B.17   Implementing Rotations

This book develops six spherical subdivision schemas. Chapter 8 showed the steps necessary to subdivide a single principal polyhedral triangle (PPT) or least common denominator triangle (LCD). The result was a set of points stored in an array *xyz*[ ][ ]. In order to cover the entire sphere, it is necessary to rotate copies of these grid points to specified locations elsewhere on the sphere. The first part of this appendix discussed the theory of rotations and listed the required rotation angles for each subdivision class and polyhedral foundation.

In this section, we present the algorithms necessary to rotate grid points to the other parts of the sphere's surface. To rotate one or more grid points, we will multiply their *xyz*-coordinates by a 4×4 transformation matrix. It is assumed that the initial position of the PPT or LCD points is the recommended position described in the text. For illustration purposes, we also assume that the PPT or LCD grid points are stored in the same *xyz*[ ][ ] of triplets we present in Appendix C:

- Define the required rotation angles (in radians). Table B.1 through Table B.9 list the required angles and the sequence they are applied to a master rotation matrix to rotate an initial set of PPT or LCD points to all other positions for each of the schemas and polyhedral frameworks presented;

- Initialize a master 4×4 transformation matrix. Initialization requires setting it to the identity matrix. Function midn() performs the initialization;

- Define the required rotation angle sine and cosine values in the appropriate transformation matrix cells. Functions xrot(), yrot(), and zrot() perform this operation and multiply the set of points by the transformation matrix;

- Define any desired translations. A translation is simply moving a set of grid points along the *x*-, *y*-, or *z*-axis, no rotation is performed. None of the schemas developed in this book require translation; we show the algorithm to demonstrate the flexibility of transformation matrices. See function mtrans();

- If a set of PPT or LCD points need to be rotated around more than one axis to achieve the required positioning, repeat the prior step for each additional axis rotation. One way to do this is to create a temporary 4×4 transformation matrix, initialize it (set it to the identity matrix), and then use xrot(), yrot(), or zrot() to define and perform each rotation one by one. A better approach, however, is to combine the separate rotations into a single 4×4 transformation matrix that accomplishes the same thing with just a single move. Matrix multiplication combines two transformation matrices into a single one capable of performing the combined operation. There is no limit to the number of matrices multiplied together to accomplish complex transformations; matrix multiplication can be more efficient than performing a series of separate ones. Function mcat() multiplies two 4×4 matrices (sometimes matrix multiplication is called *concatenation*), creating a third 4×4 matrix with the combined specifications. As explained below, if you choose to multiply matrices, you must multiply them in the reverse order of use;

• Apply the final transformation matrix to the set of PPT or LCD grid points. In most applications, a copy of the original PPT or LCD points is made and only this set of points is transformed. Transformation is performed by function xform().

B.18 Using Matrices

B.18.1 Identity

Before any transformation matrix can be used, the matrix must be initialized to the identity matrix. This sets matrix diagonal values to 1.0 and all others to 0.0. Always initialize transformation matrices before using them.

B.18.2 Specifying Angles

This book adopts the right-hand rule convention for the direction of angles. All angles are expressed in radians and rotations around the x-, y-, and z-axis. Use the following convention:

Axis of rotation	Direction of positive rotation
x	from y to x
y	from z to x
z	from x to y

B.18.3 Matrix Multiplication

Most PPT or LCD points need to be rotated around more than one axis to position them elsewhere on the sphere. Instead of performing each transformation separately, you can compute the necessary transformation matrix for each one and then combine them into a single matrix that has the combined rotational effect on the PPT or LCD point set. If you multiply matrices, you must do so in the *reverse* order in which they are needed. For example, if you need to rotate points around the x-axis and then the y-axis, you must define the y-axis rotation matrix first and then multiply it by the x-axis rotation matrix before you transform the PPT or LCD points.

Matrix multiplication is sometimes referred to as concatenation, a term that means joining together. We use the terms matrix multiplication and concatenation interchangeably in algorithm comments.

The sections that follow define the basic algorithms needed to define, multiply, and apply the rotational transformation to a set of PPT or LCD points.

B.19 Rotation Algorithms

The following C algorithms show how to initialize a transformation matrix, define transformations to rotating points around the x-, y-, and z-axis, translate points along the x-, y-, and z-axis, and then apply them to a set of points. You can define and apply these transformation matrices one at time or you can combine them into a single matrix and apply them to a set of PPT or LCD points once. The last algorithm is an example of how to use them.

B.19.1   Identity

```
//--------------------------------------------------------------
// midn
// initialize the master transformation matrix mtx[][]. Diagonal cells set
// to 1.0, all other cells set to 0.0.
//
// Identity matrix
// +-        -+
// | 1 0 0 0 |
// | 0 1 0 0 |
// | 0 0 1 0 |
// | 0 0 0 1 |
// +-        -+
//
//--------------------------------------------------------------
void midn (double mtx[4][4])
{
    int i, j;                          // general loop indices
    for (i=0;i<4;i++)                  // for each row and
    for (j=0;j<4;j++) mtx[i][j] = 0.0; // column, set element to 0.0
    for (j=0;j<4;j++) mtx[j][j] = 1.0; // set matrix diagonal to 1.0
    return;
}
```

B.19.2   Rotation Around the X-axis

```
//--------------------------------------------------------------
// xrot
// define a rotation around x-axis and concatenate
// it to the master transformation matrix, mtx[][]. Angle is expressed in
// radians.
//
// X Rotation Matrix
// +-              -+
// | 1    0     0    0 |
// | 0   COS>  SIN>  0 |
// | 0  -SIN>  COS>  0 |
// | 0    0     0    1 |
// +-              -+
//
//--------------------------------------------------------------
void xrot (double angle, double mtx[4][4])
{
    double xm[4][4],temp[4][4];    // working matrices
    midn(xm);                      // initialize working matrix to identity
    xm[1][1] = cos(angle); xm[1][2] = sin(angle);
    xm[2][1] = -sin(angle); xm[2][2] = cos(angle);
    mcat(mtx,xm,temp);             // concatenate master and x rotation
    mcpy(temp,mtx);                // copy temporary 4x4 into master matrix
    return;
}
```

B.19.3   Rotation Around the *Y*-axis

```
//---------------------------------------------------------------------
// yrot
// define a rotation around y-axis, concatenate it to the master
// transformation matrix, mtx[][].
//
// Y Rotation Matrix
// +-                  -+
// | COS> 0 -SIN>  0  |
// |  0   1   0    0  |
// | SIN> 0  COS>  0  |
// |  0   0   0    1  |
// +-                  -+
//
//---------------------------------------------------------------------
void yrot (double angle,double mtx[4][4])
{
    double ym[4][4],temp[4][4];   // working matrices
    midn(ym);                      // initialize working matrix to identity
    ym[0][0] = cos(angle); ym[0][2] = -sin(angle);
    ym[2][0] = sin(angle); ym[2][2] = cos(angle);
    mcat(mtx,ym,temp);             // concatenate master and y rotation
    mcpy(temp,mtx);                // copy temporary 4x4 into master matrix
    return;
}
```

B.19.4   Rotation Around the *Z*-axis

```
//---------------------------------------------------------------------
// zrot
// define a rotation around z-axis, concatenate it to the master
// transformation matrix, mtx[][].
//
// Z Rotation Matrix
// +-                -+
// |  COS> SIN> 0 0  |
// | -SIN> COS> 0 0  |
// |   0    0   1 0  |
// |   0    0   0 1  |
// +-                -+
//
//---------------------------------------------------------------------
void zrot (double angle, double mtx[4][4])
{
    double zm[4][4],temp[4][4];   // working matrices
    midn(zm);                      // initialize working matrix to identity
    zm[0][0] = cos(angle); zm[0][1] = sin(angle);
    zm[1][0] = -sin(angle); zm[1][1] = cos(angle);
    mcat(mtx,zm,temp);             // concatenate master and z rotation
    mcpy(temp,mtx);                // copy temporary 4x4 into master matrix
    return;
}
```

B.19.5 Translation

```
//-----------------------------------------------------------------------
// mtrans
// translate (move) a point along the x-, y- and/or z-axis and concatenate
// transformation to the master transformation matrix
//
// Translation Matrix
// +-                -+
// |  1    0    0    0 |
// |  0    1    0    0 |
// |  0    0    1    0 |
// | T(X) T(Y) T(Z)  1 |
// +-                -+
//
//-----------------------------------------------------------------------
void mtrans (double x_tr, double y_tr, double z_tr, double mtx[4][4])
{
    double xm[4][4],temp[4][4];    // working matrices
    midn(xm);                      // initialize working matrix to identity
    xm[3][0] = x_tr;               // insert the translation x,
    xm[3][1] = y_tr;               // y and
    xm[3][2] = z_tr;               // z values into transform matrix
    mcat(mtx,xm,temp);             // concatenate master and x rotation
    mcpy(temp,mtx);                // copy temporary 4x4 into master matrix
    return;
}
```

B.19.6 Matrix Multiplication (Concatenation)

```
//-----------------------------------------------------------------------
// mcat
// concatenate two 4x4 transformation matrices, m1[][] and m2[][], into
// m3[][]
//
// m1 * m2 -> m3
//-----------------------------------------------------------------------
void mcat (double m1[4][4],double m2[4][4],double m3[4][4])
{
    int i, j, k;                   // local loop indices
    for (i=0;i<4;i++) {            // for every row
       for (j=0;j<4;j++) {         // and every column
          m3[i][j] = 0.0;          // use all in that row vs all
          // in that column
          for (k=0;k<4;k++) m3[i][j] += ( m1[i][k] * m2[k][j] );
       }
    }
    return;
}
```

B.19.7 Transform Points

```
//-------------------------------------------------------------------------
// xform
//
// transform a set of points defined in an array of triplets, xyz[*][3].
// the array of triplets is defined externally, for example
// double xyz[][3];
//
//-------------------------------------------------------------------------
void xform (int from, int to, double mtx[4][4]) // from to xyz range to
// xform
{
   extern double xyz[6500][3]; // reference external PPT or LCD points
   double a[4], b[4];          // homogenous matrix working areas
   int i, j, k;                // local loop indices
   for (i=from;i<=to;i++) {    // convert every PPT or LCD point
     a[0] = xyz[i][0];         // from a triplet to a 1x4 set
     a[1] = xyz[i][1];
     a[2] = xyz[i][2];
     a[3] = 1.0;               // add homogenous scale factor
     for (j=0;j<4;j++) {       // multiply point by 4x4 transform matrix
        b[j] = 0.0;            // all elements in that row vs all
        for (k=0;k<4;k++) b[j] += a[k] * mtx[k][j];// elements in that column
     }
     xyz[i][0] = b[0];         // update point coordinates with
     xyz[i][1] = b[1];         // resulting transformed version
     xyz[i][2] = b[2];
   }
   return;
}
```

B.20 An Example

This short code segment assumes that a set of PPT or LCD grid points have already been computed and now need to rotate the set to another position on the sphere. An actual implementation would copy them and perform the transformation on the copy, leaving the initial set of points unchanged.

```
int main()
{
   double xyz[2500][3];
   double vmtx[4][4];
   .

   .

   // code to compute PPT or LCD grid points goes here
   // assume xyz[0][*] to xyz[125][*] have them
   midn(vmtx);                      // initialize master matrix to identity

   // rotate points 0 to 125 around the y axis -15 degrees and then
   // 20 degrees around the x. Note that the rotations are concatenated
   // in reverse order needed.
```

```
xrot(radian(20.),vmtx);           // concatenate x rotation to identity
yrot(radian(-15.),vmtx);          // concatenate y rotation to master
xlate(xTrans,yTrans,zTrans,vmtx); // x,y and z translations
// rotate points 1 to 125 using transformation matrix vmtx[][]
xform(0, 125,vmtx);
// PPT or LCD points 0 to 125 are now positioned
return 0;
}
```

B.21   Summary

Rotations are angular specifications around the *x*-, *y*-, and/or *z*-axes. Angles can be specified in either the left-hand rule (LHR) or right-hand rule (RHR). This book uses the right-hand rule throughout. If more than one is to be performed, they can be accumulated by multiplying individual rotation matrices together and performed as a single transformation. This can be simple (a single angle-axis rotation) or complex where a sequence of angle-axis rotations is required. Compound rotations, however, are not commutative. A rotation around one axis followed by a rotation around another generally does not result in the same position if the sequence is reversed. Rotating a point any multiple of 360° around an axis does not change its position either. Rotated geometry can be restored to its original position by applying the reverse sequence of axis rotation and the opposite angle specification. Arbitrary axis rotation allows a single point to define a new axis, which can greatly simplify angle-axis specifications. Arbitrary axis rotation still relies on rotations around the orthogonal axis, but intermediate rotations are computed automatically.

Additional Resources

Penna, Michael A. and Richard R. Patterson. *Projective Geometry and Its Applications to Computer Graphics.* Englewood Cliffs, NJ: Prentice Hall, 1986.

Van Verth, James M. and Lars M. Bishop. *Essential Mathematics for Games and Interactive Applications: A Programmer's Guide.* Amsterdam: Morgan Kaufmann, 2008.

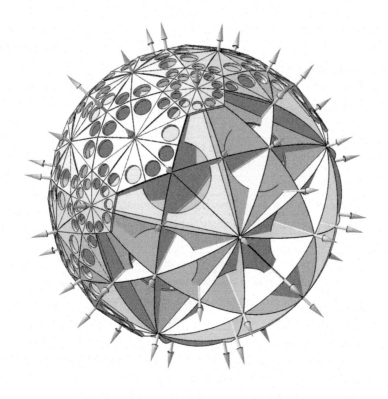

Geodesic Math

This appendix is divided into two parts. The first part presents the geodesic math necessary to subdivide the principle polyhedral triangle (PPT) or the least common denominator triangle (LCD) based on Class I, II, and III schemas presented earlier in this book. With the algorithms shown here, you can derive the coordinates of grid points, arc, and chord distances between them and all edge and face angles. Math techniques are described for each of the classes. In addition, the first part of this appendix includes some handy algorithms to solve common 3D geometric problems, such as finding the perpendicular to a plane, or the shortest distance between two nonintersecting lines in three dimensions. The second part of the appendix presents some useful math used to derive symmetrical uniqueness of vertices in the PPT as discussed in Chapter 10.

The angles that result from subdividing a PPT or LCD do not change if the PPT or LCD set of points are moved together to a different location on the surface of the sphere. However, moving them does change the coordinates of each point. The procedures developed here create one spherical PPT or LCD grid oriented either vertex-, edge-, or face-zenith. These points are sufficient for most analysis and simple displays. If you need to extend this grid to cover the rest of the sphere, there are at least two ways to achieve this. First, you can input the points into a CAD system and use its assembly features to replicate the PPT or LCD points elsewhere on the sphere. The CAD system computes the necessary coordinate transformations for you. All of the examples in Chapter 10 were created this way. Another way is to transform copies of

Geodesic Algorithms	
Subroutine	**Operation**
gcsect	calculate the two points and dihedral angle where two great circles intersect
gdihdrl	return the dihedral angle between two planes
gtricent	calculate the centroid of a triangle
stABC	calculate the surface angles of a spherical triangle
vabs	calculate the absolute value of a point (also called vector length)
vadd	add two vectors
varcv	locate a point at an arbitrary place, a geodesic arc between two other points
vcos	return the cosine of the angle between two vectors
vcrs	calculate a vector perpendicular to the plane of two other vectors
vdir	locate a point between two others at a prescribed distance
vdis	calculate the distance between two points
vdot	return the dot product (also called scalar or inner product) of two vectors
vnor	calculate the normal vector (perpendicular) to a plane
vrevs	reverse vector's sense, makes an antipode point if vector is point on sphere
vscl	scale a vector or point (moves it along infinite line through origin)
vsub	subtract one point from another
vuni	project a point to the surface of the unit sphere
vzero	initialize a range of point coordinates to (0,0,0)

Table C.1. Summary of geodesic algorithms.

the original PPT or LCD point set, the copy is rotated to a new position elsewhere on the sphere; Appendix B tells you how to achieve this.

In the appendix, we take a cookbook approach to avoid using specialized math notation. We will explain the required subdivision steps at the same time we make references to simple formulas or algorithms that implement the step. The formulas are presented as a subdivision step is explained; the algorithms are listed in their entirety later in the chapter. In addition to explaining algorithms with each step, a separate table lists them all and their function in case you need to look one up quickly. Anyone doing geodesic calculations will use a computer or calculator, thus all algorithms are written in C. Its language structure and syntax is easy to understand and convert to other computer languages or notations used by programmable calculators. Some algorithms are used by other algorithms and we include the full set. Each algorithm is fully documented and designed for clarity, not necessarily for maximum efficiency. Computer programmers can improve the data structures and efficiency of these algorithms by using computer language features, such as object classes or specific data structures (vectors, maps, lists, etc.) offered in the C++ and Java languages.[1]

The objective of this appendix is to list the algorithms and their sequence of use. Most will access a single master array of Cartesian coordinates (x, y, z). Some algorithms are performing vector operations, others are trigonometric solutions for spherical triangles, and still others are housekeeping routines. The algorithms are listed in alphabetical order in Table C.1. The actual code of each vector routine is listed at the end of this appendix.

---

[1] Vector containers and predefined algorithms are standard template library features in American National Standards Institute (ANSI) Standard C++.

C.1   Class I: Alternates and Fords

Class I subdivisions are the result of points created by the intersection of two or three great circles. The intersecting great circles are defined by reference points along the PPT's edges (and the origin of the sphere). Class I schemas differ only in the way their PPT edge reference points are located and the combination of great circles intersected to define grid points. The following section tells you how to define edge reference points and intersect great circles for each of the schemas presented in this book. Points for the Equal-arcs schema can be created by intersecting two or three great circles. Three great circles result in a superior grid, but an additional step is required and the procedure is described here.

C.1.1   Step 1: Define the PPT Apex Coordinates

Orient the deltahedron's equilateral PPT face in a vertex-zenith position symmetrical to the *xz*-plane, as shown in Figure C.1(a). The example here is a single icosahedron PPT; PPTs for the tetrahedron and octahedron will look similar. The PPT's vertex-zenith Cartesian coordinate for unit sphere will always be (0, 0, 1). The other two PPT apex coordinates are found by converting their spherical coordinates to Cartesian coordinates. See Table 8.2 for their coordinates.

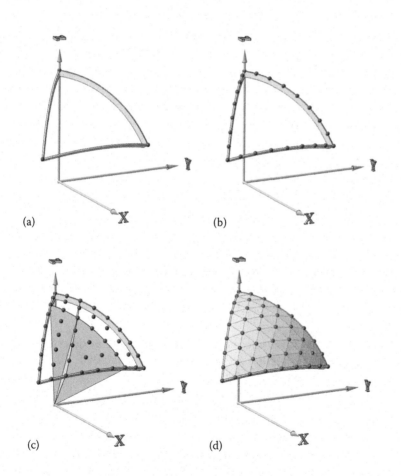

Figure C.1. Class I subdivision sequence.

C.1.2   Step 2: Define PPT Edge Reference Points

Locate $v - 1$ points between each of the PPT apex points. The total number of required reference points for all edges of the PPT, excluding the PPT's apices, will be $3(v - 1)$. Figure C.1(b) shows the PPT apex and reference points necessary for an Equal-chords 8^v icosahedron subdivision. Reference points for the other schemas will look similar. Define PPT edge reference points as follows:

- *Equal-chords.* Find the PPT's chord length with `vdis()`. Use `vdir()` to compute the coordinates of $v - 1$ evenly-spaced reference points between the vertices of each PPT edge chord. These reference points are along the polyhedron's chords; thus they are inside the sphere. Project (normalize) each reference point to the sphere's surface with `vuni()`;

- *Equal-arcs (two or three great circles).* Evenly space $v - 1$ reference points along the PPT's arc edges. Algorithm `varcv()` can create them on the surface of the sphere. They do not need to be projected;

- *Mid-arcs.* Mid-arcs is a recursive subdivision technique that uses a triangle's medial triangle to break it into four subtriangles. The triangle's three Mid-arc points define the vertices of its medial triangle. A triangle's edge length is found with `vdis()` and its midpoint is found with `vdir()`. Project each chord midpoint to the surface of the sphere with `vuni()`;

- *PPT center point.* It is useful to have a reference point in the center of the PPT's face. Some schemas for certain frequencies define a grid point at the PPT's center but not every frequency-schema combination does so. Use `gtricent()` to define the PPT's center point, given the triangle's vertex coordinates, and then use `vuni()` to project the centroid point from the plane of the PPT face to the surface of the sphere.

C.1.3   Step 3: Subdivide the PPT

Subdivision grid points within the PPT are the result of the intersection of two or three great circles, as defined in Chapter 8. A great circle is defined by three points; two are reference points along the PPT's edge, and the third is the sphere's origin. Figure C.1(c) shows how a typical grid point is defined at the intersection of two great circles, themselves defined by a pair of edge reference points (and the center of the sphere). The choice of reference points, which define great circles, depends on the schema you are creating. Chapter 8 explains how edge reference points are selected, when defining the intersection of great circles. Here, we describe how to define grid points where great circles intersect.

- *Equal-chords.* In Equal-chords, great circles defined, as described in Chapter 8, produce a highly regular set of gridpoints over the PPT. Use `gcsect()` to locate the two points where these great circles intersect. One of the two points will have a negative z component. Discard this point, since it is not on the surface of the PPT. Continue intersecting pairs of great circles until all grid points have been defined;

- *Equal-arcs.* In Equal-arcs, you have your choice of defining grid points by intersecting two or three great circles. Intersecting two great circles always produces a single grid point within the PPT and two intersections are simpler to

process than three; however, the resulting point set in the PPT is not rotationally symmetric about the PPT's centroid. Using three great circles produces superior results and the points are rotationally symmetric. Intersecting three great
circles produces three points within the PPT. Additional processing is required
to yield the single point needed. Use gtricent() to find the centroid of this
window triangle and then project the centroid point (from the plane of the
window triangle, which is inside the sphere) to the surface of the sphere with
vuni(). Use gcsect() to find the two points where a pair of great circles
intersect. One of the two points will have a negative z component. Discard this
point, since it is not within the surface of the PPT. Continue intersecting pairs
of great circles until all grid points have been defined;

- *Mid-arcs.* Mid-arcs is the easiest subdivision technique because you are simply subdividing triangles with their medial triangle. The vertices of a triangle's
medial are simply the Mid-arc points of their three edges. Mid-arcs is a recursive subdivision technique. In the first cycle, the PPT's equilateral triangle is
subdivided by its medial triangle and results in a 2^v grid. If a higher frequency
subdivision is required, the four triangles from the first cycle are each subdivided by their *medials*, resulting in a 4^v grid. Each subdivision cycle doubles the
overall subdivision frequency, thus Mid-arcs schema always produce grids with
frequencies that are powers of two. To find the three vertices of a triangle's
medial, compute the midpoint of each PPT edge chord, using vdir() and
a scale of 0.5. This midchord point is inside the sphere, so project it to the
sphere's surface, using vuni(). The three Mid-arc points of a triangle define
its medial's vertices. Repeat the process of defining medial triangles until the
desired frequency of subdivision is obtained.

C.1.4   Class I Summary

As the result of the steps followed, the PPT will be fully subdivided and
resemble the grid in the figure to the right. You can use these coordinates for analysis or display, or you can cover the entire surface of the
sphere by rotating copies of them to predefined locations elsewhere on
the sphere. See Appendix B for details.

C.2   Class II: Triacon

Unlike Classes I and III, the Class II triacon method can subdivide all five spherical
Platonics. The following steps show you how this is done. The example developed here
is an 8^v subdivision of a spherical icosahedron but the process for subdividing any of
the other four spherical Platonics are the same, only their LCD angles differ in the
initial step.

In Class I, we began by locating reference points around the edges of the PPT.
Based on these points, two or three great circles are defined and points on the surface
of the PPT are derived from their intersections. All other spherical information, such
as arcs and chords or triangle angles, are computed from the coordinates of these PPT
grid points. The algorithms used in Class I schemas are mostly based on coordinate
geometry.

In Class II, the sequence or steps are just the opposite of Class I. Subdivision starts
by solving the parts of the spherical polyhedron's LCD right triangle. Instead of using

coordinate geometry, the geodesic math is exclusively trigonometric solutions to right triangle and these types of triangles are the easiest to solve. The coordinates of grid points are solved by using spherical trigonometry. Once the LCD's grid points are defined, it is trivial to define the coordinates of the other three LCDs that make up the edge-zenith triacon diamond. You can use these grid points as is, input them into a Computer Aided Design (CAD) system, or cover the rest of the sphere. Appendix B describes how this is done.

C.2.1   Step 1: Position and Define the Triacon LCD

The first step in triacon subdivision is to position the spherical polyhedron edge-zenith. The triacon diamond is symmetrical to this edge; the LCD we subdivide is one fourth of a diamond and located in its positive (x, y, z) quadrant. Figure C.2 shows our example spherical icosahedron and triacon diamond in this position; the LCD portion we subdivide is accented.

- *LCD principal parts.* Label the LCD spherical parts, a right triangle, as shown in Figure C.2, and find sides a, b, and c, given its three surface angles A, B, and C. The basic relationships are shown in Figure C.2.

It's easy to establish the icosahedron's LCD surface angles A, B, and C; we can do this by inspection. They are also listed in Table 8.4. Through symmetry and knowing that the sum of the surface angles around any point on the sphere is always 360°, the LCD angles for a spherical icosahedron are $A = 60°$ and $B = 36°$ and, by definition $C = 90°$. We now use the formulas in Figure C.2 and surface angles A, B, and C to find sides (central angles) a, b, and c. They are 31.7175°, 20.9052°, and 37.3774°, respectively.

We also need the arc and chord factors for the LCD's sides as well as the full icosahedron's full polyhedral edge and face meridian arc length. The arc factors for any of the LDC's sides are simply the radian value of their central angle, or $2\pi\theta/360$, where θ is the arc's central angle expressed in degrees. The chord factor between two points on the unit sphere, is the straight-line distance between them and is simply $2r \sin(\theta/2)$, where r is the radius of the sphere and θ is the central angle between the two points. Given LCD sides, the polyhedron's edge and face medians are simply

$$\text{PPT polyhedron edge} = 2a, \quad \text{and}$$
$$\text{PPT face median} = b + c.$$

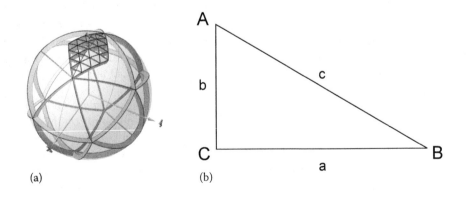

(a)   (b)

Figure C.2. Orientation of triacon LCD triangle.

C.2.2 Step 2: Subdivide the Triacon LCD

- *Define nested right triangles.* The triacon schema subdivides the LCD by nesting a series of right spherical triangles. Figure C.3 shows how an eight frequency grid is developed. The first subdivision step is to divide half the polyhedron's edge (arc) into v/2 equal arc segments. Each segment forms all or part of the base of a series of right spherical triangles. The result is (v/2) – 1 nested spherical right triangles within the LCD right triangle. All of these right triangles share angle *B*. In this example, there are three nested right triangles within the overall LCD right triangle. We will label their angles and sides the same way we labeled the overall LCD triangle. Side *a* of each nested right triangle is an integral number of polyhedron edge segments. In Figure C.3, it appears that each of the nested right triangles has the same angle *A*, but this is not the case. There are no similar triangles on a sphere; each nested right triangle has a slightly different value for *A*, but angles *B* and *C* remain constant and are 36° and 90°, respectively. Given *a*, *B*, and *C* for each nested triangle, we find parts *A*, *b*, and *c* using these three relationships:

$$\cos A = \cos a \times \sin B,$$
$$\cos b = \cos B / \sin A, \quad \text{and}$$
$$\cos c = \cos a \times \cos b.$$

- *Subdivide LCD sides.* With all the nested triangle parts solved, divide LCD side *b* into arc segments equal to the *b* sides of each of the nested right triangles, as shown in Figure C.3(b). The *b* sides for each nested right triangle were found in the previous step. Each side *b* is slightly different and the overall LCD side *b* will be subdivided into v/2 arc segments;

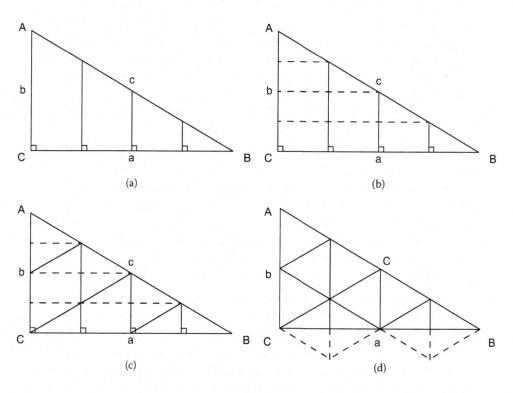

Figure C.3. LCD subdivision sequence.

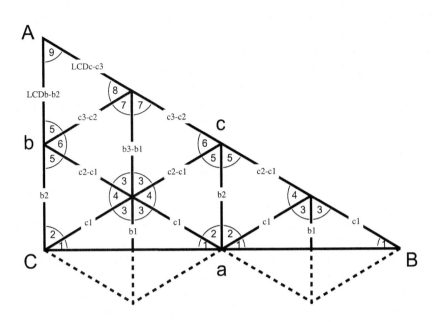

Figure C.4. LCD 8ᵛ subdivision summary.

- *Complete the triangulation.* Triangulate the grid with diagonals by alternating nested triangle *c* arcs across the LCD. Each row of *c* arcs is the same length. The result is a series of triangles, each with their own *a*, *b*, and *c* arcs, as shown in Figure C.3(c).

Diagonals are constructed with nested triangle *c* arcs to alternate points, as shown in Figure C.3(d). When completed, all interior tessellation vertices have three-way intersections of two diagonals and a perpendicular to side *a*. Sides *a* and *c* of the overall LCD triangle will be subdivided v/2 times, while *b* is subdivided v/4 times.

Solving the LCD edge perpendiculars above solves all six parts (three angles, three sides) of each nested right triangle within the LCD. Subdividing the LCD sides, striking diagonals, and completing the triangulation replicate these various arcs across the LCD to create the three-way grid. The elegance of the triacon schema lies in the fact that parts *A*, *b*, and *c* of the nested triangles are replicated across the LCD to define the angles and sides of the other triangles in their respective rows. Figure C.4 shows how the replication works.

Starting at the smallest nested right triangle, the one immediately next to LCD vertex *B*, notice how its angles 1 and 3 and side *b*1 carry across the LCD to all other triangles in its row. Notice that angle 2 is the complement of angle 1. Other angles are found this way; they are either complements or supplements of one or more nested triangle parts. For example, angle 4 is the supplement of two angle 3s. Some diagonal *c* arcs are found by factoring out *c* arcs from another nested triangle. Figure C.4 shows how this is done.

C.2.3 Step 3: Define Grid Points

Vertex references greatly facilitate geodesic work, and you have complete freedom to name chords, triangles, surfaces, and central angles any way that is convenient to your application. For illustration purposes, we number triacon diamond quadrants 1 through 4 and name grid points with numbered *a*-columns and *b*-rows. This is the convention used in Figure C.5. Other conventions might be more convenient for calculator or PC programming. The polyhedral edge, not part of the final grid, is shown

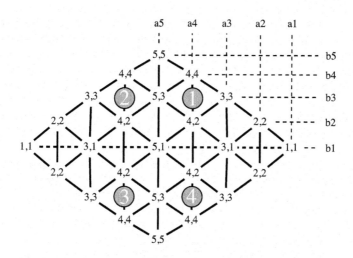

Figure C.5. Triacon grid vertex references.

dashed in the figure. Points 1,1 in all quadrants are the vertices of the subdivided poly-hedron. Points 5,5 are points in the middle of the icosahedron face.

Class I and III schemas start by computing the coordinates of the PPT's vertices and end by finding the central and surface angles of all the grid triangles. Class II is just the opposite. We start with the surface angles of the LCD and end by finding the coordinates of the grid points. It is easy to find the Cartesian coordinates of grid points in the triacon diamond. Only two arc angles, *a* and *b*, are needed for each point and we already have their values when we solved the parts of each nested right triangle in the previous steps. Note that angles *a* and *b* of each point correspond to the φ and λ in the coordinate (ρ, φ, λ) system described in Chapter 4:

$$x = \sin a \times \cos b,$$
$$y = \sin b,$$
$$z = \cos a \times \cos b.$$

LCD Vertex Angles	
Vertex	**Angle°**
a1	31.71748
a2	23.78811
a3	15.85874
a4	7.92937
a5	0
b1	0
b2	5.72354
b3	11.22946
b4	16.33348
b5	20.90516

Icosahedron 8ᵛ LCD Vertices			
a,b	**X**	**Y**	**Z**
1,1	0.52573	0	0.85065
3,1	0.27327	0	0.96194
5,1	**0.00000**	**0.00000**	**1.00000**
2,2	0.40134	0.09973	0.91048
4,2	0.13726	0.09973	0.9855
3,3	0.26804	0.19474	0.94352
5,3	0.00000	0.19474	0.98085
4,4	0.13238	0.28123	0.95046
5,5	0.00000	0.35682	0.93417

Table C.2. Icosahedron 8ᵛ LCD vertices, angles, and coordinates.

Angle *a* is measured along the *xy*-plane and is positive if it is on the +*x*-axis side of the plane; otherwise, it is negative. Angle *b*, measured on the *yz*-plane, is positive, when it is on the +*y*-axis side and negative if on the −*y* side. We simply find the cumulative *a* and *b* values for each vertex in the LCD grid and apply the above formula. Using the vertex reference system suggested in Figure C.5, the *a* and *b* vertex angles listed in Table C.2 and the formulas above, we find the coordinates of each LCD grid point. Figure C.2 also lists the coordinates of every vertex in the LCD. Notice that vertex 5,1 (highlighted in the table) is at the zenith of the sphere with coordinates (0, 0, 1), as expected.

To find vertex coordinates for grid points in quadrants 2, 3, and 4, simply change the signs of the coordinates in quadrant 1. The *z*-coordinate for every point in the edge-zenith diamond will always be positive. The sign changes for each quadrant are shown in Table C.3.

Triacon LCD Coordinate Signs			
Quadrant	X	Y	Z
1	+	+	+
2	+	+	+
3	+	−	+
4	−	−	+

Table C.3. Triacon quadrant coordinate signs.

C.2.4   Class II Summary

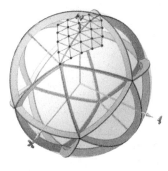

After following the previous steps, the triacon LCD is fully subdivided and its coordinates reflected to define grid points in the other three diamond quadrants. The fully subdivided diamond will resemble the grid in the figure to the left. You can leave the diamond's grid as is for analysis or display, or you can cover the entire sphere by rotating copies of the coordinates to predefined locations elsewhere on the sphere. Appendix B explains how this is done. You can also use these grid points with any of the coordinate geometry algorithms listed in this appendix to find centers of triangles, face normals, distances between arbitrary pairs of points, and so on.

C.3   Class III: Skew

This section describes how to create a Class III skew subdivision. Unlike Class I and III schemas where 3D grid points are created by intersecting great circles or by solving spherical right triangles, Class III defines points in two dimensions and then projects them onto the surface of the sphere. Skew grids are not parallel or perpendicular to the edges of their base polyhedron; thus, we do not have the convenience of locating reference points or defining right triangles along the edges of the foundation polyhedron. In projection, we essentially develop a flat triangular grid that meets the subdivision specification and then we project it onto the surface of the sphere, much like a movie screen, except that our screen is spherical.

The steps that follow describe how to pre-subdivide a flat, 2D, equilateral PPT with a skewed grid. A simple coordinate system based on a triangular grid (also called an *isometric grid*) creates the points needed for any (b,c) subdivision specification. Once the PPT's 2D grid points are defined, the entire set or points is centered over the Cartesian origin, scaled to fit the face of the spherical polyhedron we are subdividing, and then projected onto the surface of the unit sphere. The projected PPT and its grid are oriented face-zenith. The edges of the equilateral PPT are not retained as subdivision grid members. Projection subdivides a single PPT. Its grid can be used as is, input to a CAD system for further development, or copied and rotated to cover the remainder of the sphere's surface.

The projection method described here can also create Class I and II grids as well, depending on the subdivision grid's (b,c) specification. If b and c are zero, these steps create a Class I grid. Likewise, if $b = c$ and neither is zero, a Class II grid results. Class I subdivisions made by projection are the same as Class I Equal-chords discussed earlier in the book. Class II subdivisions made by projection resemble the triacon, but do not minimize the number of parts (chords, arcs, face shapes) as triacon does.

C.3.1   Step 1: Define the PPT Grid

The PPT grid points are initially defined on the xy-plane, thus the z-coordinate of each grid point is zero. The triangular grid axis system was detailed in Section 8.7.6, illustrated in Figure 8.43. It is an example of an 8^v ($b = 2$, $c = 6$) grid. The trigrid coordinates of all points within the PPT are found with these formulas:

$$x = \cos 60°(c + b),$$
$$y = \sin 60° c, \quad \text{and}$$
$$z = 0.$$

The PPT's apex points will be $(0,0)$, (b,c) and (c,b). Define the remainder of the points within the PPT with the same triangular grid axis system.

C.3.2   Step 2: Position PPT for Projection

The orientation of the PPT and its point set relative to the Cartesian x- and y-axis depends on the (b,c) specification used to make the grid in step one. In this step, the PPT and its point set is reoriented, centered, and scaled to a standard position suitable for projection.

- *PPT edge perpendicular to the x-axis.* One vertex of the PPT is already at the origin of the coordinate system. In this step, the entire PPT and its point set are rotated around this vertex to position the PPT side opposite it perpendicular to the x-axis.[2] The rotation angle will never exceed 30°. To find the exact rotation angle needed to place the PPT edge perpendicular to the x-axis, use vdir () to find its midpoint. Then use vcos () to find the angle between its mid-edge point and the x-axis. Use zrot () to rotate the PPT's point set counterclockwise (right hand rule) around the origin (z-axis). The command zrot () is listed in Appendix C;

- *Center the PPT on the z-axis.* The PPT must be centered over the origin (z-axis). Use gtricent () to find the PPT's centroid. Since the PPT edges are already symmetrical to the x-axis, the result of the rotation in the previous step, the PPT's centroid x-coordinate is the distance the PPT's center is from the

---

[2] If you are using the projection method to make a Class II grid ($b = c$), the edge opposite the PPT apex at the origin is already perpendicular to the x-axis, and you do not need to rotate the trigrid point set.

Deltahedron Unit Sphere Characteristics		
Deltahedron	**Chord Length**	**Inradius**
Tetrahedron	1.63299316	0.33333333
Octahedron	1.41421356	0.57735027
Icosahedron	1.05146222	0.79465447

Table C.4. Deltahedron chord lengths and inradius.

origin (the centroid's y- and z-coordinates are zero). To center the PPT over the origin, subtract the centroid's x-coordinate from the x-coordinate of all points in the PPT point set. The PPT point set is now centered on the Cartesian origin with one of the PPT's edges perpendicular to the x-axis;

- *Scale the trigrid PPT.* The PPT edge points were defined by a unit triangular grid, thus it is not the same size as the edge chord of the deltahedron we are subdividing. To correct this, the PPT and its points set must be rescaled. The scale needed is the ratio of the PPT's edge length to the polyhedral edge length (inscribed in a unit sphere). Use vdis() and any two vertices of the PPT to find the PPT's edge length. Table C.4 lists the chord lengths of the deltahedra inscribed in a unit sphere. Use the scale and algorithm vscl() to scale the PPT. After scaling the edges of the PPT equilateral triangle are now equal to the chord lengths of the deltahedron we are subdividing.

To review, the PPT equilateral triangle and its point set is positioned with one apex coincident to the origin, and the PPT edge opposite that vertex is perpendicular to the x-axis. The PPT's centroid is found and the entire PPT point set is translated in the $-x$ direction to center it over the origin. The PPT and its points set are scaled, so the PPT's edges are the same length as the chord length of the deltahedron we are subdividing.

C.3.3   Step 3: Project Trigrid Points

The PPT point set is on the xy-plane and centered at the origin (z-axis). The z-coordinate of every point is zero. They must be moved off the xy-plane before they can be projected. Translate the PPT point set along the $+z$-axis, a distance equal to the unit insphere radius of the deltahedron being subdividing. This is accomplished by simply adding the inradius distance to the z-coordinate of every point in the PPT set. Table C.4 lists the inradius for each deltahedron. Once translated, the apex points of the trigrid PPT will be on the surface of the unit sphere and tangent to the vertices of the face-zenith deltahedron being subdividing. The trigrid points are on the plane of the PPT grid and, thus, are inside the sphere. Use vuni() to project the grid points onto the surface of the sphere. A single face of the base deltahedron is now subdivided.

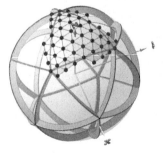

C.3.4   Class III Summary

At the conclusion of step three, the icosahedron's Class III {3, $\}_{2,6}$ PPT will be fully subdivided and look like the figure to the left. Notice that a number of triangles straddle the polyhedral edge. Thus, to define all of their vertices, you must copy the grid points within the PPT and rotate them to define the points in the adjacent polyhedral faces. Appendix B explains how this is done. The next section describes how to find the chord and arc factors as well as the central and surface angles of any triangle.

C.4   Characteristics of Triangles

Almost every Class I, II, or III spherical application requires supporting information regarding the breakdown. It is easy to extract chord and arc factors or the central and surface angles of grid members and triangles from the coordinates of the grid vertices. The procedures that follow are based on grid points located on the surface of the unit sphere.

- *Chord factors.* The chord factor, or chord length on a unit sphere, is simply the straight line distance between two grid points; use vdis(). A chord factor is typically found for chords between neighboring points, but there are no computational reasons why a chord factor between any arbitrary pair of points cannot be computed. This might be needed in some applications, for example, the edge of an opening spanning two non-neighboring points;

- *Arc factors and central angles.* Arc factors are simply the radian measure of the central angle between two points (measured from the origin). Like chord factors, arc factors are typically the geodesic arc distance between neighboring points but an arc factor can be computed the same way between any two points on the sphere. Use vcos() to find the cosine of the arc factor. The arccosine is the central angle;

- *Surface and dihedral angles.* The spherical surface angle between any two great circles is the same as the dihedral angle between their planes. Use gdihdrl() to find the angle (in decimal degrees). Three points define each plane. For great circles, one of these points will be the origin of the sphere. Use gdihdrl() to find the dihedral angle between two adjacent planar triangles (with vertices on the unit sphere). If the six point input to gdihdrl() are the vertices of two adjacent triangles, the dihedral angle between them is returned.

C.5   Storing Grid Points

The computer programs that follow assume all PPT or LCD grid points are stored in a single double precision array of triplets called *xyz*[SIZE][3] where the minimum size is determined by the subdivision frequency "v" plus working area. The formula for SIZE is

$$SIZE = \frac{(v+1)(v+2)}{2}.$$

This 2D array is easy to visualize. The Cartesian *xyz*-coordinates of the fifth point in a grid, for example, would be *xyz*[4][0], *xyz*[4][1], and *xyz*[4][2], respectively. Note that in C, as in many other computer languages, array elements are indexed by their offset from the beginning of the array. Thus, the first array element index starts at zero [0].

Some algorithms need temporary locations to hold coordinates. Those algorithms use a simple integer array, *r*[], that contain indices to available unused elements in the master *xyz*-coordinate array. The actual index contained in them is not important to understanding the algorithm. How the temporary location is used is documented.

C.5.1   gcsect(): Intersection Points of Two Great Circles

The command gsect() returns two points, *p*0 and *p*1, and the dihedral angle between two intersecting great circles. Two great circles are defined by points v1,v2 and v3,v4 and the origin of the sphere. The command gsect() is based on normals to the planes of the great circles. See Figure 4.41 for an explanation of how normals of two planes define another plane perpendicular to their line of intersection. Note, also, that the dihedral angle returned is also the surface angle at their intersection.

```
int gcsect (int v1,int v2, int v3,int v4, int p0,int p1, double *dihedral)
{
    // find normals to both great circles and then normal to the normals
    vzero(r[0],r[0]);            // r[0] is origin of sphere
    vnor(v1,r[0],v2,r[1]);       // r[1] normal to first gc's plane
    vnor(v3,r[0],v4,r[2]);       // r[2] normal to second gc's plane
    // find normal to plane created by above normals, the resulting normal
    // is parallel to the line of intersection of the two gc planes
    vnor(r[1],r[0],r[2],r[3]); // r[3] normal to normals,
    // normal 1 / origin / normal 2
    // check that the normal to normals is non-zero. A zero normal means
    // the input points are from coincident great circles
    if (vabs(r[3])==0.0) {       // test for coincident great circles
        vzero(p0,p0);            // initialize the return n to
        vzero(p1,p1);            // zero and
        *dihedral = 0.0;         // initialize dihedral angle to zero
        rput(count,r);           // return working registers to free pool
        return 9;                // signal error with return code
    }
    vuni (r[3],p0);              // unitize the normal, this is intersection 1
    vrevs(p0,p1);                // intersection 2 is antipodal of 1
    // find dihedral, return smallest angle
    // Note that cos function for vector dot products can return an
    // angle 0>angle<180. To insure angle is always the smallest dihedral
    // take absolute value of cos. Result is always the supplement.
    *dihedral = acos( fabs(vdot(r[1],r[2]) / (vabs(r[1]) * vabs(r[2]))) );
    return 0;                    // two valid points found, return
}
```

C.5.2   gdihdrl(): Dihedral Angle between Two Planes

The command gdihdrl () is a general function for finding the dihedral angle (also surface angle if points are on intersecting great circles) between any two planes. Plane 1 is defined by points *p*1, *p*2, and *p*3. Plane 2 is defined by points *p*4, *p*5, and *p*6.

```
double gdihdrl (int p1, int p2, int p3, int p4, int p5, int p6)
{
    extern double xyz[6500][3];          // xyz coordinate space
    double dihedral = 0.0;               // dihedral angle initialized
    vnor(p1,p2,p3,r[0]);                 // find normal to first plane
    vnor(p4,p5,p6,r[1]);                 // find normal to second plane
    // compute angle between normals
    dihedral = degree(acos(vdot(r[0],r[1]) / (vabs(r[0])*vabs(r[1]))));
    if (dihedral <= 90.0) dihedral = 180.0 - dihedral; // check for smallest
    return dihedral;
}
```

C.5.3   gtricent(): Centroid of a Triangle

The command gtricent () returns the centroid point *v*4 of a triangle defined by points *v*1, *v*2, and *v*3. Point *v*4 is on the plane of the triangle.

```
void gtricent (int v1, int v2, int v3, int v4)
{
    extern double xyz[SIZE][3];           // xyz coordinate space
    xyz[v4][0] = (xyz[v1][0] + xyz[v2][0] + xyz[v3][0]) / 3.0;
    xyz[v4][1] = (xyz[v1][1] + xyz[v2][1] + xyz[v3][1]) / 3.0;
    xyz[v4][2] = (xyz[v1][2] + xyz[v2][2] + xyz[v3][2]) / 3.0;
    return;
}
```

C.5.4 stABC(): Surface Angles of a Spherical Triangle

The command stABC() returns the three surface angles of spherical triangle and the triangle's spherical excess (its area in steradians) given its three vertices on the surface of the unit sphere. The command gdihdrl() computes the dihedral angle.

```
void stABC(int v1, int v2, int v3, double *A, double *B, double *C,
           double *Excess)
{
    *A = radian(gdihdrl(v2,v1,v3));
    *B = radian(gdihdrl(v1,v2,v3));
    *C = radian(gdihdrl(v2,v3,v1));
    *Excess = (*A + *B + *C) - PI;
    return;
}
```

C.5.5 vabs(): Length of a Vector

The command vabs () returns the absolute value (also called vector length) of vector v1.

```
double vabs (int v1)                     // find absolute value (lengths) of v1
{
    extern double xyz[SIZE][3];
    double sum = 0.0;                     // init sum or squares
    for (int i=0;i<3;i++) sum += xyz[v1][i] * xyz[v1][i];// sum the squares
    sum = sqrt(sum);                     // keep summing the squares
    return sum;                          // return the final total
}
```

C.5.6 vadd(): Add Two Vectors

The command vadd () adds vector v1 and v2 placing sum in v3.

```
void vadd (int v1, int v2, int v3)
{
    extern double xyz[6500][3]; // xyz coordinate space
    int i; // general loop index
    double temp[3];
    for (i=0;i<3;i++) temp[i] = xyz[v1][i] + xyz[v2][i];
    for (i=0;i<3;i++) xyz[v3][i] = temp[i];
    return;
}
```

C.5.7   varcv(): Locate a Point on a Geodesic Arc between Two Other Points

The command `varcv()` is a very useful algorithm for placing a point on a geodesic arc between two other points on the surface of the sphere. The point is placed a percentage of the distance from v1 towards v2. If the percent is 0.5, v3 is placed on the arc midway between v1 and v2. If the percent is 1.0, v3 will have the same value as v2. The command `varcv()` returns the central angle between v1 and v3 in radians. No error checking is performed.

```
double varcv(int v1, int v2, double scaleDistance, int v3)
{
    extern double xyz[SIZE][3];          // xyz coordinate space
    double alpha, beta, angle;
    double arcangle;
    vzero(r[2],r[2]);                    // create an origin point
    angle = acos(vcos(v1,r[2],v2));      // find great circle v1-v2 arc
    alpha = sin((1.-scaleDistance)*angle)/sin(angle);
    beta = sin(scaleDistance*angle)/sin(angle);
    vscl(v1,alpha,r[0]);                 // scale v1
    vscl(v2,beta, r[1]);                 // scale v2
    vadd(r[0],r[1],v3);                  // addition is desired point,
                                         // returned in v3
    vzero(r[2],r[2]);                    // create an origin point
    arcangle=acos(vcos(v1,r[2],v3));     // return v1 to v3 arc angle (in
                                         // radians)
    return arcangle;                     // return central angle
}
```

C.5.8   vcos(): Cosine of Angle between Two Vectors

The command `vcos()` returns the cosine of the angle between two vectors, v1 and v3, measured from v2. If v1 and v3 are points on the surface of a unit radius sphere and v2 is the sphere's origin, the cosine is the central radian angle. This subroutine can be used in many situations.

```
double vcos (int v1, int v2, int v3)
{
    extern double xyz[SIZE][3];          // xyz coordinate space
    double cosine;                       // cosine of angle between two
                                         // vectors
    vsub(v2,v1,r[0]);                    // r[0] free v2-v1 edge vector
    vsub(v2,v3,r[1]);                    // r[1] free v2-v3 edge vector
    cosine = vdot(r[0],r[1]) / (vabs(r[0]) * vabs(r[1]));
    // divide dot product by
    // product of absolutes
    return cosine;
}
```

C.5.9   vcrs(): Cross Product of Two Vectors

The command `vcrs()` computes the cross product of vectors v1 and v2 and returns a vector, v3, that is perpendicular to their plane. v3's magnitude is also returned. The cross product of two vectors returns a normal vector (one that is 90°) to the plane defined by the two vectors.

```
double vcrs (int v1, int v2, int v3)
{
    extern double xyz[SIZE][3];          // xyz coordinate space
    // column diagonals multiplied
    xyz[v3][0] = xyz[v1][1] * xyz[v2][2] - xyz[v1][2] * xyz[v2][1];// v3x
    xyz[v3][1] = xyz[v1][2] * xyz[v2][0] - xyz[v1][0] * xyz[v2][2];// v3y
    xyz[v3][2] = xyz[v1][0] * xyz[v2][1] - xyz[v1][1] * xyz[v2][0];// v3z
    return (vabs(v3));
}
```

C.5.10   vdir(): Point on a Parametric Line

The command vdir() defines a point, v3, on an infinite line defined by points v1 and v2 at a scalar distance from v1, which is sometimes called the anchor point. If the scalar value is 0 or 1, v3 will equal v1 or v3, respectively. If scalar value is 0.5, v3 will be a point midway between v1 and v2. Negative scales place v3 away from v1 opposite the direction from v2. This algorithm is useful in Equal-chords schemas that space reference points along a chord. Reference points are inside the sphere and need to be projected to its surface. See vuni() to project them.

```
void vdir (int v1, int v2, double scalar, int v3)
{
    extern double xyz[SIZE][3];// xyz coordinate space
    vsub(v2,v1,r[0]);              // translate v2 by v1 relative to origin
    vscl(r[0],scalar,r[1]);       // scale v2, put in temp registers at 3
    vadd(v1,r[1],v3);             // translate scaled v2 back relative to v1
    return;
}
```

C.5.11   vdis(): Distance between Two Points

The command vdis() returns the distance between points v1 and v2.
```
double vdis (int v1, int v2)
{
    extern double xyz[SIZE][3];  // xyz coordinate space
    double dist;
    vsub(v2,v1,r[0]);              // find difference between n
    dist = vabs(r[0]);           // return the absolute value
    return dist;
}
```

C.5.12   vdot(): Dot Product

The command vdot() returns the dot product, also called the scalar or inner product, of two vectors.
```
double vdot (int v1, int v2)
{
    extern double xyz[SIZE][3];          // xyz coordinate space
    double sum = 0.0;                    // temp summation variable
    for (int i=0;i<3;i++) sum += xyz[v1][i] * xyz[v2][i];
    return sum;                          // return sum of squares
}
```

C.5.13   vnor(): Normal Vector to a Plane

The command vnor finds the normal vector, v4, to a plane defined by points v1, v2, and v3. v4's magnitude is returned.

```
double vnor (int v1, int v2, int v3, int v4)
{
    extern double xyz[SIZE][3];        // xyz coordinate space
    vsub(v1,v2,r[0]);                  // define t1 vector on the plane
    vsub(v3,v2,r[1]);                  // define t2 vector on the plane
    vcrs(r[1],r[0],v4);                // find cross product t2 X t1
    return vabs(v4)/2.0;               // area is half the abs value of
                                       // normal vector
}
```

C.5.14   vrevs(): Reverse a Vector's Sense

With the command vrevs(), the reverse sense of v1 is returned in v2 placing. If v1 is a vector, it now points the opposite direction and its magnitude remains the same. If v1 is a point, it is relocated to a new position on the opposite side of the origin along an infinite line between its old position and the origin of the sphere. If v1 is on the surface of the unit sphere, v2 will be its antipode.

```
void vrevs (int v1, int v2)
{
    extern double xyz[SIZE][3];        // xyz coordinate space
    for (int i=0;i<3;i++) xyz[v2][i] = xyz[v1][i] * -1.0; // reverse sense
    return;
}
```

C.5.15   vscl(): Scale a Vector or Point

With the command vscl(), the vector or point v1 is scaled along an infinite line through the origin and returned in v3. The scale factor can be any real (positive value) number. The command vscl() is very useful in scaling up or down all points on a sphere or simply to locate a new point in the same direction.

```
void vscl (int v1, double scalar, int v3)
{
    extern double xyz[SIZE][3];            // xyz coordinate space
    int i;                                 // general loop index
    for (i=0;i<3;i++) xyz[v3][i] = xyz[v1][i] * fabs(scalar);
    if (scalar<0) for (i=0;i<3;i++) xyz[v3][i] = xyz[v3][i] * -1;
    return;
}
```

C.5.16   vsub(): Subtract a Vector or Point

The command vsub() subtracts v2 from v1. The difference is returned in v3.

```
void vsub (int v1, int v2, int v3)
{
    extern double xyz[SIZE][3];        // xyz coordinate space
    for (int i=0;i<3;i++) xyz[v3][i] = xyz[v1][i] - xyz[v2][i];
    return;
}
```

C.5.17   vuni(): Normalize a Vector

The command `vuni ()` normalizes the vector or point v1 and returns its position on the surface of the unit sphere in v2. This is a highly used algorithm.

```
void vuni (int v1, int v2)
{
    extern double xyz[SIZE][3];          // xyz coordinate space
    int i;                               // general loop index
    double len = 0.0;                    // init sum or squares
    for (i=0;i<3;i++) len += sq(xyz[v1][i]);// sum the squares
    len = sqrt(len);                     // find length of vector
    for (i=0;i<3;i++) xyz[v2][i] = xyz[v1][i] / len;
    return;
}
```

C.5.18   vzero(): Initialize a Vector

The command `vzero ()` initializes the vector or point v1 to zero and returns the value in v2.

```
void vzero (int v1, int v2)
{
    extern double xyz[SIZE][3];             // xyz coordinate space
    for (int i=v1;i<=v2;i++) {              // for every vector row element
      for (j=0;j<3;j++) xyz[i][j] = 0.0; // set its coordinates to zero
    }
    return;
}
```

C.6   Symmetrical Uniqueness

In Chapter 10, Section 10.2 we introduced the concept of identifying vertices in the **BC** grid by their symmetry group. Each vertex that resides in section S^0 forms a unique symmetry group. For Class I and II section S^0 will be a single Schwarz triangle and for Class III section S^0 will be a dual Schwarz triangle. For computational purposes it useful to have a methodology for determining if a vertex in the **BC** grid resides in section S^0 and therefore defines a unique symmetrical position in the overall geometry. This section will detail how to perform this symmetry assessment for any vertex of the **BC** grid. The process will only rely on the 2D geometry of the **BC** grid, prior to projection, because it represents all that is required geometrically to make that determination.

The corners vertices of the PPT within the **BC** grid are labeled p_0, p_1, and p_2 as shown in Figure C.6 along with an outline of what is included in section S^0. As you can see some of the edges of section S^0 are included and some excluded. All the vertices of

Symmetry Algorithms	
Subroutine	**Operation**
hspace	determine position of point in relation to a 2D line
section0	determine if 2D point in BC grid is in section zero (S^0) for Class III config

Table C.5. Summary of symmetry algorithms

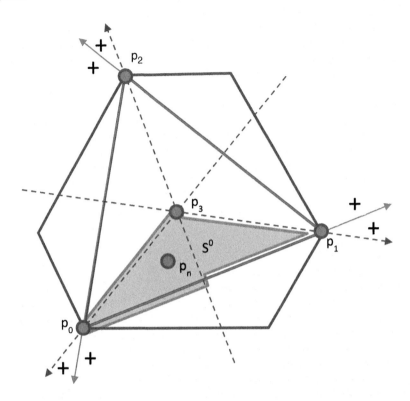

Figure C.6. BC Grid, PPT Triangle, and Unique Section S^0

the **BC** grid that fall in section S^0 are symmetrically unique. Now we will look at how to exactly determine what is considered *included* in section S^0 and by introducing the concept of 2D half-spaces.

The concept of half-space in 2D space is that a line divides the plane into two halves. The line on the plane is defined by two points and any other point will have a distinct relationship with that line. Every other point will either be exactly on the line or on either side. The following equation is used to determine the relation of point

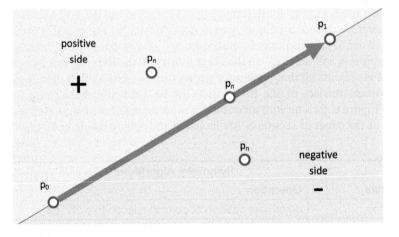

Figure C.7. Half-space defined by 2D line and three possible locations relative to line.

p_n to the line defined by p_0 and p_1. The value of *side* is a real number but we are only concerned with the sign, not the magnitude, of the *side* value. The direction of the line changes which side is positive or negative following the right hand rule (RHR). Figure C.7 illustrates the line defined by line p_0 and p_1 and the possible positions of every other point p_n.

$$side = \left(p_1^x - p_0^x \right) * \left(p_n^y - p_0^y \right) - \left(p_n^x - p_0^x \right) * \left(p_1^y - p_0^y \right)$$

The PPT triangle inside the **BC** grid has 4 points that define the half-spaces required to determine if a vertex is included in section S^0; the 3 corners and the center of the PPT. The *hspace( )* command computes the sign of a point in relation to a 2D line. The *section0* command uses multiple *hspace( )* commands to report if the point in question is properly inside section S^0 and therefore symmetrically unique.

C.6.1   Storing BC Grid Points

First consider the **BC** grid as outlined in Chapter 10 (Section 10.2.2). The number of vertices in the **BC** grid is directly related to the grid's (b,c) values as shown in the following equation. Consecutive rows will have a different number of vertices based on the class of the grid. The computer programs that follow assume all BC grid points stored in a single double precision array of triplets called bc_xyz[SIZE][3].

$$SIZE = (b+c+1)^2 - \frac{b^2 + c^2 + b + c}{2}$$

C.6.2   hspace(): Which Side of a 2D Line Does Point Lie

The command hspace () returns the position of point v2 in relation to line defined from v0 to v1. If the point v2 is exactly on the line the function returns zero. If v2 is on the right side of the line (right hand rule) then the function will return 1 else the function will return –1. Notice that very small values are checked and treated as zero.

```
int hspace (int v0, int v1, int v2 )
{
    extern double bc_xyz[SIZE][3];        // xyz coordinate space
    double sign;
    sign = (xyz[v1][0] - xyz[v0][0]) * (xyz[v2][1] - xyz[v0][1]) -
           (xyz[v2][0] - xyz[v0][0]) * (xyz[v1][1] - xyz[v0][1]);
    if ( fabs ( sign ) < 0.000000000001 )
        return 0; // exactly on the line
    return sign > 0?  1 : -1;
}
```

C.6.3   section0(): Determines If Point in BC Grid Is in Section Zero (S0) for Class III

The command section0 () returns true if the 2D coordinates of point v4 is in section zero, S^0, of the PPT triangle defined by three 2D points (v0,v1,v2) and its center (v3).

```
int section0 (int v0, int v1, int v2, int v3, int v4)
{
   extern double bc_xyz[SIZE][3];        // xyz coordinate space
   if ( hspace ( v3, v0, v4 ) >= 0 )
      if ( hspace ( v3, v2, v4 ) >= 0 && hspace ( v0, v1, v4 ) >= 0 )
         return TRUE;
      else if ( hspace ( v3, v0, v4 ) < 0 && hspace ( v0, v1, v4 ) > 0 )
         return TRUE;
      return FALSE;
}
```

Additional Resources

Clinton, Joseph D. *Advanced Structural Geometry Studies Part I—Polyhedral Subdivision Concepts for Structural Applications.* National Aeronautics and Space Administration contractor report NASA CR-1734, Carbondale, IL: Southern Illinois University, 1971.

____. "Geodesic Math." *Domebook 2* ed. Lloyd Kahn et al. Bolinas, CA: Pacific Domes, 1971. 106–113.

Kenner, Hugh. *Geodesic Math and How to Use It.* Berkeley: University of California Press, 1976.

Kitrick, Christopher J. "A Unified Approach to Class I, II and III Geodesic Domes." *International Journal of Space Structures* 5.3-4 (1990): 223–246.

Messer, Peter W. "Mathematical Formulas for Geodesic Domes." Appendix to *Spherical Models* by Magnus J. Wenninger. New York: Dover, 1999. 145–149.

Bibliography

This is a comprehensive bibliography that details all citations in the text as well as expands references to include the history of spherical subdivision and spherical applications in science, technology, engineering, entertainment, and the arts. Many references are technical and provide a more rigorous background to the challenges of spherical work and different approaches to subdivision. Readers will find this expansive resource an easy way to identify classic resources, current research and key personalities involved in pioneering work.

Abbott, Edwin A. 1884. *Flatland: A Romance of Many Dimensions*. London: Seeley & Co.

Aguilar, Rodolfo Jesus. 1964. "A Study of the Stability of Framed, Triangulated Geodesic Domes Under the Action of Concentrated Loads." PhD, Department of Civil Engineering, North Carolina State University.

Ahern, William W., and Brigit T. Mathe. 1977. Spherical Structural Arrangement. US Patent 4, 026,078, issued May 31, 1977.

Ahern, William W., and William H. Wainwright. 1968. Spherical Structural Arrangement. US Patent 3,392,495, issued July 16, 1968.

Al Mamum, Mohammad Abdullah. 2015. A Short Historical Review of Kepler Conjecture. Department of Materials and Metallurgical Engineering (Bangladesh University of Engineering & Technology).

Alexander, James. 2004. "Loxodromes – A Rhumb Way to Go." *Mathematics Magazine* 77 (5): 349–356.

Aoyama, Steven. 1985. Golf Ball. US Patent 4,560,168, issued December 24, 1985.

____. 1990. Golf Ball. US Patent 4,948,143, issued August 14, 1990.

____. 1990. Golf Ball Dimple. US Patent 6,162,136, issued December 19, 2000.

___. 1995. Three Parting Line Quadrilaterals Golf Ball Dimple Pattern. US Patent 5,415,410, issued May 16, 1995.

___. 1999. Golf Ball Dimple Pattern. US Patent 5,957,786, issued September 28, 1999.

___. 2007. Golf Ball Dimples. US Patent 7,207,905, issued April 24, 2007.

Aoyama, Steven, and Douglas E. Jones. 2005. Golf Ball with Improved Flight Performance. US Patent 6,916,255, issued July 12, 2005.

___. 2005. Golf Ball with Improved Flight Performance. US Patent 6,923,736, issued August 2, 2005.

Aoyama, Steven, and William E. Morgan. 2007. Golf Ball Dimple Pattern with Overlapping Dimples. US Patent 7,258,632, issued 21 Aug. 2007.

Aste, Tomaso, and Swnia Weaire. 2008. *The Pursuit of Perfect Packing*. second ed. New York: Taylor & Francis.

Baer, Steve. 1970. *Zome Primer: Elements of Zonohedra Geometry; Two- and Three-Dimensional Growths of Stars with Five-Fold Symmetry*. Albuquerque, NM: Zomeworks Corp.

Bagchi, Bhaskar. 1997. "How to Stay Away from Each Other in a Spherical Universe." *Resonance* 2 (September): 18–26.

Ball, Walter William Rouse, and H. S. M. Coxeter. 1987. *Mathematical Recreations and Essays*. New York: Dover Publications.

Bartholdi, John, and Paul Goldsman. 2001. "Continuous Indexing of Hierarchical Subdivisions of the Globe." *International Journal of Geographical Information Science* 15 (6): 489–522.

Baverel, Olivier. 2000. "Nexorades: a Family of Interwoven Space Structures." Ph. D., University of Surrey.

Baverel, Olivier, C. Dothe, and J. F. Caron. 2006. "Nexorade: A Structure for 'Free Form' Architecture." Adaptables 2006: International Conference on Adaptable Building Structures, Eindhoven, The Netherlands, July 3–5, 2006.

Baverel, Olivier, and O. Popovic Larsen. 2011. "A Review of Woven Structures with Focus on Reciprocal Systems — Nexorades." *International Journal of Space Structures* 26 (4): 281–288.

Baverel, Olivier, and Hoshyar Nooshin. 2007. "Nexorades Based on Regular Polyhedra." *Nexus Network Journal* 9: 281–299.

Baverel, Olivier, Hoshyar Nooshin, Y. Kueoiq, and G. A. R. Parke. 2000. "Nexorades." *International Journal of Space Structures* 15 (2): 115–120.

Baverel, Olivier and M. Sidani. 1998. "The Multi-Reciprocal Grid System." Proceedings of the International Conference on Lightweight Structures in Civil Engineering, Warsaw, Poland

Berman, Joel D., and Kit Hanes. 1970. "Volumes of Polyhedra Inscribed in the Unit Sphere in *E3*." *Mathematische Annalen* 188 (1): 78–84.

Blenkinsop, Tom. 1999. "Pedagogy of Stereographic Projection." *Journal of African Earth Sciences* 29 (4): 897–902.

Bohlen, J. C. 1974. *Trigonometric Relationships for Geodesic Domes with Special Reference to the Dodecahedron*. Western Forest Products Laboratory, Information Report VP-X-121, Vancouver, BC.

Bourke, Paul. 1996. "Distributing Points on a Sphere." paulbourke.net/geometry/spherepoints/.

Bowditch, Nathaniel. 2002. *The American Practical Navigator: An Epitome of Navigation*. 2002 ed. Vol. 9. Bethesda, MD: U.S. National Imagery and Mapping Agency.

Boyt, Dick. 1991. *Rotegrity – An Alternate Method of Spanning Space with a Compound Curve*. Neosho, MO: Crowder College.

Brand, Stuart, and James Baldwin. 1968. Whole Earth Catalog. Menlo Park, CA: Portola Institute.

Briddell, Charles "Don". 2012. "Multiple Certainties: An Implication of Field Structure Theory (FST)." 19th Natural Philosophy Alliance Proceedings, Albuquerque, NM.

Brodsky, Isaac. 2018. "H3: Uber's Hexagonal Hierarchical Spatial Index." [Web page]. Uber Engineering.eng.uber.com/h3/.

___. 2019. "Geospatial Analytics with H3+kepler.gl." Spatial Data Science Conference 2019, New York City: Columbia University.

Bromberg, Manuel. 2004. Re: Founding of Skybreak Carolina Corp., early 1950s geodesic dome experiments and R. Buckminster Fuller guest lectures at North Carolina State School of Design.

___. 2008. Re: Duncan Stuart's 1951 experiments with geodesic dome geometry and construction.

Brook, David Louis Sterrett. 2005. "Henry Leveke Kamphoefner, The Modernist, Dean of the North Carolina State University School of Design 1948–1972." Master's Thesis, Department of Adult and Community College Education, North Carolina State University.

Brückner, Max. 1900. *Vielecke und Vielflache: Theorie und Geschichte (Polygons and Polyhedra: Theory and History)*. Leipzig: Teubner.

Bullard, James Atkins, and Arthur Kiernan. 1922. *Plane and Spherical Trigonometry with Stereographic Projections*. Boston: D.C. Heath.

Bunger, Donald J., and Joseph F. Stiefel. 1991. Golf Ball. US Patent 5,060,953, issued October 19, 1991.

Burt, Michael. 2011. "The Periodic Table of the Polyhedral Universe." *International Journal of Space Structures* 26 (2): 75–93.

Cahill, Bernard J. S. 1909. "An Account of a New Land Map of the World." *Scottish Geographical Magazine* 9: 449–469.

___. 1913. Geographical Globe. US Patent 1,081,207, issued December 9, 1913.

___. 1913. Map of the World. US Patent 1,054,276, issued February, 25, 1913.

Calter, Paul A. 2006. *Squaring the Circle – Geometry in Art and Architecture*. Emeryville, CA: Key College Pub.

Caspar, Donald L. D. 1993. "Deltahedral Views of Fullerene Polymorphism." *Philosophical Transactions of the Royal Society of London. Series A: Mathematical, Physical & Engineering Sciences* 343: 133–144.

Caspar, Donald L. D., and Aaron Klug. 1962. "Physical Principles in the Construction of Regular Viruses." *Cold Spring Harbor Symposia on Quantitative Biology* 27: 1–24.

Cessna, Joseph B., and Thomas R. Bewley. 2009. "Honeycomb-Structured Computational Interconnects and their Scalable Extension to Spherical Domains." Proceedings of the 11th International Workshop on System Level Interconnect Prediction (SLIP'09), San Francisco.

Ch, Adam, and Paul Tannery. 1996. "Progymnasmata de Solidorum Elementis." *Oeuvres de Descartes* 10: 265–276.

Chemical Rubber Company. 1953. *CRC Standard Mathematical Tables*. Boca Raton, FL: CRC Press.

Chu, Hsiao-yun, and Roberto G. Trujillo. 2009. *New Views on R. Buckminster Fuller*. Stanford, CA: Stanford University Press.

Clinton, Joseph D. 1970. "Chord Factors and Angles." In *Domebook One*, ed. Lloyd Kahn, 50–52. Los Gatos, CA: Pacific Domes.

___. 1971. Advanced Structural Geometry Studies: Part I – Polyhedral Subdivision Concepts for Structural Applications. ed. Carbondale Southern Illinois University, Ill. Washington, DC: National Aeronautics and Space Administration.

___. 1971. Advanced Structural Geometry Studies: Part II – A Geometric Transformation Concept for Expanding Ridge Structures. ed. Carbondale Southern Illinois University, Ill. Washington, DC: National Aeronautics and Space Administration.

___. 1971. "Geodesic Math." In *Domebook 2*, ed. Lloyd Kahn, 106–113. Bolinas, CA: Pacific Domes.

___. 1990. "Lowest Common Frequency: b2 + bc + c2." *International Journal of Space Structures* 5 (3-4): 213–222.

___. 1999. *Spheres, Polyhedra and Circles*. Clinton International Design Consultants (CIDC), West Chester, PA

___. 2002. "A Group of Spherical Tessellations Having Edges of Equal Length." 5th International Conference on Space Structures, London.

___. 2002. "A Limited and Biased View of Historical Insights for Tessellating a Sphere." *Space Structures* 5.

___. 2004. A Method of Tessellating a Surface. US Patent 2004/0257433, issued December 23, 2004.

___. 2004. "A Search in Progress: Polyhedra from Intersecting Cylinders." BRIDGES Mathematical Collections in Art, Music and Science.

Clinton, Joseph D., and Jay Salsbury. n.d. *Geodesic Math with Computer Aided Design Illustrations*. mr-fusion.hellblazer.com/pdfs/geodesicmath.pdf.

Cohen, Albert, Phillip Davis, and Joseph F. Orabona. 1961. Radome Structural Devices. US Patent 2,978,704, issued April 4, 1961.

Collidge, Julian Lowell. 1971. *A Treatise on the Circle and the Sphere*. Bronx, NY: Chelsea Pub. Co.

Colorado State University. 2001. "Geodesic Climate Model Uses Different Mapping Technique, Coordinates and Supercomputing to Improve Predictions." *ScienceDaily.*

Congleton, Carol A. 1980. *Navigational Applications of Plane and Spherical Trigonometry*. Centreville, MD: Cornell Maritime Press.

Conway, John H., Heidi Burgiel, and Chaim Goodman-Strauss. 2008. *The Symmetries of Things*. Wellesley, MA: AK Peters.

Conway, John H., and N. J. A. Sloane. 1993. *Sphere Packings, Lattices and Groups*. Second ed. Vol. 290. *Volume A Series of Comprehensive Studies in Mathematics*. New York: Springer-Verlag.

Coxeter, H. S. M. 1936. "The Partition of a Sphere According to the Icosahedral Group." *Scripta Math* 4: 156–157.

___. 1962. "The Problem of Packing a Number of Equal Non-Overlapping Circles on a Sphere." *Transactions of the New York Academy of Sciences* 24: 320–331.

___. 1969. *Introduction to Geometry*. second ed. New York: Wiley.

___. 1970. *Twisted Honeycombs*. Vol. 4. *Regional Conference Series in Mathematics*. Providence, RI: Published for the Conference Board of the Mathematical Sciences by the American Mathematical Society.

___. 1972. "Virus Macromolecules and Geodesic Domes." In *A Spectrum of Mathematics*, edited by J. C. Butcher, 98–107. Auckland: Auckland University Press and Oxford University Press.

___. 1973. *Regular Polytopes*. third ed. New York: Dover Publications.

___. 1987. *Projective Geometry*. second ed. New York: Springer-Verlag.

___. 1991. *Regular Complex Polytopes*. second ed. New York: Cambridge University Press.

___. 1998. *Non-Euclidean Geometry*. sixth ed. *Spectrum Series*. Washington, D.C.: Mathematical Association of America.

___. 2012. "Polyhedral Numbers." edited by Robert S. Cohen, John J. Stachel, and Marx W. Wartofsky, vol. 11. Berlin: Springer Science & Business Media.

Coxeter, H. S. M., Michele Emmer, Roger Penrose, and Marianne L. Teuber. 1985. "M. C. Escher: Art and Science." Proceedings of the International Congress on M.C. Escher, Rome, Italy, 26–28 March.

Coxeter, H. S. M., M. S. Longuet-Higgins, and J. C. P. Miller. 1954. "Uniform Polyhedra." *Philosophical Transactions of the Royal Society of London. Series A: Mathematical, Physical & Engineering Sciences* 246 (916): 401–450.

Coxeter, H. S. M., and László Fejes Tóth. 1963. "The Total Length of the Edges of a Non-Euclidean Polyhedra with Triangular Faces." *Oxford Journals Mathematics & Physical Sciences, the Quarterly Journal of Mathematics* 14 (1): 273–284.

Coxeter, H. S. M., and Patrick Du Val. 1982. *The Fifty-Nine Icosahedra*. New York: Springer-Verlag.

Critchlow, Keith. 1970. *Order in Space: A Design Source Book*. New York: Viking Press.

Critchlow, Keith, and Rod Bull. 1982. *Time Stands Still – New Light on Megalithic Science*. New York: St. Martin's Press.

Cromwell, Peter R. 1997. *Polyhedra*. Cambridge, U.K.: Cambridge University Press.

Cundy, H. Martyn, and A. P. Rollett. 1961. *Mathematical Models*. second ed. Oxford, U.K.: Clarendon Press.

Dawson, Robert J. MacG. 2005. "Some New Tilings of the Sphere with Congruent Triangles." Department of Mathematics and Computing Science, Saint Mary's University. www.mi.sanu.ac.rs/vismath/bridges2005/dawson/index.html.

Di Carlo, Biagio. 2008. "The Wooden Roofs of Leonardo and New Structural Research." *Nexus Network Journal* 10 (1): 27–39.

___. 2015. *Reciprocal Frames*. Raleigh, NC: Lulu.com Editions.

___. 2017. La Scienza del Design – Nelle Strutture Poliedriche, Geodetiche, Reciproche e Tensegrali. Raleigh. lulu.com Editions.

Dorst, Leo, Daniel Fontijne, and Stephen Mann. 2007. *Geometric Algebra for Computer Science: An Object-Oriented Approach to Geometry. Morgan Kaufmann Series in Computer Graphics*. Amsterdam: Elsevier.

Doskas, Gary. 2011. *Spherical Harmony – A Journey of Geometric Discovery*. LuLu Marketplace: Hedron Designs.

Dunham, William. 1994. *The Mathematical Universe – An Alphabetical Journey Through the Great Proofs, Problems, and Personalities*. New York: Wiley & Sons.

Dutton, Geoffrey. 1984. "Geodesic Modeling of Planetary Relief." *Cartographica* 21 (2-3): 118–138.

___. 1991. "Polyhedral Hierarchical Tessellations: The Shape of GIS to Come." *Geographical Information Systems* 1 (3): 49–55.

___. 1996. "Encoding and Handling Geospatial Data with Hierarchical Triangular Meshes." 7th Symposium on Spatial Data Handling, Delft, Netherlands.

___. 1999. *A Hierarchical Coordinate System for Geoprocessing and Cartography*. Vol. 79. *Lecture Notes in Earth Sciences*. New York: Springer.

___. 2000. "Universal Geospatial Data Exchange via Global Hierarchical Coordinates." International Conference on Discrete Global Grids, Santa Barbara, CA.

Easton, Robert, and Lloyd Kahn. 1970. *Domebook One*. Los Gatos, CA: Pacific Domes.

Edmondson, Amy C. 1986. "Geodesic Reports: The Deresonated Tensegrity Dome." *Synergetica; Journal of Synergetics* 1 (4).

___. 1987. *A Fuller Explanation – The Synergetic Geometry of R. Buckminster Fuller*. *Design Science Collection*. Boston: Birkhäuser (reprinted Pueblo, CO: Back-in-Action Book Series 2007).

Eisenberg, Michael, et al. 2013. "Turtle Temari." *Bridges Enschede 2013: Mathematics, Music, Art, Architecture, Culture*. The Bridges Organization, Enschede, The Netherlands.

Emerson, Brent D., George R. Wallace, Matthew B. Stanczak, and Sanjay Kattappa. 2003. Distance Golf Ball-DDH Steel Distance. US Patent 6,572,494, issued June 3, 2003.

Euclid. 1956. *The Thirteen Books of Euclid's Elements*. ed. Thomas L. Heath. New York: Dover Publications.

Faux, I. D., and M. J. Pratt. 1979. *Computational Geometry for Design and Manufacturing*. New York: Halsted Press.

Fearnley, Christopher J. 2009. "Exploring Foldable Great Circle Geometries." Eight Interdisciplinary Conference of the International Society of the Arts, Mathematics, and Architecture (ISAMA), Albany, NY: University of Albany.

___. 2011. "CJ Fearnley's List of Buckminster Fuller Resources on the Internet." www.cjfearnley.com/buckyrefs.html.

Fejes Tóth, László. 1949. "On the Densest Packing of Spherical Caps." *American Mathematical Monthly* 56: 330–331.

___. 1964. *Regular Figures*. New York: Macmillan.

___. 1985. "Stable Packing of Circles on the Sphere." *Structural Topology* 11: 9–14.

Fekete, György. 1990. "Rendering and Managing Spherical data with Sphere Quadtrees." *IEEE Visualization, Proceedings of the 1st Conference on Visualization* (1): 176–186.

Fernández-Serrano, Martino Peña, and José Calvo López. 2019. "Projecting Stars, Triangles and Concrete – The Early History of Geodesic Domes, from Walter Bauersfeld to Richard Buckminster Fuller." *Architectura*, 47: 92–114.

Fischbeck, Richard D. 2008. Geodesic Structure. US Patent 7389612.

Fisher, Irving. 1943. "A World Map on a Regular icosahedron by Gnomonic Projection." *Geographical Review* 33 (4): 605–619.

___. 1948. Global Map. US Patent 2,436,860, issued March 2, 1948.

Fisher, N. I., Toby Lewis, and B. J. J. Embleton. 1987. *Statistical Analysis of Spherical Data*. New York: Cambridge University Press.

Fitzgibbon, James W. 1957. Letter to: R. A. Lehr, Union Tank Car, Subject: Geodesic dome construction progress. Raleigh, NC: Synergetics, Inc.

___. 1957. Letter to: R. A. Lehr, Union Tank Car, Subject: Geodesic dome construction progress. Raleigh, NC: Synergetics, Inc.

___. 1957. Letter to: William H. Wainwright, Geometrics, Inc., Subject: Availability of computer reference geometry tables for geodesic domes. Raleigh, NC: Synergetics, Inc.

___. 1957. Memo to: R. Buckminster Fuller, Subject: Current work assignments. Raleigh, NC: Synergetics, Inc.

Fowler, Patrick W., and D. E. Manolopoulos. 1996. *An Atlas of Fullerenes*. Oxford England: Clarendon Press.

Fowler, Patrick W., Tibor Tarnai, and Zsolt Gáspár. 2002. "From Circle Packing to Covering on a Sphere with Antipodal Constraints." *Philosophical Transactions of the Royal Society of London. Series A: Mathematical, Physical & Engineering Sciences* 458 (2025): 2275–2287.

Fowler, Patrick W., Tibor Tarnai, and S. Kabai. 2005. "Packing Regular Triplets of Circles on a Sphere." *Philosophical Transactions of the Royal Society of London. Series A: Mathematical, Physical & Engineering Sciences* 461: 2355–2367.

Freeman, Ira, and Mae Freeman. 1946. *Fun with Figures – Easy Experiments for Young People*. New York: Random House.

Friedman, Michael, and Joachim Krausse. 2016. "Folding and Geometry: Buckminster Fuller's Provocation of Thinking." In *On Folding: Towards a New Field of Interdisciplinary Research*, edited by Wolfgang Schäffner & Michael Friedman, 139–174. Transcript Verlag.

Fuller, R. Buckminster. 1937. Motor Vehicle. US Patent 2,101,057, issued December 7, 1937.

___. 1940. Prefabricated Bathroom. US Patent 2,220,482, issued November 5, 1940.

___. 1946. Cartography. US Patent 2,393,676, issued January 29, 1946.

___. 1948. 3 Way Geodesic Grid B Calculations on Present. Department of Special Collections and University Archives, Stanford University Libraries.

___. 1950. Project Noah's Ark #2 Discovering New Man Advantage. Stanford, CA: Department of Special Collections and University Archives, Stanford University Libraries.

___. 1954. Building Construction. US Patent 2,682,235, issued June 29, 1954.

___. 1959. Geodesic Tent. US Patent 2,914,074, issued November 24, 1959.

___. 1959. Self-Strutted Geodesic Plydome. US Patent 2,905,113, issued September 22, 1959.

___. 1961. Synergetic Building Construction. US Patent 2,986,241, issued May 30, 1961.

___. 1962. Tensile-Integrity Structures. US Patent 3,063,521, issued November 13, 1962.

___. 1965. Geodesic Structures. US Patent 3,197,927, issued August 3, 1965.

___. 1965. Laminar Geodesic Dome. US Patent 3,203,144, issued August 31, 1965.

___. 1967. Octahedral Building Truss. US Patent 3,354,591, issued November 28, 1967.

___. 1971. *Utopia or Oblivion: The Prospects for Humanity*. Toronto: Bantam Books.

___. 1975. Non-Symmetrical Tension-Integrity Structures. US Patent 3,866,366, issued February 18, 1975.

Fuller, R. Buckminster, and E. J. Applewhite. 1975. *Synergetics: Explorations in the Geometry of Thinking*. New York: Macmillan.

___. 1986. *Synergetics Dictionary – The Mind of Buckminster Fuller with an Introduction and Appendices*. Vol. 4. New York: Garland.

Fuller, R. Buckminster, and Carl Solway Gallery. 1981. *Buckminster Fuller: Inventions, Twelve Around One*. Cincinnati, OH: Carl Solway Gallery.

Fuller, R. Buckminster, and Shoji Sadao. 1992. Fuller Projection Dymaxion Air-ocean World. Los Angeles: Buckminster Fuller Institute.

Fuller, R. Buckminster, and James Ward. 1983. *Inventions*. New York: St. Martin's Press.

___. 1985. *The Artifacts of R. Buckminster Fuller: A Comprehensive Collection of His Designs and Drawings*. Vol. 4. New York: Garland.

Fuller, R. Buckminster, and Thomas Tse Kwai Zung. 2002. *Buckminster Fuller: Anthology for the New Millennium*. New York: St. Martin's Griffin Press.

Gabriel, J. François, ed. 1997. *Beyond the Cube: The Architecture of Space Frames and Polyhedra*. New York: John Wiley.

Gailiunas, Paul. 2006. *Twisted Domes*. (Newcastle, England).

Gasson, Peter C. 1983. *Geometry of Spatial Forms – Analysis, Synthesis, Concept Formulation and Space Vision for CAD*. Edited by Ellis Harwood. *Mathematics and its Applications*.

Geodesics, Inc. 1955. *Largest Rigid Plastic Structure in History – Press Release*. Cambridge, MA.

Gheorghiu, Adrian, and Virgil Dragomir. 1978. *Geometry of Structural Forms*. Applied Science Publishers Ltd.

Giuffre, Carl, and Lee Stemkoski. 2020. "Virtual Temari: Artistically Inspired Mathematics." *Journal of Humanistic Mathematics* 10 (2): 237–257.

Gobush, William. 1988. Golf Ball. US Patent 4,765,626, issued August 23, 1988.

___. 1988. Golf Ball. US Patent 4,772,026, issued September 20, 1988.

___. 1989. Multiple Dimple Golf Ball. US Patent 4,804,189, issued February 14, 1989.

___. 1990. Multiple Dimple Golf Ball. US Patent 4,960,283, issued October 2, 1990.

___. 1991. Multiple Dimple Golf Ball. US Patent 5,060,954, issued October 29, 1991.

Goldberg, Michael. 1935. "The Isoperimetric Problem for Polyhedra." *Tohoku Mathematical Journal* 40: 226–236.

___. 1937. "A Class of Multi-Symmetric Polyhedra." *Tohoku Mathematics Journal* 43: 104–108.

___. 1967. "Viruses and a Mathematical Problem." *Journal of Molecular Biology* 24 (2): 337–338.

Graver, Jack E. 2004. "Encoding Fullerenes and Geodesic Domes." *Society for Industrial and Applied Mathematics* 17 (4): 596–614.

Gray, Robert W. 2009. "Great Circle and LCD Triangle Information." The Projects of R. W. Gray. www.rwgrayprojects.com/rbfnotes/greatc/greatc1.html.

Grünbaum, Branko, and G. C. Shephard. 1987. *Tilings and Patterns*. W. H. Freeman.

Hales, Thomas C. 1998. "An Overview of the Kepler Conjecture." *Metric Geometry*.

Hales, Thomas C., et al. 2017. "A Formal Proof of Kepler Conjecture." *Forum of Mathematics, Pi* 5 e2.

Hardin, R. H., N. J. A. Sloane, and W. D. Smith. 2008. "Tables of Spherical Codes with Icosahedral Symmetry." www.research.att.com/~njas/icosahedral.codes.

Hart, George W. 1998. "Icosahedral Constructions." Bridges – Mathematical Connections in Art, Music and Science. Winfield, KA: Southwestern College, July 28–30, 1998.

___. 2008. "The Encyclopedia of Polyhedra." www.georgehart.com/virtual-polyhedra/vp.html.

___. 2008. "Pavilion of Polyhedreality." www.georgehart.com/pavilion.html.

___. 2019. "Max Brückner's Wunderkammer of Paper Polyhedra." Bridges Linz 2019, Linz, Austria.

Hart, George W., and Henri Picciotto. 2001. *Zome Geometry: Hands-on Learning with Zome Models*. Emeryville, CA: Key Curriculum Press.

Heartney, Eleanor, and Kenneth D. Snelson. 2009. *Kenneth Snelson – Forces Made Visible*. Lenox, MA: Hard Press Editions.

Heikes, Ross, Joon-Hee Jung, C. S. Konor, and D. Randall. 2007. "Continuing Development of Models Based on Spherical Geodesic Grids." PDE Sphere 2007.

Henderson, David W., and Eduarda Moura. 1996. *Experiencing Geometry: On Plane and Sphere*. Englewood Cliffs, NJ: Prentice Hall.

Henderson, David W., and Daina Taimiṇa. 2005. *Experiencing Geometry: Euclidean and Non-Euclidean with History*. third ed. Upper Saddle River, NJ: Pearson Prentice Hall.

Hoberman, Charles. 1990. Radial Expansion/Retraction Truss Structures. US Patent 5,024,031, issued January 18, 1990.

___. 1990. Reversibly Expandable Doubly-Curved Truss Structure. US Patent 4,942,700, issued July 24, 1990.

___. 2002. Connection to Make Foldable Structures. US Patent 2002/0083675 A1.

Holden, Alan. 1991. *Shapes, Space and Symmetry*. New York: Dover Publications.

Hoschek, Josef, and Dieter Lasser. 1993. *Fundamentals of Computer-Aided Geometric Design*. Wellesley, MA: AK Peters.

Hotchkiss, John F., and John Stuart Martin. 1997. *500 Years of Golf Balls: History & Collector's Guide*. Dubuque, IA: Antique Trader Books.

Howard, T. C. 1958. *Possible Ways the Random Geometric Grid Developed by Lincoln Laboratories may be Covered by Patent 2,682,235 Owned by Inventor R. Buckminster Fuller*. Geodesics, Inc. (Raleigh, NC: Stanford University Libraries Department of Special collections and University Archives).

Hume, Andrew. November 1986. *Exact Descriptions of Regular and Semi-Regular Polyhedra and Their Duals*. AT&T Bell Laboratories, Computing Science Technical Report No. 130 (Murray Hill, NJ).

Hunt, Jarvis. 1924. Golf Ball. US Patent 1,517,514, issued April 2, 1924.

Huybers, Pieter. 1993. "Computer-Aided Design of Polyhedral Building Structures." *Design Studies* 14 (1): 19–39.

___. 1997. "The Polyhedral World." In *Beyond the Cube: The Architecture of Space Frames and Polyhedra*, edited by J. François Gabriel, 243–279. New York: John Wiley & Sons, Inc.

___. 2002. "The Morphology of Building Structures." International Conference on Computational Science, Amsterdam.

Hwang, In H. 1996. Golf Ball. US Patent 5,564,708, issued October 15, 1996.

___. 1996. Golf Balls. US Patent 5,575,477, issued November 19, 1996.

Ihara, Keisuke. 1989. Golf Ball. US Patent 4,844,472, issued July 4, 1989.

Kabai, Sándor. 2006. "Rhombic Structures: Use of Mathematica in Combination with Solid Models for Studying Space Structures." 2006 Wolfram Technology Conference, Champaign, IL.

Kahn, Lloyd. 1974. *Domebook Two*. Bolinas, CA: Shelter Publications.

___, ed. 1989. *Refried Domes*. Bolinas, CA: Shelter Publications.

___. 1989. "The Wonder of Jeana." In *Refried Domes*, 12–13. Bolinas, CA: Shelter Publications, Inc.

Kaiser Aluminum & Chemical Sales Inc. 1958. *Kaiser Aluminum Dome; Engineering Data for Mean Plan Diameter 125 Feet; A-80-11.5.* Chicago.

Kajikawa, Yasushi. 1990. "Growing Icosahedra." *Scientific American (Japanese edition)*, 2: 1–12.

Kang, Hye Sook. 2007. Dimple Pattern on Golf Ball. US Patent 7,278,933, issued October 9, 2007.

Kasashima, Atsuki. 2003. Golf Ball. US Patent 2003/0171167, issued September 11, 2003.

___. 2004. US Patent. US Patent 2004/0121858, issued June 24, 2004.

___. 2005. Golf Ball. US Patent 6,971,962, issued December 6, 2005.

___. 2008. Golf Ball. US Patent 7,354,358, issued April 8, 2008.

Kells, Lyman M., Willis F. Kern, and James R. Bland. 1943. *Plane and Spherical Trigonometry*. New York: McGraw-Hill Book Co., Inc.

Kennedy III, Thomas J. 2003. Golf Ball with Elevated Dimple Portions. US Patent 6,626,772, issued September 30, 2003.

Kenner, Hugh. 1973. *Bucky: A Guided Tour of Buckminster Fuller*. New York: Morrow.

___. 2003. *Geodesic Math and How to Use It*. second ed. Berkeley, CA: University of California Press.

Kepler, Johannes, and C. Hardie (Trans.). 1966. *Strena seu de nive sexangula*. Oxford, UK: Clarendon Press.

Keyes, Gene. 2009. "B. J. S. Cahill Butterfly Map – Octahedral Map of the World." www.genekeyes.com/B.J.S._CAHILL_RESOURCE.html.

Kimerling, A. Jon, Kevin Sahr, Denis White, and Lian Song. 1999. "Comparing Geometrical Properties of Global Grids." *Cartography and Geographic Information Science* 26 (4): 271–288.

Kitrick, Christopher J. 1980. Tensegrity Module Structure and Method of Interconnecting the Modules. US Patent 4,207,714, issued June 17, 1980.

___. 1984. "Nonlinear Analysis of Normal and Inverted Geodesic Domes Under the Action of Concentrated Loads." University of Cincinnati.

___. 1985. "Geodesic Domes." Structural Topology 11, University of Montreal, 15–20.

___. 1990. "A Unified Approach to Class I, II & III Geodesic Domes." International Journal of Space Structures 5 (3 & 4): 223–246.

___. 1993. "Dimpled Single-layer Geodesic Shells." Space Structures 4, Surrey, Guildford, UK, September 5–10, 1993.

___. 2013. "Geodesic Tessellations with Maximal Identical Equilateral Triangles." Technical Paper International Association for Shell and Spatial Structures (IASS), Wroclaw, Poland.

___. 2013. "Quasi-regular Geodesic Tessellations with Chained Triangles." International Association for Shell and Spatial Structures (IASS). Poland: Wroclaw University of Technology.

___. 2015. "Equal Edge Hexagonal Spherical Tessellations." Proceedings of the International Association for Shell and Spatial Structures (IASS) Symposium 2015, Amsterdam, the Netherlands, August 17–20, 2015.

___. 2015. "Plane Hexagonal Geodesic Tessellations." International Association for Shell and Spatial Structures (IASS), Amsterdam, the Netherlands.

___. 2016. "Plane Snub Tessellations." Spatial Structures in the 21st Century, Tokyo, Japan, September 26–30, 2016.

___. 2017. "Icosahedral Roundest Polyhedra." Interfaces: Architecture Engineering Science, Hamburg, Germany, September 25–28, 2017.

___. July 2019. Dymaxion Sky-Ocean World 1980 Grip-Kitrick Edition. La Jolla, CA: ResearchGate.

___. 2019. "Non-Linear Solution to Spherical Coverings and Application to Forms." International Association for Shell and Spatial Structures (IASS), Barcelona, Spain, October 7–10, 2019.

Kitrick, Christopher J., and Robert E. Grip. 1990. "Helical Forms of Face Bonded Tetrahedra." *Topologie Structurale* 16: 31–48.

Knebel, Klaus, J. Sanchez-Alfarez, and S. Zimmerman. 2002. "The Structural Making of the Eden Domes." *Space Structures* 1: 245–253.

Kovács, Flórián, Tibor Tarnai, and S. D. Guest. 2004. "Double–Link Expandohedra: A Mechanical Model for Expansion of a Virus." *Philosophical Transactions of the Royal Society of London. Series A: Mathematical, Physical & Engineering Sciences* 460 (2051): 3191–3202.

Krausse, Joachim (editor), and Claude Lichtenstein. 2001. *Your Private Sky: R. Buckminster Fuller*. Lars Müller Publishers.

Kroto, Harold. 1997. "Symmetry, Space, Stars and C80." *Reviews of Modern Physics* 69 (3): 703–722.

Kunszt, Peter Z., Alexander S. Szalay, and Aniruddha R. Thakar. 2000. "The Hierarchical Triangular Mesh." Proceedings of the MPA/ESO/MPE Workshop, Garching, Germany, July 31– August 4, 2002.

Lalvani, Haresh. 1976. "Transpolyhedra and Explosion-Implosion Principles for Dual Transformations." IASS World Congress on Space Enclosures, Montreal.

___. 1977. *Transpolyhedra: Dual Transformations by Explosion-Implosion*. Papers in Theoretical Morphology 1: 1–66.

___. 1981. "Multi-Dimensional Periodic Arrangements of Transforming Structures." Ph.D., University of Michigan.

___. 1982. "Structures on Hyper-Structures: Multi-Dimensional Periodic Arrangements of Transforming Space Structures." *Papers in Theoretical Morphology*, Lalvani, 112–114.

___. 1990. "Continuous Transformations of Subdivided Periodic Surfaces." *Space Structures* 5 (3-4): 255–279.

___. 1996. "Origins of Tensegrity: Views of Emmerich, Fuller and Snelson." *International Journal of Space Structures* 11 (1-2): 27–55.

___. 1996. Space Structures with Non-Periodic Subdivisions of Polygonal Faces. US Patent 5,524,396, issued June 11, 1996.

___. 1997. Building Systems with Non-Regular Polyhedra Based on Subdivisions of Zonohedra. US Patent 5,623,790, issued April 29, 1997.

___. 2008. Sports Ball. US Patent 2008/0268989, issued October 30, 2008.

Lane, Henry C. 1955. Final Report Study of Shelter Logistics for Marine Corps Aviation. edited by *U. S. Marine Corps Aviation Logistics and Material Branch*. Washington, D.C.: U. S. Marine Corps.

Larsen, Olga Popovic. 2008. *Reciprocal Frame Architecture*. Architecture Press, an imprint of Elsevier.

Latypov, Nurakhmed Nurislamovich, and Nurulla Nurislamovich Latypov. 2003. System for Placing a Subject into Virtual Reality. US Patent 6,563,489, issued May 13, 2003.

Lavallee, Gerald A., and Edward F. Mendrala. 1994. Golf Ball. US Patent 5,356,150, issued October 18, 1994.

Leighton, Henry Leland Chapman. 1943. *Solid Geometry and Spherical Trigonometry*. New York, NY: D. Van Nostrand.

Lemos, Lane D., and Matthew B. Stanczak. 2000. Golf Ball with Secondary Depressions. US Patent 6,010,442, issued January 4, 2000.

Leon, Bruno. 2004. Letter to Edward Popko discussing the first wooden lattice geodesic dome prototype built in Raleigh, NC in 1951. Sta. Fe, NM

___. 2004. Re: Early history of Marine Corps geodesic dome projects at Geodesics, Inc. in the 1950s.

Lettvin, Jonathan D. 2003. "Diffuse – A Computer Program for Distributing n Points on a Sphere." www.wasp.uwa.edu.au/~pbourke/geometry/spherepoints.

Leytem, Charles. 1996. "Hidden Symmetries in the Snub Dodecahedron." *European Journal of Combinatronics* 17 (5): 451–460.

Li, Yue, Xi-Qiao Feng, Yan-Ping Cao, and Huajian Gao. 2010. "Constructing Tensegrity Structures from One-Bar Elementary Cells." *Philosophical Transactions of the Royal Society of London. Series A: Mathematical, Physical & Engineering Sciences* 466 (2113): 45–61. rspa.royalsocietypublishing.org/content/466/2113/45.abstract.

LIFE. 1953. *Tape, Plastic and Aluminum – Ford Builds a 'Geodesic Dome'*. LIFE, 8 June, 67–70.

Lim, Dong-keun. 1995. Dimple Pattern and the Placement Structure on the Spherical Surface of the Golf Ball. US Patent 5,441,276, issued August 15, 1995.

Lisle, Richard J., and Peter R. Leyshon. 2004. *Stereographic Projection Techniques for Geologists and Civil Engineers*. second ed. New York: Cambridge University Press.

Livio, Mario. 2002. *The Golden Ratio: The Story of Phi, The World's Most Astonishing Number*. New York: Broadway Books.

Loeb, Arthur L. 1976. *Space Structures: Their Harmony and Counterpoint. Advanced Book Program*. Reading, MA: Addison Wesley Pub. Co.

___. 1997. "Buckminster Fuller and the Relevant Pattern." In *Beyond the Cube: The Architecture of Space Frames and Polyhedra*, edited by J. François Gabriel, 147–159. New York: John Wiley & Sons, Inc.

López-Pérez, Daniel, and R. Buckminster Fuller. 2019. *R. Buckminster Fuller: Pattern-Thinking*. Edited by Lars Müller Fuller. Baden, Switzerland: Lars Müller Publishers.

Lorance, Loretta. 2009. *Becoming Bucky Fuller*. Cambridge, MA: MIT Press.

Machin, Brian E. 1993. Golf Balls with Isodiametrical Dimples. US Patent 5,377,989, issued January 3, 1995.

Mackey, Gary T. 1991. Golf Ball. US Patent 5,046,742, issued September 10, 1991.

MacLean, Kenneth James Michael. 2007. *A Geometric Analysis of the Platonic Solids and Other Semi-Regular Polyhedra: With an Introduction to the Phi Ratio - for Teachers, Researchers and the Generally Curious*. Ann Arbor, MI: Loving Healing Press.

Maehara, Kazuto, Keisuke Ihara, and Atsuki Kasashima. 2001. Dimpled Golf Ball and Dimple Distributing Method. US Patent 6,254,496, issued July 3, 2001.

Makai, Endre Jr., and Tibor Tarnai. 1973. On Polyhedra with Approximately Equal Edges. Budapest, Hungary: Mathematics Institute of the Hungarian Academy of Sciences.

Makai Jr., Endre. 1975. "On Some Geometrical Problems of Single-Layered Spherical Grids with Triangular Network." *II International Conf. on Space Structures*, Guildford, England: University of Surrey.

Makai Jr., Endre, and Tibor Tarnai. 1976. "Morphology of Spherical Grids." *Acta technica Academiae Scientiarum Hungaricae* 83: 247–283.

Maor, Eli. 1994. *e: The Story of a Number*. Princeton, NJ: Princeton University Press.

___. 1998. *Trigonometric Delights*. Princeton, NJ: Princeton University Press.

Marks, Robert W., R. Buckminster Fuller. 1960. *The Dymaxion World of Buckminster Fuller*. New York: Reinhold Pub. Corp.

Marks, Robert W., and R. Buckminster Fuller. 1973. *The Dymaxion World of Buckminster Fuller*. Garden City, NY: Anchor Books.

Martin, Frank S. 1955. Golf Balls. US Patent 2,728,576, issued December 27, 1955.

Martin, Frank S., and Thaddeus A. Pietraszek. 1978. Golf Ball. US Patent 4,090,716, issued May 23, 1978.

Martin, John Stuart. 1968. *The Curious History of the Golf Ball, Mankind's Most Fascinating Sphere*. New York: Horizon Press.

Matsko, Vincent J. 2009. "Polyhedra and Geodesic Structures." Illinois Mathematics and Science Academy. staff.imsa.edu/~vmatsko/.

McAtee, Cammie. 2017. *Montreal's Geodesic Dreams: Jeffrey Lindsay and the Fuller Research Foundation Canadian Division*. Edited by Michelangelo Sabatino. *Canadian Modern*. Halifax, Nova Scotia, Canada: Dalhousie Architectural Press.

McGuire, Kenneth Stephen. 2002. Golf Ball with Non-Circular Shaped Dimples. US Patent 6,409,615, issued June 25, 2002.

McHale, John. 1962. *R. Buckminster Fuller. Makers of Contemporary Architecture.* New York: G. Braziller.

Melvin, Terence. 1989. Golf Ball. US Patent 4,880,241, issued November 14, 1989.

Melvin, Terence, Joseph K. Pieroni, Frank S. Martin, and James F. Little. 1979. Two-Piece Solid Golf Ball. US Patent 4,141,559, issued February 27, 1979.

Messer, Peter W. 1999. "Mathematical Formulas for Geodesic Domes." In *Spherical Models*, edited by Magnus J. Wenninger, 145–149. New York: Dover.

Mikloweit, Ulrich. 2020. "Exploring Noble Polyhedra with the Program Stella4D." Bridges 2020, Helsinki, Finland.

Miyazaki, Koji. 1979. *A History of Polyhedra in Japan.* Kobe, Japan: Kobe University.

___. 1983. *An Adventure in Multidimensional Space: The Art and Geometry of Polygons, Polyhedra and Polytopes.* New York: John Wiley & Sons.

___. 2016. *Maruzen Encyclopedia of Polyhedra.* Tokyo: Maruzen Publishing.

Mizuta, Ai (Mayumi). 2020. *TEMARI-Jiwari and Color Magic.* eBook. Tokyo: Shibucho Japan Temari Association.

Mizuta, Ai (Mayumi), and Genny Thompson (translator). 2020. *Temari – The Geometric Shapes & Needlework.* eBook. Tokyo: Shibucho, Japan Temari Association.

Morell, Joseph. 1990. Golf Ball. US Patent 4,973,057, issued November 27, 1990.

Morgan, Gregory J. 2003. "Historical Review: Viruses, Crystals and Geodesic Domes." *Trends in Biochemical Sciences* 28 (2): 86–90.

Morgan, George W. 1985. *Geodesic & Geolatic Domes & Space Structures – Geometric Design Methods.* San Jose, CA, USA: Sci-Tech Publications.

Morgan, William E. 2005. Dimple Pattern for Golf Balls. US Patent 6,849,007, issued February 1, 2005.

___. 2007. Golf Ball with Spherical Polygonal Dimples. US Patent 7,309,298, issued December 18, 2007.

Morrison, James E. 2007. *The Astrolabe.* Rehoboth Beach, DE: Janus.

Motro, René. 2003. *Tensegrity – Structural Systems for the Future.* London: Kogan Page Science.

___. 2009. *An Anthology of Structural Morphology.* Hackensack, NJ: World Scientific.

Museum of Modern Art. 1960. *Three Structures by Buckminster Fuller in the Garden of the Museum of Modern Art.* New York: Museum of Modern Art.

Nardacci, Nicholas M. 2005. Dimpled Golf Ball and Dimple Distributing Method. US Patent 2005/0176525, issued August 11, 2005.

Naruo, Takeshi, and Taketo Mizota. 1914. "The Influence of Golf Ball Dimples on Aerodynamic Characteristics." *Procedia Engineering* 72, 780–785, The 2014 Conference of the International Sports Engineering Association.

National Oceanic & Atmospheric Administration. 2008. Offshore Aquaculture in the United States: Economic Considerations, Implications & Opportunities. Silver Spring, MD: US Dept. of Commerce.

Nesbit, Dennis R., and Joseph F. Stiefel. 1991. Golf Ball. US Patent 5,044,638, issued September 3, 1991.

Nomura, June, and Keisuke Ihara. 1989. Golf Ball. US Patent 4,869,512, issued September 26, 1989.

Nooshin, Hoshyar. 1979. *FORMEX Formulation of Double-Layered Grids*. Guildford, England: University of Surrey.

Nooshin, Hoshyar, Peter L. Disney, and O. C. Champion. 1997. "Computer-Aided Processing of Polyhedric Configurations." In *Beyond the Cube: The Architecture of Space Frames and Polyhedra*, edited by J. François Gabriel, 343–384. New York: John Wiley & Sons, Inc.

Nooshin, Hoshyar, Peter L. Disney, and Chiaki Yamamoto. 1993. *Formain*. Brentwood, Essex, UK: Multi-Science Publishing Co. Ltd.

NPD Group Inc. 2009. 2008 Video Game Software Sales Across Top Global Markets Experience Double-Digit Growth (press release).

Ogg, Steven S. 2002. Golf Ball with Multiple Sets of Dimples. US Patent 6,482,119, issued November 19, 2002.

___. 2002. Golf Ball with Pyramidal Protrusions. US Patent 6,471,605, issued October 29, 2002.

___. 2005. Aerodynamic Surface Geometry for a Golf Ball. US Patent 6,958,020, issued October 25, 2005.

___. 2007. Aerodynamic Surface Geometry for a Golf Ball. US Patent 7,198,577, issued April 3, 2007.

Oka, Kengo. 1992. Golf Ball. US Patent 5,145,180, issued September 8, 1992.

Oka, Kengo, and Shinji Ohshima. 1994. Golf Ball. US Patent 5,338,039, issued August 16, 1994.

Organization, Food and Agriculture. 2018. Fishery and Aquaculture Statistics 2018. edited by FAO. New York, USA: Fisheries and Aquaculture Department, edited by FAO.

___. 2020. Global Fishery and Aquaculture Production Statistics. edited by Fisheries Division (NFI). Rome: Food and Agriculture Organization of the United Nations.

___. 2020. The State of World Fisheries and Aquaculture. Rome: Food and Agriculture.

Otero, Cesar, and Reinaldo Togores. 2002. "Computational Geometry and Spatial Meshes." International Conference Computational Science ICCS 2002 Part II, Amsterdam.

Page, Stephen H. 2007. Containment Pens for Finfish Aquaculture. US Patent 7,284,501, issued October 23, 2007.

Pantke, Conrad. 1930. Space-Covering Structure. US Patent 1,773,851, issued August 26, 1930.

Pauline, Sperry. 1928. *Short Course in Spherical Trigonometry*. Edited by Johnson's Mathematics Series. GA: Atlanta: Johnson Publishing Co.

Pavlov, G. N. 1990. "Determination of Parameters of Crystal Latticed Surfaces Composed of Hexagonal Plane Facets." *International Journal of Space Structures* 5, (3 & 4): 169–185.

Pearce, Peter. 1978. *Structure in Nature is a Strategy for Design*. Cambridge, MA and London, England: MIT Press.

Pearce, Peter, and Susan Pearce. 1978. *Polyhedra Primer*. New York: Van Nostrand Reinhold.

Pearson, Frederick. 1984. *Map Projection Methods*. Blacksburg, VA: Sigma Scientific, Inc. Computer Science Corporation.

Peck, Charles E. 1995. *A Taxonomy of Fundamental Polyhedra and Tessellations*. Winkworks.

Pedersen, Jean. 1981. "Some Isonemal Fabrics on Polyhedral Surfaces." New York: Springer, *The Geometric Vein* 99–122.

___. 1983. "Geometry: The Unity of Theory and Practice." *The Mathematical Intelligencer* 5 (4): 37–49.

Pocklington, Terence W. 1996. Golf Ball. US Patent 5,536,013, issued July 16, 1996.

___. 1996. Golf Ball Dimple Construction. US Patent 5,547,197, issued August 20, 1996.

Pooley, Richard, and Ron Ward. 1962. *The Energetic-Synergetic Geometry of R. Buckminster Fuller and Its Applications to Geodesic Structures*. Vol. 1. Carbondale, IL: Design Department, Southern Illinois University.

Popko, Edward S. 1968. *Geodesics*. Detroit: University of Detroit Press.

___. 1969. *SPHERES – FORTRAN IV Programs for Spherical Subdivision*. Detroit: University of Detroit.

Posamentier, Alfred S., and Ingmar Lehmann. 2007. *The (Fabulous) Fibonacci Numbers*. New York: Prometheus Books.

Press, Cambridge University. 2017. "Mathematicians deliver formal proof of Kepler Conjecture." *ScienceDaily*. phys.org/news/2017-06-mathematicians-formal-proof-kepler-conjecture.html..

Puderbaugh, Homer L. 1964. "Projections for a Geodesic Sphere." *Architectural Science Review* 7 (March): 19–26.

Pugh, Anthony. 1976. *An Introduction to Tensegrity. The Dome Series*. Berkeley, CA: University of California Press.

___. 1976. *Polyhedra – A Visual Approach*. Berkeley, CA: University of California Press.

Radin, Charles. 1999. *Miles of Tiles*. Vol. 1. *Student Mathematical Library*. Providence, RI: American Mathematical Society.

___. 2006. "Review of The Pursuit of Perfect Packing." *The Mathematical Association of America* 113: 88–90.

Randall, David A., Todd D. Ringler, Ross P. Heikes, Phil Jones, and John Baumgardner. 2002. "Climate Modeling with Spherical Geodesic Grids." *Computing in Science and Engineering IEEE Computational Science and Engineering* 4 (5): 32–41.

Reimann, David A. 2017. "Transforming Squares to Strips in Expanded Polyhedral Forms." Bridges Waterloo 2017: Mathematics, Art, Music, Architecture, Education, Culture, July 27–31, 2017.

Riblet, Byron C. 1908. Game-Ball. US Patent 906,932, issued December 15, 1908.

Richeson, David S. 2008. *Euler's Gem: The Polyhedron Formula and the Birth of Topology*. Princeton, NJ: Princeton University Press.

Richter, Donald L. 1962. Building Construction. US Patent 3,026,651, issued March 27, 1962.

___. 1962. Structural Unit. US Patent 3,058,550, issued October 16, 1962.

Ringler, Todd D., Ross P. Heikes, and David A. Randall. 2000. "Modeling the Atmospheric General Circulation Using a Spherical Geodesic Grid: A New Class of Dynamical Cores." *American Meteorological Society* 128 (7): 2471–2489.

Roberts, Siobhan. 2006. *King of Infinite Space – Donald Coxeter, the Man Who Saved Geometry*. New York: Walker & Company.

Roelofs, Rinus. 2008. "Two- and Three-Dimensional Constructions Based on Leonardo Grids." *Nexus Network Journal* 10 (1): 7–16.

___. 2010. "About Weaving and Helical Holes." Bridges Pécs 2010: Mathematics, Music, Art, architecture, Culture, Pécs, Hungary.

Sadao, Shoji. 1974. Geodesic Pentagon and Hexagon Structure. US Patent 3,810,336, issued May 14, 1974.

___. 2011. *Buckminster Fuller and Isamu Noguchi: Best of Friends*. Milan: 5 Continents Editions.

Sadao, Shoji, and Buckminser Fuller. 1954. Dymaxion Airocean World – The Raleigh Edition of Fuller Projection. In *Student Publications of the School of Design*. Raleigh: North Carolina State University.

Sadourny, Robert, Akio Arakawa, and Yale Mintz. 1968. "Integration of the Nondivergent Barotopic Vorticity Equation with Icosahedral-Hexagonal Grid for the Sphere." *Monthly Weather Review* 96 (6): 351–356.

Saff, E. B., and A. B. J. Kuijlaars. 1997. "Distributing Many Points on the Sphere." *The Mathematical Intelligencer* 12 (1): 5–11.

Sahr, Kevin, Denis White, and A. Jon Kimerling. 2003. "Geodesic Discrete Global Grid Systems." *Cartography and Geographic Information Science* 30 (2): 121–134.

Sajima, Takahiro. 2007. Golf Ball. US Patent 7,223,183, issued May 29, 2007.

___. 2008. Golf Ball. US Patent 7,331,879, issued February 19, 2008.

Sato, Katsunori. 2007. Golf Ball. US Patent 7,252,601, issued August 7, 2007.

Sato, Katsunori, and Atsuki Kasashima. 2007. Golf Ball. US Patent 7,160,212, issued January 9, 2007.

___. 2007. Golf Ball. US Patent 7,252,601, issued August 7, 2007.

Scheel, H. 1969. "Geodesic Surface Division." *The Canadian Architect*, 14 (5): 61–66.

Schläfli, Ludwig. 1950. *Gesammelte Mathematische Abhandlungen.* 3 vols. Vol. 1–3. Verlag Birkhäuser.

Schneider, Michael S. 1995. *A Beginner's Guide to Constructing the Universe.* New York: Harper Perennial.

Sharp, Abraham. 1717. *Geometry Improved.* London: A. S. Philomath.

Shaw, Michael. 1990. Golf Balls. US Patent 4,960,282, issued October 2, 1990.

Shaw, Michael, and Robert C. Haines. 1979. Golf Balls. US Patent 4,142,727, issued March 6, 1979.

Shaw, Michael, and Brian F. Machin. 1988. Golf Balls. US Patent 4,722,529, issued February 2, 1988.

Sherwood, Anton. 2007. "How can I arrange N points evenly on a sphere?" A collection of links for sphere arrangement problems. bendwavy.org/sphere.htm.

Shimosaka, Hirotaka, Keisuke Ihara, Yutaka Masutani, Michio Inoue, and Atsuki Kasashima. 1999. Golf Ball. US Patent 5,902,193, issued May 11, 1999.

Šiber, Antonio. 2020. "Icosadeltahedral Geometry of Geodesic Domes, Fullerenes and Viruses: A Tutorial on the T-Number" *Symmetry* 12 (4): 556–578, Symmetry in Molecular Biology.

Sieden, Lloyd Steven. 2000. *Buckminster Fuller's Universe.* Cambridge, MA: Perseus Pub.

Simonds, Vincent J., and Thomas J. Bergin. 2006. Aerodynamic Surface Geometry for a Golf Ball. US Patent 7,338,392, issued March 4, 2008.

___. 2008. Aerodynamic Surface Geometry for a Golf Ball. US Patent 7,338,392, issued March 4, 2008.

Skilling, J. 1975. "The Complete Set of Uniform Polyhedra." *Philosophical Transactions of the Royal Society of London. Series A: Mathematical, Physical & Engineering Sciences* 278: 111–135.

Smart, William Marshall. 1942. *Astronomical Navigation – A Handbook for Aviators*. New York: Longmans.

___. 1990. *Textbook on Spherical Astronomy*. Edited by R. M. Green. sixth ed. New York: Cambridge University Press.

Snelson, Kenneth D. 1965. Continuous Tension, Discontinuous Compression Structures. US Patent 3,169,611, issued February 16, 1965.

___. 1968. Model for Atomic Forms. US Patent 3,276,148, issued October 4, 1966.

___. 1978. Model for Atomic Forms. US Patent 4,099,339, issued July 11, 1978.

___. 2000. Magnetic Geometric Building System. US Patent 6,017,220, issued January 25, 2000.

___. 2002. "Circles, Spheres and Atoms." *Symmetry: Culture and Science* 13 (1): 1–18.

Snyder, John Parr. 1987. *Map Projections – A Working Manual*. U.S. Geological Survey. Washington: U. S. Government Printing Off..

Sobel, Dava. 1995. *Longitude: The True Story of a Lone Genius Who Solved the Greatest Scientific Problem of His Time*. New York: Walker.

Sobel, Dava, and William J. H. Andrewes. 1998. *The Illustrated Longitude: The True Story of the Lone Genius Who Solved the Greatest Scientific Problem of His Time*. New York: Walker.

Solheim, Karsten. 1987. Golf Ball. US Patent 4,653,758, issued 31 Mar. 1987.

Song, Lian, Jon Kimerling, and Kevin Salr. 2002. *Developing an Equal Area Global Grid by Small Circle Subdivision*. Santa Barbara, CA: University of California, National Center for Geographic Information & Analysis.

Song, Peng, Chi-Wing Fu, Prashant Goswami, Jianmin Zheng, Niloy J. Mitra, and Daniel Cohen-Or. 2013. "Reciprocal Frame Structures Made Easy." *ACM Transactions on Graphics* 32 (4): 10.

Sperry, Pauline. 1928. *Short Course in Spherical Trigonometry*. Mathematics Series. Atlanta, GA: Johnson Publishing Co.

Spunt, Leonard. 1976. "Modular Dome Structures." IASS World Congress on Space Enclosures, Montreal.

___. 1976. Modular Dome Structures. US Patent 3,959,937, issued June 1, 1976.

State University Colorado. 2001. "Geodesics Climate Model Uses Different Mapping Techniques, Coordinates and Supercomputing to Improve Predictions." Science-Daily. www.sciencedaily.com/releases/2001/09/010926071704.htm.

Stewart, Ian. 1997. "Crystallography of a Golf Ball." *Scientific American*, (February), 276: 96–98.

Stiefel, Joseph F. 1999. Golf Ball and Method of Forming Dimples Thereon. US Patent 5,890,975, issued April 6, 1999.

Stiefel, Joseph F., and Dennis Nesbitt. 1999. Tetrahedral Dimple Pattern Golf Ball. US Patent 5,890,974, issued April 6, 1999.

Stuart, Duncan R. 1951. Topological Data. Department of Special Collections and University Archives, Stanford University Libraries (Raleigh: Fuller Research Foundation).

___. 1952. *Letter to: Buckminster Fuller, Subject: A New Approach to Subdividing an Icosahedron.* Raleigh, NC: Skybreak Carolina Corp.

___. 1952. *A Report on the Triacon Gridding System for Spherical Surface.* Skybreak Carolina Corp. Raleigh, NC: Stanford University Libraries, Department of Special Collections and University Archives.

___. 1953. *Spherical Truss for Ford Motor Co. by R. B. Fuller.* Raleigh, NC: Skybreak Carolina Inc., Geodesic geometry calculations, Department of Special Collections and University Archives, Stanford University Libraries.

___. 1962. "Polyhedra." *Student Publications of the School of Design North Carolina State University* 3 (1).

___. 1963. "On the Orderly Subdivision of Spheres." *The Student Publications of the School of Design, North Carolina State University* 5: 23–33.

___. 1963. "Polyhedral and Mosaic Transformations." *The Student Publications of the School of Design, North Carolina State University (monograph)* 12 (1): 2–28.

___. 1963. "The Transformation of Polyhedra: An Experiment in Animated Film Making." *Journal of Architectural Education* 18 (3): 41–43.

Stuart, Duncan R., and Fred Eichenberger. 1968. *The Mass Production of Unique Items.* Raleigh, NC: Design Research Laboratory of the School of Design, North Carolina State University.

Sullivan, Michael J. 2000. Golf Ball with Elevated Dimple Portions. US Patent 6,139,448, issued October 31, 2000.

___. 2003. Golf Ball Dimples. US Patent 6,569,038, issued May 27, 2003.

___. 2004. Golf Ball dimples. US Patent 6,709,349, issued March 23, 2004.

Sullivan, Michael J., Steven Aoyama, and Edmund A. Hebert. 2009. Golf Ball Surface Patterns Comprising Multiple Channels. US Patent 2009/0017941, issued January 15, 2009.

Sultan, Cornel, Martin Corless, and Robert E. Skelton. 2002. "Symmetrical Reconfiguration of Tensegrity Structures." *International Journal of Solids and Structures* 39: 2215–2234.

Sutton, Daud. 2002. *Platonic & Archimedean Solids. Wooden Books.* New York: Walker & Co.

Szalay, Alexander S., Jim Gray, George Fekete, Peter Z. Kunszt, Peter Kukol, and Ani Thakar. 2005. *Indexing the Sphere with the Hierarchical Triangular Mesh.* Microsoft Corp. Research Advanced Technology Division, Technical Report MSR-TR-2005-123 (Redmond, WA).

Szpiro, George. 2003. *Kepler's Conjecture – How Some of the Greatest Minds in History Helped Solve One of the Oldest Math Problems in the World.* Hoboken, NJ: John Wiley & Sons.

Tammes, P. M. L. 1930. "On the Origin of Number and Arrangement of the Places of Exit on the Surface of Pollen-Grains." *Recueil des Travaux Botaniques Néerlandais* 27: 1–84.

Tarnai, Tibor. 1974. "Spherical Grids of Triangular Network." *Acta Technica Academiae Scientiarum Hungaricae* 76: 307–336.

___. 1980. "Problems Concerning Spherical Polyhedra and Structural Rigidity." *Structural Topology* 4: 61–66.

___. 1983. "Geodesic Dome with Three Different Bar Lengths." *Structural Topology* 8: 23–24.

___. 1984. "Spherical Circle-Packing in Nature, Practice and Theory." *Structural Topology* 9: 39–58.

___. 1987. *Geodesic Domes with Skew Networks.* Hungarian Institute of Building Science.

___. 1987. *Spherical Grid Structures. Geometric Essays on Geodesic Domes.* Budapest: Hungarian Institute for Building Science.

___. 1992. "Dense Sphere Packing on a Sphere." *First International Seminar on Structural Morphology.* France: Montpellier - Le Grand Motte.

___. 1993. "Geodesic Domes and Fullerenes." *Philosophical Transactions of the Royal Society of London. Series A: Mathematical, Physical & Engineering Sciences* 343: 1667–1673.

___. 1993. "Geodesic Domes and Golf Balls." *Space Structures* 4, University of Surrey, Guildford, UK. 2: 1176–1184.

___. 2002. "Polymorphism in Multi-Symmetric Close Packings of Equal Spheres on a Spherical Surface." *Structural Chemistry* 13 (3-4): 289–295.

___. 2009. "Ivory Shells and Polyhedra." *International Association for Shell and Spatial Structures (IASS) Symposium*. Spain: Universidad Politecnica de Valencia.

___. 2011. "Geodesic Domes: Natural and Man-Made." *International Journal of Space Structures* 26 (3): 215–227.

Tarnai, Tibor, P. W. Fowler, and S. Kabai. 2003. "Packing of Regular Tetrahedral Quartets of Circles on a Sphere." *Philosophical Transactions of the Royal Society of London. Series A: Mathematical, Physical & Engineering Sciences* 459: 2847–2859.

Tarnai, Tibor, and Zsolt Gáspár. 1985. "Covering the Sphere with Equal Circles." *Colloquium on Intuitive Geometry*, Hungary: Balatonszéplak.

___. 2001. "Packing of Equal Regular Pentagons on a Sphere." *Philosophical Transactions of the Royal Society of London. Series A: Mathematical, Physical & Engineering Sciences* 457: 1043–1058.

Tarnai, Tibor, Zsolt Gáspár, and A. Lengyel. 2013. "From Spherical Circle Coverings to the Roundest Polyhedra." *Philosophical Magazine* 93 (31-33): 3980–3992.

Tarnai, Tibor, Zsolt Gáspár, and L. Szalai. 1995. "Pentagon Packing Models for "All-Pentamer" Virus Structures." *Biophysical Journal* 69 (2): 612–618.

Tarnai, Tibor, F. Kovács, P. W. Fowler, and S. D. Guest. 2012. "Wrapping the Cube and Other Polyhedra." *Proceedings of the Royal Society (London)* 468: 2652–2666.

Tarnai, Tibor, and Andras Lengyel. 2012. "Dimple Patterns of Golf Balls and Node Arrangements of Geodesic Domes." *International Association for Shell and Spatial Structures (IASS)* 53 (3): 185–194.

Tarnai, Tibor, and Koji Miyazaki. 2003. "Circle Packings and the Sacred Lotus." *Leonardo* 36 (2): 145–150.

Tarnai, Tibor, and Magnus J. Wenninger. 1990. "Spherical Circle Coverings and Geodesic Domes." *Structural Topology* 7: 5–22.

Taylor, William W. 1908. Golf Ball. US Patent 878,254, issued February 4, 1908.

___. 1918. Golf Ball. US Patent 1,286,834, issued December 3, 1918.

___. 1990. Golf Ball. US Patent 4,921,255, issued May 1, 1990.

Termes, Dick. 2020. *Parts of the Whole: 1-6 Point Perspective on the Sphere*. Spearfish, SD.

Thieman, Sharon. 2018. "Power of Wrapping." *Master of Fine Arts, College of Arts & Letters*. Florida Atlantic University.

Thompson, D'Arcy Wentworth. 1992. *On Growth and Form*. New York: Dover.

Tickoo, Sham, and Vivek Singh. 2008. *CATIA V5R18 for Designers*. Schererville, IN: CADCIM Technologies.

Todhunter, I. 1871. *Spherical Trigonometry, for the Use of Colleges and Schools*. London: Macmillan and Co.

Tomlow, Jos, ed. 1997. *Polyhedra from Pythagoras to Alexander Graham Bell*. Edited by J. Francois Gabriel, *Beyond the Cube: The Architecture of Space Frames and Polyhedra*. New York: John Wiley & Sons, Inc.

Turner, Merle B. 1986. *Celestial for the Cruising Navigator*. Centreville, MD: Cornell Maritime Press.

Twarock, Reidun. 2006. "Mathematical Virology: A Novel Approach to the Structure and Assembly of Viruses." *Philosophical Transactions of the Royal Society of London. Series A: Mathematical, Physical & Engineering Sciences* 364: 3357–3373.

U. S. Department of Commerce. 1949. *Table of Sines and Cosines to Fifteen Decimal Places at Hundredths of a Degree*. edited by Applied Mathematics Laboratories Nat. Bureau of Standards. Washington, DC: U.S. Government Printing Office.

U. S. Department of Labor. 2011. "CPI Inflation Calculator." data.bls.gov/cgi-bin/cpicalc.pl.

Unger, Kirby. 1991. "The Invention Behind the Invention." *Synergetica Journal, Buckminster Fuller Institute* 1 (1).

University of Minnesota Science and Technology Center. 2008. "The Geometry Center – A Project of the National Science Foundation." www.geom.uiuc.edu/.

University of New South Wales. 2008. "Distributing Points on the Sphere." School of Mathematics and Statistics. www.maths.unsw.edu.au/about/distributing-points-sphere.

Van Brummelen, Glen. 2017. *Heavenly Mathematics – The Forgotten Art of Spherical Trigonometry*. *Popular Mathematic/History of Science*. Princeton: Princeton University Press.

Veilleux, Thomas A., Vincent J. Simonds, and Kevin J. Shannon. 2007. Golf Ball Dimple Pattern. US Patent 7,179,178, issued February 20, 2007.

Verheyen, Hugo F. 1989. "The Complete Set of Jitterbug Transformers and the Analysis of their Motion." *Computers & Mathematics with Applications* 17 (13): 203–225.

___. 1996. *Symmetry Orbits*. Boston: Birkhäuser.

Vince, John. 1985. *Computer Graphics for Graphic Designers*. Video bookshelf. White Plains, NY: Knowledge Industry Publications.

___. 1992. *Computer Graphics*. London: Design Council.

___. 2005. *Geometry for Computer Graphics – Formula, Examples and Proofs*. London: Springer.

___. 2006. *Mathematics for Computer Graphics*. second ed. London: Springer.

___. 2008. *Geometric Algebra for Computer Graphics*. London: Springer.

Watkins, Christopher D., and Vincent P. Mallette. 1996. *Stereogram Programming Techniques*. Rockland, MA: Charles River Media.

Webb, Robert. 2000. "Stella: Polyhedron Navigator." *Symmetry: Culture and Science* 11 (1–4): 231–268.

___. 2008. "Great Stella Manual." Software 3D. www.software3d.com/StellaManual.php.

Wenninger, Magnus J. 1971. *Polyhedron Models*. Cambridge [Eng.]: University Press.

___. 1975. *Polyhedron Models for the Classroom*. second ed. Reston, VA: National Council of Teachers of Mathematics.

___. 1978. "Geodesic Domes by Euclidean Construction." *The Mathematics Teacher* 17 (7): 582–587.

___. 1979. *Spherical Models*. New York: Cambridge University Press.

___. 1983. *Dual Models*. New York: Cambridge University Press.

___. 1990. *Artistic Tessellation Patterns on the Spherical Surface*. International Journal of Space Structures 5 (3 & 4): 247–253.

___. 1999. *Spherical Models*. Mineola, NY: Dover.

___. 2002. *Symmetrical Patterns on a Sphere*. Collegeville, MN: St. John's Abbey.

___. 2002. *Symmetrical Patterns on a Sphere*, essay #5 in Part I, *Symmetry 2000*, Ed. Istvan Hargittai and Torvard C. Laurent, Wenner-Gren International Series 80: 41–51. London: Portland Press.

Wertz, James Richard, and Wiley J. Larson. 1999. *Space Mission Analysis and Design*. third ed. *Space Technology Library*. El Segundo, CA: Microcosm.

Wester, True. 1984. *Structural Order in Space: The Plate-Lattice Dualism*. Copenhagen: Plate Laboratory, Royal Academy of Arts School of Architecture.

Weyl, Hermann. 1952. *Symmetry*. Princeton, NJ: Princeton University Press.

White, Denis. 2000. "Global Grids from Recursive Diamond Subdivisions of the Surface of an Octahedron or Icosahedron." *Environmental Monitoring and Assessment* 64 (1): 93–103.

White, Denis, A. Jon Kimerling, and W. Scott Overton. 1992. "Cartographic and Geometric Components of a Global Sampling Design for Environmental Monitoring." *Cartography and Geographic Information Systems* 10 (1): 5–22.

White, Denis, A. Jon Kimerling, Kevin Sahr, and Lian Song. 1992. "Comparing Area and Shape Distortion of Polyhedral-Based Recursive Partitions of the Sphere." *International Journal of Geographical Information Science* 12 (8): 805–827.

White, John. 1920. Golf Ball. US Patent 1,418,220, issued October 2, 1920.

White, Mark P. 2003. Rafiki Model and Map to the Genetic Code. US Patent 2003/0190657, issued October 9, 2003.

___. 2011. "Rafiki." Rafiki, Inc. www.codefun.com/.

Whittaker, E. J. W. 1984. *The Stereographic Projection*. Cardiff, Wales: Published for the International Union of Crystallography by University College Cardiff Press.

Wigley, Mark. 2015. *Buckminster Fuller Inc. Architecture in the Age of RADIO*. Zurich: Lars Müller Publishers.

Williams, Robert Edward. 1968. *Handbook of Structure*. Huntington Beach, CA: Douglas Advanced Research Laboratories.

___. 1979. *The Geometrical Foundation of Natural Structure a Sourcebook of Design*. New York: Dover.

Wilson, David P. 2016. "Protruding Features of Viral Capsids are Clustered on Icosahedral Great Circles." *PLoS ONE Academia Sinica* 11 (4): 1–22.

Winfield, Douglas C., and Steven Aoyama. 2003. Pentagonal Hexecontahedron Dimple Pattern on Golf Ball. US Patent 6,527,653, issued March 4, 2003.

Wodehouse, Roger P. 1935. *Pollen Grains: Their Structure, Identification and Significance in Science and Medicine. Publications in the Agricultural and Botanical Sciences*. New York: McGraw-Hill.

Woo, Young-Kyu. 1993. Golf Ball. US Patent 5,192,078, issued March 9, 1993.

Xiaochong, Tong, Ben Jin, Zhang Yongsheng, and Meng Li. 2002. *The Expression of Spherical Entities and Generating of Voronoi Diagrams Based on Truncated Icosahedron DGG*. Zhengzhou, China: Institute of Surveying and Mapping.

Yackel, Carolyn A. 2009. "Marking a Physical Sphere with a Projected Platonic Solid." Bridges Banff 2009: Mathematics, Music, Art, Architecture, Culture. Banff, Alberta, Canada: The Banff Centre.

___. 2011. "Teaching Temari: Geometrically Embroidered Spheres in the Classroom." Bridges Coimbra 2011: Mathematics, Music, Art, Architecture. Coimbra, Portugal: University of Coimbra.

Yagley, Michael S., Steven S. Ogg, Pijush K. Dewanjee, David M. Bartels, and Geoffrey P. M. Goodman. 2003. Golf Ball. US Patent 6,634,965, issued October 21, 2003.

Yamada, Kaname. 1988. Golf Ball. US Patent 4,720,111, issued January 19, 1988.

___. 1990. Golf Ball. US Patent 4,946,167, issued August 7, 1990.

Zumkeller, Roland et al. 2015. "A Formal Proof of Kepler Conjecture." Forum of Mathematics, Pi 5.

Zung, Thomas Tse Kwai, and R. Buckminster Fuller. 2002. *Buckminster Fuller: Anthology for the New Millennium*. New York: St. Martin's Griffin Press.

Index

About the Authors

Edward S. Popko is a graduate of the University of Detroit's School of Architecture and has both masters and PhD degrees from Massachusetts Institute of Technology. He is a registered architect and former Fulbright Scholar. In the 1960s, he was an apprentice in Buckminster Fuller's affiliate office, Geometrics, Inc., in Cambridge, Massachusetts, and later authored *Geodesics*, a primer on geodesic domes, and *Transitions*, a documentary of urban settlements in developing countries.

For the past 30 years, he has held research, product development, and marketing positions for Computer Aided Design (CAD) and Geographic Information Systems technology within IBM. He has managed software development teams in computer mapping and CAD solutions for architects, industrial buildings, high-tech plants, and shipbuilding.

He is retired from IBM and lives with his wife, Geraldine, in Woodstock, New York.[1] They have a daughter Ellen and a son Gerald.

---

[1] Edward S. Popko holds a model of a Class III subdivision of a spherical icosahedron $\{3,5+\}_{5,2}$ made by Father Magnus Wenninger.

Chris J. Kitrick is a graduate of the University of Cincinnati with a bachelor's in Architecture and a masters in Structural Engineering. In the late 1970s, he interned for three years at Buckminster Fuller's Professor Emeritus office on the campus of the University of Philadelphia. During that tenure he was involved in Fuller's *Synergetics II* book, multiple dome developments, tensegrity research, and the first computerized version edition of the Dymaxion Airocean map.

For the past 36 years, he has been involved with graphics software and hardware development from mainframe to mobile devices enabling the rapid advancement of visual technology. Independently, he has authored and presented numerous technical papers on spherical geometry at international conferences on space structures.

He is still working and resides in La Jolla, California, with his wife Tomoko.[2] They have two sons Francis and Ian, and one daughter, Eileen.

---

[2] Chris J. Kitrick is holding a model of quasi-regular geodesic tessellation of the tetrahedron. See (Kitrick 2011).